Design and
Construction Failures

Design and Construction Failures

Lessons from Forensic Investigations

Dov Kaminetzky, M.S., P.E.
Feld, Kaminetzky & Cohen, P.C.
New York, New York

McGraw-Hill, Inc.

New York   St. Louis   San Francisco   Auckland   Bogotá
Caracas   Hamburg   Lisbon   London   Madrid
Mexico   Milan   Montreal   New Delhi   Paris
San Juan   São Paulo   Singapore
Sydney   Tokyo   Toronto

Library of Congress Cataloging-in-Publication Data

Kaminetzky, Dov.
    Design and construction failures: lessons from forensic investigations
    / Dov Kaminetzky.
        p.   cm.
    Includes index.
    ISBN 0-07-033565-6
    1. Structural failures.   2. Structural design.   I. Title.
    TA656.K36   1991
    624.1'7—dc20   90-46691

    2 3 4 5 6 7 8 9 0   DOC/DOC   9 6 5 4 3 2 1

ISBN   0-07-033565-6

*The sponsoring editor for this book was Joel Stein, the editing supervisor
was Stephen M. Smith, the designer was Naomi Auerbach, and the
production supervisor was Suzanne W. Babeuf. It was set in Century
Schoolbook by Progressive Typographers.*

Printed and bound by R. R. Donnelley & Sons Company.

Contents

Preface

This book is a compilation of more than 40 years' personal experience in the investigation and analysis of distressed and failed buildings and structures.

What distinguishes the case histories of failures included in this book, many with disasterous results, is that they are derived from the first-hand experiences of an engineer who followed the failure or collapse — from the first agitated phone call, through detailed on-site investigations and repair of the structure, and often to a dramatic conclusion in a court of law. These are not stories collected from the scrapbooks of others.

This book is directed principally to my peers in the engineering profession, and to architects and contractors. This triumverate shares the responsibilities for design and construction integrity — and the burdens for failures as well. Owners, inspectors, insurers, administrators, and attorneys will also find much of interest and value in this book. It is my earnest hope that educators and students will use this book as a learning tool, thereby minimizing the incidence of failures.

The numerous photographs and detail drawings that support the descriptive text are intended to contribute to an easy understanding of the technical discussions. The "lessons" that conclude each case history serve as practical summaries which should be used to avoid failures.

This book benefits from careful review by my professional colleagues, and I owe each my sincere appreciation. Chief among them are my partners, Harold R. Cohen, P.E., and Martin Fradua, P.E. I am especially indebted to Kenneth M. Cushman, Esq., of the law firm of Pepper, Hamilton & Scheetz, Philadelphia, and Robert A. Rubin, Esq., of the office of Postner & Rubin, New York City, for reviewing the sections touching on law.

However, I take full responsibility for the accuracy and integrity of this work. In every instance I have striven to be objective. Criticism, when offered, has been intended to be constructive. If I have stepped on anyone's toes, it was inadvertent. To the best of my ability, I have presented the facts as I obtained them personally or as they were given to me by parties I considered above reproach.

I must single out for praise the late Jacob Feld, Ph.D., whom I was privileged to be closely associated with for 23 years. Dr. Feld's book, *Construction Failure,* published in 1968, remains a landmark work in the field of forensic engineering.

Credit must go to my entire office staff, without whose contributions this book could not have been completed; and special thanks to my secretary Charlene Stanley, and to Sherry La Fond, who typed the entire manuscript. I am also indebted to Stewart Halsey Ross for his important contributions as editor, coordinator, and motivator. He worked closely at my side during the final phases of writing this book and his help was invaluable. I would like to express my appreciation, too, to Joel Stein, senior editor for civil engineering for McGraw-Hill, for his patience and guidance, and to the entire McGraw-Hill production staff, who worked diligently to design and assemble this book.

My special gratitude goes to my wife, Harriette, for her devoted encouragement over the many months involved in putting this book together.

Dov Kaminetzky, M.S., P.E.

Design and Construction Failures

Failures

1.1 General

> Large scale structural failure is a nightmare that haunts the construction industry. The financial devastation, the demolished reputations and the loss of life that could result from a collapse have troubled the sleep of probably every architect, engineer, contractor or owner at some time.
>
> This frightening quality of failures almost guarantees that they will continue to happen. Fear, embarrassment and the gag of interminable lawsuits have kept information on failures from traveling quickly enough, what little of it ever gets into general circulation at all.
>
> The way to dispel a nightmare is to attack it with hard fact, with eyes open wide and the mind alert . . . availability of complete and accurate information could be the first step towards shaking the dread of collapse.[1]

These remarks underscore the problems our firm has confronted during more than five decades of investigating both major and minor construction failures.

The question that has preoccupied me, personally, for over 35 years of daily intense involvement in failure investigations is the "why" of repeated construction failures. The striking fact is that construction failures do not discriminate. We see failures result from the efforts of engineering firms of the highest caliber and reputation. Some of these firms have, in fact, investigated failures by others, only to find themselves involved in major failures later on.

Often, failure investigators reach erroneous conclusions as to the cause of failure. Only later do they discover that they have committed gross errors.

There is no way to understand failures without full comprehension of *failures in general,* and there is no way to understand construction failures without understanding *construction.*

1.1.1 Failure

Failure is a human act and is defined as: omission of occurrence or performance; lack of success; nonperformance; insufficiency; loss of strength; and cessation of proper functioning or performance.

The four essential elements of a construction project are:

1. Concept
2. Design
3. Performance
4. Use

For a *successful* project these elements must include:

1. Knowledge (training and education)
2. Competence (experience)
3. Care (control)

If any of these three is lacking, no compensating measure of excellence in the other two will prevent a failure of the facility.

President Herbert Hoover, who was an engineer before he became a politician, said:

> The great liability of the engineer . . . compared to men of other professions . . . is that his works are out in the open where all can see them.
>
> His acts . . . step by step . . . are in hard substance.
>
> He cannot bury his mistakes in the grave like the doctors.
>
> He cannot argue them into thin air . . . or blame the judge . . . like the lawyers.
>
> He cannot, like the architect, cover his failures with trees and vines.
>
> He cannot, like the politicians, screen his shortcomings by blaming his opponents . . . and hope the people will forget. The engineer simply cannot deny he did it.
>
> If his works do not work . . . he is damned.

Failures occur far more frequently than we realize or admit. Only a few have significant consequences. We miss the train; the steak is turned to charcoal; the bill is not paid on time; the canary flies out the window; the ball misses the basket; the lawyer loses the case; the patient dies; the building collapses.

Of these failures, only the last three or four might get into the newspapers, but the last one will surely receive dramatic attention from the media.

Everybody wants to *prevent* once and for all the occurrence of failures. The simple fact still remains, however, that *there is no way* failures may be entirely eliminated.

After every major construction collapse there is a public outcry about construction failures, and there is a rush to "eliminate construction failures and uproot this evil."

During August 1982, Representative Albert Gore, Jr., of Tennessee, chairman of the House Subcommittee on Investigations and Oversight of the House Committee on Science and Technology, conducted hearings with "industry experts" to study the serious problem of structural failures. This subcommittee was to determine (1) the underlying causes of structural failures and (2) what part, if any, shortcomings in technology or engineering training played in these failures. At the hearings, Mr. Gore made the following statement:

> We are not here to investigate the cause of any particular failure or to place blame on any part of the construction industry. Rather, our purpose [is] to obtain a better understanding of the actions that either take place, or fail to take place during the course of a building project which eventually results in failure. If we can do that, the industry may be able to determine measures which would help reduce the number of these occurrences.[2]

The Subcommittee issued a final report identifying *the six significant factors* that it found to be critical in preventing structural failures:

1. Communications and organization in the construction industry
2. Inspection of construction by the structural engineer
3. General quality of design
4. Structural connection design details and shop drawings
5. Selection of architects and engineers
6. Timely dissemination of technical data

Yet, this prestigious subcommittee failed to identify the one most critical element necessary to reduce construction failures. As a matter of fact, for a committee charged with seeking to determine *shortcomings in engineering training,* its report failed to mention engineering training and education among its "six critical factors," "five moderately significant factors," and even the "eleven least significant factors."

Subsequently, in November 1983, an engineering conference was convened at Santa Barbara, California, to discuss building structural failures, their causes and prevention. We have studied the Santa Barbara Science Foundation's conference and its famous "five recommendations":[3]

1. Use of ductility, continuity, and redundancy in design to improve *structural integrity* of buildings.
2. Certification by the design professional that all safety requirements

have been met before a certificate of occupancy is issued. The designer's responsibility should include assuring the *structural safety* of the design, details, checking shop drawings, and field supervision.

3. *Peer review* by an independent professional for all buildings above a certain threshold.

4. Development of a document which sets forth the *definition and assignment of responsibilities.*

5. Development by the insurance industry of a *unified risk insurance* policy covering all members of the construction team under a single umbrella.

Unfortunately, these proposals, which will help somewhat to reduce the occurrence of failure, were only peripheral. Surprisingly, this conference showed a lack of understanding of the causes of construction failures, and therefore was itself a failure by *not identifying the one most important factor which could reduce the level of failures — education and training.*

In 1986, Congress appropriated funds to the National Bureau of Standards (NBS; renamed the National Institute of Standards and Technology in 1988) to investigate construction failures. This latest "governmental investment" has made little progress in eliminating failures or reducing their number. NBS has not been a factor in influencing changes in building codes, construction methods or controls, and has not in other ways improved the construction process with its failure investigations.

We also witnessed the allocation of public funds (a National Science Foundation grant) to the University of Maryland for the Architecture and Engineering Performance Information Center (AEPIC), which is intended to reduce the incidence of failures by making data on past construction failures available to the public (for a fee).

Other than serving a few students and attorneys involved in failure litigation, this latest data bank will have little impact in helping reduce construction failures. Design engineers, construction engineers, inspectors, and contractors cannot be expected to flock to the AEPIC library to study how to properly design and build successful structures.

Also highly questionable are the sources of information available to AEPIC. Since there are often various opposing opinions as to the cause or causes of failures, a "dry" report cannot be objective, and we all know how inaccurate media reporting sometimes can be.

In 1985, a roundtable of 40 engineers, contractors, lawyers, and educators met in Clearwater Beach, Florida, to discuss construction failures. The participants issued recommendations which they "thought would help."

They called on the American Society of Civil Engineers (ASCE) to develop procedures for cooperative failure investigations and guidelines spelling out how civil engineers should take the lead in quality assurance and quality control.

This workshop, which was yet another effort to spur action to reduce structural failures, once more failed to come up with the obvious, most important recommendation, which is improved education and training.

Here again we have a failure of the failure experts, and their own lack of knowledge. Their carelessness led them, too, down the path to certain failure.

Engineers and constructors in general do not have a monopoly on failures. Parents and teachers fail in education. Couples fail in marriages. As individuals, we fail to arrive at our offices on time. We overdraw our bank accounts. Failure is very definitely a human trait. We erroneously refer to material failure, machine failure, weld failure, and bolt failure, which, in fact, are all human failures.

Materials do not fail. They follow the laws of nature and physics perfectly. Humans who produced the materials failed to manufacture them properly. The engineer who selected the wrong material failed. The welder who produced low-quality welds failed. The producer of the steel used to manufacture the bolt, the engineer who designed the bolt, the designer who selected the bolt — all in combination or each individually may have failed. *All failures are caused by human errors.*

The headlines that frequently appear in *Engineering News Record* (*ENR*) make statements like "Bad beams shut five schools," "Weight suspect in collapse," "Methane blamed for tunnel blast," "Poor materials, roof-top pool blamed for collapse," "Poor Weld Caused Collapse," "Radio Tower Kills Youth," and many others (Fig. 1.1).

"Acts of God" are very rare, indeed. Once we recognize this reality, we will be on our way to a reduction in the number — but not the elimination — of construction failures.

Errors are unintended deviations from correct and acceptable practice, and thus are avoidable. There are three basic types of human errors:

1. Errors of knowledge (ignorance)
2. Errors of performance (carelessness and negligence)
3. Errors of intent (greed)

Ignorance is often the result of insufficient education, training, and experience. Sometimes ignorance is evidenced when we, without sufficient basis, extrapolate from limited experience and knowledge into a wide and large field. Lack of communication is another form of ignorance. When we design "blind work," there is no reliable way of confirming the adequacy and quality of construction, and therefore we have built ignorance into our structure. Typical examples are filling concealed cavities with concrete

Figure 1.1 Blame is typically misplaced. Failures are attributed to materials and connections, rather than to the people who designed and built the structure.

(drilled caissons and pipe piles), and confirming the adequacy of rock surfaces for support under water.

Ignorance is also evidenced when one uses bold new designs on large-scale projects without thorough preparation, study, mock-ups, and testing.

Carelessness and negligence include errors in calculations and detailing, incorrect reading of drawings and specifications, and defective construction and workmanship. These are errors of execution, and are the result of lack of care.

Greed, on the other hand, is an error of intent which is done with full knowledge. The use of worn-out equipment and tools to reduce costs, taking unwarranted risks and shortcuts, substituting deficient materials and accepting work of poor or marginal quality to accelerate construction schedules all fit in this category.

T. H. McKaig, a failure investigator himself, recognized all these facts and determined as early as 1962 that construction failures are the result of ignorance, carelessness, negligence, and greed. During the last 35 years, I have tried unsuccessfully to find even *one* failure from a cause other than these four. There are none! Additionally, what emerges is that we can do something about only one of these — ignorance.

Ignorance may be attacked by *education and training*. The engineer's own experience is clearly a very important educational tool. However, this must be supplemented by the experiences of others. There is the famous old saying: "A wise man learns from the mistakes of others. Nobody lives long enough to make them all himself."

While education may have some secondary effect on the other three (carelessness, negligence, and greed), it is my belief, however, that we can do very little in these areas. I will go even to the extent of making the following statement: *Education and training are the only effective ways to minimize failures.*

All other solutions which have been advocated by others are only peripheral. They will help somewhat, but will not attack the core of the problem, which is a low level of education of our engineers and architects and inadequate training of our labor forces, including job superintendents and crew supervisors.

Learning from past experiences and the ability to identify potential problems may be achieved by *bringing this knowledge to engineering students during their college education, not waiting for them to search for it in libraries when they are professionals.*

The training of an engineer, architect, or contractor should provide an understanding not only of the best solutions that may be adopted, but also of practices that should be avoided. The study of construction failures and their causes ought to be part of the curriculum of our schools alongside the study of normal construction. This knowledge combined with care (quality control) will greatly reduce the number of failures that presently confront us.

The development and use of new materials and construction practices should, of course, be encouraged, but their potential problems must be recognized. Their use on a large scale should be avoided until a good track record has been established.

1.1.2 Errors, mistakes, and blunders

Errors. Deviation from the true value, lack of precision, variation in measurement because of a lack of human and mechanical perfection — this is the realm of errors. Errors are divided into two classes: *accidental* and *systematic*. The latter are easier to deal with and will be discussed first.

Systematic errors are errors which are always of approximately the same magnitude. A form of systematic error is using instruments with inaccurate calibrations. A steel rule may be too short when it is cold and too long when it is hot. By observing the temperature of the rule at each reading the systematic error can be eliminated. This is actually done with steel tapes when accurate measurements are required.

Accidental errors, on the other hand, will be *distributed at random* in accordance with the laws of probability.

The method of "least squares" is frequently employed to determine the most probable value of a quantity from a set of physical measurements after the systematic errors have been removed by observations.

Accidental or chance errors are a result of combinations of random influences which may be changed from observation to observation. It should be noted, however, that to misread a scale or instrument is not an error. It is a *blunder.*

Errors are always with us; we can reduce their effect by following scientific rules of observation; we can anticipate and make allowance for them; *but we can never eliminate them!*

Mistakes.   Mistakes result from *lack of judgment,* caused by a misconception or misapprehension — that is, by conceiving or understanding wrongly. Lack of judgment may be divided into two categories: mistakes due to acceptance of wrong data and mistakes due to lack of experience.

If data were lacking as to the probable load that a beam would have to support, it would be impossible to design the beam properly. To attempt such a design would be a mistake (faulty judgment), since all of the data are not known.

To *assume* we know what we do not, often leads to failure and is a mistake. To use an untried material, for instance, for our construction program when we do not know all its properties is a mistake. Similarly, to substitute *intuition* when we lack experience is a mistake. Experience has been defined as "all that is perceived, understood, and remembered." It must be remembered that intuition works only part of the time even for those of genius, and should be relied on only when experience is impossible to obtain.

Blunders.   Blunders are the result of *lack of care.* A blunder may be the result of a gross disregard of the facts, proceeding with information so sketchy as to foredoom the project, or lack of experience with no attempt to rectify this condition.

A blunder is frequently the result of many failings. Checking, taking time to be sure, requesting expert advice, i.e., information and guidance based on experience, can and will minimize blunders.

1.1.3   Fatalities

According to the National Safety Council (NSC), an average of about 2500 persons have been killed and 218,000 injured yearly in construction accidents in the 10-year period 1977 – 1987 in the United States. Figures by the

Department of Labor's Bureau of Labor Statistics differ slightly, but nevertheless show a large number of construction fatalities. Only about 10 percent are related to construction collapses. The balance is attributed to accidents during construction resulting from falls from heights; falling objects; failures or overturning of cranes, hoists, or other equipment; and trench excavation collapses. Falling from ladders, platforms, and floor edges is especially critical, where guardrails and toeboards are inadequate or nonexisting (see checklist in Chap. 10). Sewer excavations are also constant offenders, especially shallow, unbraced trenches.

Accident Facts, data published by the NSC, also shows that the probability of accidental death in construction is four times greater than the average for all industries. This statistic clearly points out that construction is hazardous work.

One of the most effective methods of minimizing these accidents is *advance analysis.* This method was developed from a study of reports of thousands of accidents and phases of construction operations during which they occurred. From this study a series of steps for determining accident-producing conditions and their location was produced, taking into consideration the necessary safe flow of operations. The analysis consists of two stages:

1. Study of the job during the planning phase to discover accident-producing conditions

2. Execution of the needed corrective measures at locations identified by the study

Research of hundreds of accidents has produced four distinct accident-prone situations:

1. When temporary structures fail — forms, scaffolds, ramps, ladders, cofferdams, and sheeted and unsheeted cuts, for example.

2. When workers and material or workers and equipment come into contact. In particular, accidents are likely to occur where movement and dynamics are present: spaces congested with workers, areas where workers and equipment cross paths or where high-speed free-swinging loads move, material storage areas, and truck routing lanes.

3. When operations involve inherent engineering hazards — explosives, injurious gases, toxic fumes, or dusts. In operations involving explosives, we see cases of premature explosion or misfire. In confined areas without adequate ventilation in construction activities such as tunneling, pulmonary injuries may result from the use of gasoline-powered equipment. Similarly, during rock-drilling operations, excessive accumulation of dust may become dangerous to the health of workers.

4. When individuals follow unsafe practices. Using impact tools improp-

erly, leaving protruding nails, and working without hard hats are just a few examples. Education and training can sharply reduce accidents in this area.

Safety engineers at the Liberty Mutual Insurance Company determined many years ago, after intensive studies, that over 60 percent of construction accidents are controllable. They therefore proposed an eight-point plan designed to formulate a safety program. They recommended:

1. Sustained management support and direction
2. Adequate safety organization
3. Active participation by superintendents and crew supervisors
4. Effective employee education
5. Control of accident hazards by:
    a. Evaluation of construction jobs
    b. Establishment of safe methods and good practices
    c. Safeguarding equipment and machines
    d. Control of exposure to occupational disease
    e. Provision of personal protective devices
6. On-the-job medical programs
7. Inspection on construction jobs
8. Investigation of accidents

There is an urgent need to develop an effective program to reduce the occurrence of accidents on construction jobs. This cannot be accomplished by leaving the problem to chance. *It must be adopted by law at all construction sites.* Because contractors have the responsibility for methods and means of construction, they should be required to develop such safety programs.

1.1.4 Life of structures

During humanity's long history, many structures have survived thousands of years. The Egyptian pyramids and the Roman aqueducts are monuments that show that structures may last for millennia, provided that they are properly designed and constructed.

The United States is a young country. Most of the structures here are comparatively new, and have been built since the turn of this century. The practice has been to build and demolish and rebuild again in a continuous cycle.

We are not used to asking: "What is the life span of the structure?", nor do we construct our buildings, roads, or bridges for a predetermined life span.

One of the reasons why it is difficult to estimate the life expectancy of a structure is that it depends greatly on the maintenance the structure will receive.

The life cost of a building is the total sum of the initial cost (first cost) and the costs of maintenance, operation, and periodic remodeling. To minimize the life cost it is necessary to minimize the maintenance that will have to be performed during the life of the structure, which often runs counter to a reduction of first cost.

Most of our present-day structures are estimated to have an average life span of about 50 years. The exceptions are those structures which experience "premature deaths" — those that fail, collapse, deteriorate prematurely, burn down, or meet with a natural disaster such as a hurricane or an earthquake.

It must be recognized, however, that the concept of design for infinite life is as unrealistic as that of design for zero probability of failure.

1.1.5 Safety

Two principal criteria that must be considered in choosing a failure probability (safety factor) are public safety and economy. *Safety* can be defined as the state of being safe or freedom from risk of injury or danger.

In relation to construction, public safety depends on the control and performance of each component involved in the design, construction, maintenance, and use of a structure. Any deviation or nonconformance of these factors may expose the structure to damage or total collapse. Absolute safety is generally not feasible and is obtainable only when *all* the following conditions are met: the designer prepares flawless plans, the contractor executes them perfectly, the owner maintains the structure continuously so that there is no degradation, the structure is used as it was designed, without overloads, and there is no reduction of strength by structural modification (alteration) or by removal of support by creating new conditions at adjacent construction or demolition.

We cannot design for zero probability of failure. If structural failures have been mostly the result of rare occurrences of low strength or unusually high loads, then it makes a lot of sense to increase the factor of safety for a very little cost. Statistical estimates have shown that often, for an increase of 10 percent in cost, the factor of safety may be increased by 100 percent. However, failures will be affected only to a small degree by altering the factors of safety. Higher factors of safety will reduce the number of failures resulting from the occurrences of extremely high loads or low strengths. Most structural failures in service, however, result from *gross errors or omissions* by persons (designers or constructors) or building codes (present state of the art). Therefore they may be effectively controlled only by proper training and education leading to better communication, inspection, and review and are only insignificantly affected by increase of factors of safety.

Latest statistical studies of failure reports in the United States, Canada, United Kingdom, and Europe[4-13] show that human errors are the only factor to be considered. The same results could have been arrived at simply by studying human behavior.

Studies performed between 1979 and 1985 by various researchers have identified errors as they relate to construction stages, i.e., design, construction, and maintenance (Table 1.1). We must recognize that all these statistical results are based on voluntarily obtained and incomplete data. The sampling is not necessarily random, and the determinations are based on a particular investigator's opinion.

The great variation among these data attests to their overall inaccuracy. Our own experience tells us that there is a near even distribution of errors between the design and construction phases.

An encouraging statistical study conducted by David E. Allen of the National Research Council of Canada has shown that structural failures of completed structures have a rather low probability of causing death when compared to many other activities. Some of his results are listed for comparison, showing the death rate per billion (10^9) for 1 hour of exposure for various activities:

Structural failure (of completed structure)	<1
Fires	3
Failure of temporary structures	35
Accidents during construction	218
Coal mining	400
Automobile travel	1,040
Air travel	2,400
Alpine rock climbing	40,000
Professional boxing	70,000

TABLE 1.1   Incidences of Error by Phase

Reference	Design, %	Construction, %	Maintenance, %	Total, %
Allen[4]	55	49	—	104
Fraczek[5]	55	53	—	108
Yamamoto and Ang[6]	36	43	21	100
Matousek[7,8]	45	49	6	100
CEB 157[9]	50	40	8	98
Melchers et al.[10]	55	24	21	100
Rackwitz and Hillemeier[11]	46	30	23	99
AEPIC[12]	67	33	—	100
Summary				
Range	36–67	24–53	6–23	
Average	51	40	16	

1.1.6 Factor of safety

The concept of factor of safety (often referred to as "factor of ignorance") was introduced more than a century ago.

By definition,

$$\text{Safety factor} = \frac{\text{assumed strength}}{\text{assumed load}}$$

The safety of the structure is thus assured during its useful life as long as this ratio is greater than unity.

When the actual load is greater than actual strength, the result is failure. Such was the case when a warehouse on Duane Street in New York City collapsed after it had been overloaded with boxes of merchandise. The floor was not designed for the heavy load imposed on it (Fig. 1.2).

Figure 1.2 Overloading of residential floor with stacked boxes resulted in the typical V-shaped collapse.

The factor of safety is, in fact, a ratio of two assumptions — assumed strength and assumed load. If either of these assumptions or both prove to be wrong, there may be a failure.

While the factor of safety is merely an assumption, it is fixed and regulated by law, and is set at a level considered high enough to allow for a considerable error in assumption. This factor is set at different levels for different materials. Structural steel design, for instance, requires a factor of 1.65 for tensile stresses; timber design requires a factor which goes up to 4.0; for pullout of metal embedments from concrete, a factor of safety of 4.0 is used.

The choice of a factor of safety is dependent not only on the uncertainties concerning material strengths and quality control of the construction, but also on psychological effects created by such obvious flaws as cracking and deterioration, on a desire to avoid public criticism if failure occurs, and most of all on economic considerations when the cost of extra strength is evaluated. The factor of safety is therefore determined so as to achieve an optimal balance between safety and economy.

Factors of safety should also take into consideration the sensitivity of the particular structural element within the entire structure and also the level of warning which may be expected prior to failure of certain materials. The consequence of the failure of a major column in a structure may be by far more devastating than the failure of a typical floor beam, which is likely to have only a local effect.

Factors of safety must also relate to a particular material and be accurately defined as to the specific kind of failure they apply to, e.g.:

1. Shear failure
2. Bending failure
3. Buckling failure
4. Overturning

The definition of the factor of safety must be specific as to the level of failure which is used, e.g., yield of structural steel or total fracture for steel elements; or the appearance of the first crack in concrete or complete separation for concrete structures.

Similarly, a factor of safety against a failure without adequate warning (such as punching shear of concrete slabs) should be greater than a factor of safety for a structure which is expected to offer ample advance warning (such as wood beams, which usually make splitting noises prior to failure).

It must be recognized, however, that factors of safety do not take human errors into account but only stochastic (involving chance and probability) variability of material strengths and loading.

1.1.7   Loading

The assumed load consists of two basic components:

1. Dead loads
2. Superimposed loads (live loads, wind loads, earthquakes, and the like)

Dead loads can be estimated accurately by computing the weights of the various components. Since we know the unit weights of materials, the weight of the structure itself and all the permanent components which it supports can be calculated.

The so-called superimposed loads are temporary loads which are more difficult to assess. Live loads, for example, involve weights such as furniture and its contents, storage, people, vehicles, and equipment.

1.1.8   Strength

The assumed strength is computed from available material strength data, test reports, or certification by producers.

The geometry of the cross sections and the frame coordinates of the structure have to be precisely established. These may be designed and assumed for a new structure and measured for an existing facility.

We should not lose sight of the fact that the strength of the structure changes with time. Practically, there is a reduction of strength of all materials over time. The only exception is concrete, which actually gains strength with age, to a certain degree, but later loses strength from long-term exposure to the environment.

1.1.9   Load tests*

Whenever there is a question as to the safety of a structure, there are several avenues open to the engineer to evaluate its strength and adequacy.

One of the obvious methods is performing an analysis based on the assumed geometry and properties of the components of the structure. Sometimes we resort to small-scale model tests.

If there are still unanswered questions remaining regarding the safety of the structure, the local department of buildings or, at the department's option, the engineer in charge may order a full-scale, in-place load test. This load test is performed for the total design load multiplied by a load factor. For reinforced concrete structures the total test load is $(1.4DL + 1.7LL)0.85$, where DL is dead load and LL is live load. The exact testing procedures are described in Sec. 20 of the American Concrete Institute (ACI) code 318-89. This code gives a detailed method of testing, load increments, and acceptability criteria.

---

* Dov Kaminetzky, "Verification of Structural Adequacy," American Concrete Institute (ACI), Detroit, MI, 1985, "Rehabilitation, Renovation, and Preservation of Concrete and Masonry Structures," SP-85-6.

Figure 1.3 Full-scale, in-place load test, with water pool, to confirm the structural adequacy of a precast-concrete garage deck.

The load itself is often applied by water contained in special pools (Fig. 1.3) or by means of lead or steel weights, or sometimes concrete blocks.

The load is applied usually in four increments and the corresponding deflections are measured with extensometers, often attached to long rods (Fig. 1.4). It is important that readings of deformations be taken continuously and load-versus-deformation charts be plotted (Figs. 1.5 and 1.6). If at any time it is found that the deformations are not directly proportional

Figure 1.4 Extensometers attached to vertical rods measure slab deflections with great accuracy during full-scale load tests.

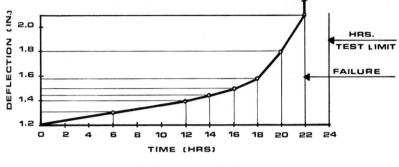

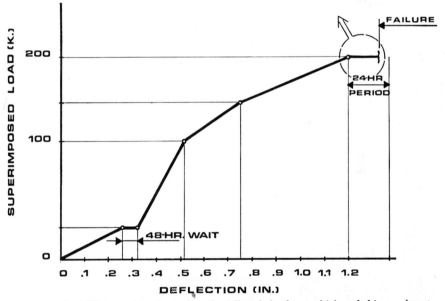

Figure 1.5 Load/deformation charts describe full-scale load test which ended in an abrupt failure 22 hours after start.

to the applied loads, failure has already occurred. In such a case, the load test should be immediately terminated.

Load tests must be conducted only by engineers experienced in such an activity. Personnel not directly involved in the testing should be excluded from the test site.

Of great importance are the safety precautions that must be taken to avoid casualties or damage during load testing. Such precautions include constructing temporary shoring towers several inches below the tested structure. In the event of failure, the failed structure will come to rest on

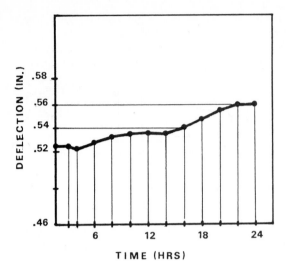

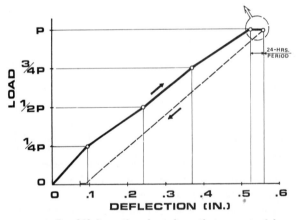

Figure 1.6   Load/deformation chart shows that concrete slab successfully passed full-scale load test.

the top of shores. This precaution was not taken by a testing laboratory in Dearborn, Michigan on March 7, 1963, when two men were killed during a load test. The *Detroit News* headline read: "Two Testing Experts Dead in Dearborn Motel Collapse." The news story continued:[14]

> Two experienced testing experts were killed yesterday while conducting a weight test on concrete planks being used in the construction of a motel . . . .
>
> The technicians were checking stress gauges on a pole attached between the floor and the roof.

. . . The owner of the motel said the tests had been ordered after cracks were discovered in the precast planks . . . .

. . . [The] head of Dearborn's building and safety engineering department said that his department agreed to the test only if a qualified company, such as the firm hired, was used. . . .

Unfortunately, this "qualified" firm did not insist on protective shoring below the tested roof.

1.1.10   History

Humans have been concerned with safety since early in their history. The earliest building code that we are aware of was the code of Hammurabi, the King of Babylonia, which was written about 2200 B.C., nearly 4000 years ago. This code was very stringent in respect to shoddy construction which resulted in loss of human life or property loss. The code went to the extent of demanding the life of the builder of a collapsed house if the owner of the house had been killed during the collapse (Figs. 1.7 and 1.8):

> If a builder build a house for a man and do not make its construction firm and the house which he has built collapse and cause the death of the owner of the house — that builder shall be put to death.

Collapses of structures have been reported throughout history, are occurring in our lifetime, and, no doubt, will continue to occur forever, as long as human beings build.

My three favorite biblical failures are the collapse of the house in Gaza by Samson, the collapse of the walls of Jericho, and the failure during the construction of the Tower of Babel (Babylonia).

Samson.   It is written in the *Book of Judges,* Chap. 16:

> 27. . . . and all the lords of the Philistines were there; and there were upon the roof about three thousand men and women, that beheld while Samson made sport . . . .
>
> 29. And Samson took fast hold of the two middle pillars upon which the house rested . . . .
>
> 30. And Samson said "Let me die with the Philistines" and he bent with all his might; and the house fell . . . .

The moral of this biblical story is that when a failure occurs, the cause for the failure is not necessarily the one that appears obvious. One must always investigate thoroughly to find the correct cause.

In the particular story of Samson we are all led to believe that Samson's strength applied laterally to the two columns caused them to buckle and fail (Fig. 1.9). Yet verse 27 reveals that 3000 men and women were on the roof at the time of the collapse. This fact opens the case again to the

A. If a builder build a house for a man and do not make its construction firm and the house which he has built collapse and cause the death of the owner of the house — that builder shall be put to death

B. If it cause the death of the son of the owner of the house — they shall put to death a son of that builder.

C. If it cause the death of a slave of the owner of the house — he shall give to the owner of the house a slave of equal value.

D. If it destroy property, he shall restore whatever it destroyed, and because he did not make the house which he built firm and it collapsed, he shall rebuild the house which collapsed at his own expense.

E. If a builder build a house for a man and do not make its construction meet the requirements and a wall fall in, that builder shall strengthen the wall at his own expense.

Translated by R.F. Harper.
"Code of Hammurabi" p.83 – seq.

Jacob Feld 1922.

Figure 1.7 The world's oldest building code, the Code of Hammurabi, promulgated literally an eye for an eye and a son for a son.

Figure 1.8 The Code of Hammurabi as it appears on a limestone relief in the British Museum, London.

question: Was it Samson's strength that caused the collapse, or was it the overloading of the roof?

The Walls of Jericho. This biblical story describes the conquest of the city of Jericho by Joshua. His victory was achieved by the use of horns to blow the walls down. In short, in modern engineering terms, it was a collapse caused by vibrations of the walls.

It is written in the *Book of Joshua,* Chap. 6:

19. Now Jericho was straitly shut up because of the children of Israel: none went out, and none came in . . . .

20. So the people shouted, and the priests blew with the horns. And it came to pass, when the people heard the sound of the horn, that the people shouted with a great shout, and the wall fell down flat . . . .

Figure 1.9 Which was the real cause for the collapse of "the house in Gaza" — Samson pushing apart the two middle pillars or the overloaded roof with 3000 Philistines standing on it?

On the other hand, Dr. Jacob Feld, my late partner and a noted forensic engineer, wrote after his visit to Jericho in Israel, where he examined the ruins, that the foundations of the walls had first been undermined by Joshua's spies. This fact is referred to in *Joshua,* Chap. 2: "And Joshua the son of Nun sent out of Shittim two spies secretly, saying: 'Go view the land, and Jericho'."

In an interview with the *Jerusalem Post,* Dr. Feld said that he had first studied the problem of the walls of Jericho in 1931 when he joined Major Tulloch, an engineer officer in General Allenby's army who stayed to settle in Jericho and excavated the site. Some 50 feet (15 m) down they found shards, which were dated to approximately Joshua's times, and their position definitely showed a human-induced failure to have been the probable cause of the collapse (Fig. 1.10):

> The lower stones, which were the larger ones and must have been the foundation of the city wall, were tipped downwards and outward, as though somebody had deliberately undermined the wall from the outside. He [Dr. Feld] believed that Joshua calculated correctly that by undermining the wall in this manner he could bring it down without a fight. . . .

The Tower of Babel.   The failure which occurred during construction in Babylonia did not involve a physical collapse, but was rather a failure of communications. The various parties engaged in the construction of this tower literally did not speak the same language and could not complete the project (Fig. 1.11).

The story itself is described in the *Book of Genesis,* Chap. 11:

1. And the whole earth was of one language and of one speech.
2. And it came to pass, as they journeyed east, that they found a plain in the land of Shinar; and they dwelt there.
3. And they said one to another: "Come, let us make brick, and burn them thoroughly." [*Author's comment:* Even then, builders recognized the importance of proper burning of brick, which some brick producers today are not aware of.] . . . And they had brick for stone, and slime had they for mortar.
4. And they said: "Come, let us build us a city, and a tower, with its top in heaven . . . ."
6. And the Lord said: ". . .
7. Come, let us go down, and there confound their language, that they may not understand one another's speech." . . .
8. And they left off to build the city.
9. Therefore was the name of it called Babel; because the Lord did there confound the language of all the earth.

Which all goes to prove that lack of communication is a certain path to failure.

Figure 1.10 Did the walls of Jericho crash down because of vibration or foundation undermining by spies?

1.2 Ingredients of Failure

1.2.1 Collapse

When all the built-in resistances in a structure are no longer available, the unfortunate result is a total collapse.

Many statistical surveys have been conducted relating to collapses. With varying degrees of accuracy, they all point out that total collapses

The Tower of Babylon.
from a 15th century French manuscript.

La Tour de Babel.
manuscrit du 15 siècle. France.

מגדל בבל
מתוך כתב־יד צרפתי, סוף המאה ה־15

Figure 1.11   The Tower of Babel never "reached the sky." Construction stopped as a result of poor communication.

occur more frequently during construction than during the useful life of the structure. Ratios of 20 or 30 to 1 are often cited. The 30-to-1 ratio is even more acceptable when the first year of life of the structure is included with the construction period.

However, distress and other nonperformance failure cases occur more frequently after construction and during the use of the structure.

Collapses during construction are often caused by:

1. Formwork failure

2. Inadequate temporary bracing

3. Overloading and impact during construction

1.2.2 Progressive collapse

Progressive collapses are usually very severe since they take the form of swift, "domino effect" failures. Such failures may be avoided by using the *alternate path theory* or may be mitigated by introducing *blocks* by designing separation joints. The three most common occurrences of this type of collapse are:

1. *High-rise concrete flat-plate* structures (during construction or earthquake)

2. *Formwork* for concrete structures

3. High-rise structures constructed with *precast-concrete elements*

Recently, two more types of construction have joined this group: *lift-slab* and *hung-ceiling* systems.

The most well-known failures in these categories are the Commonwealth Avenue collapse in Boston (Fig. 1.12), the Bailey's Crossroads collapse in Virginia (Fig. 1.13), and the Ronan Point explosion-collapse in London (Fig. 1.14).

The Ronan Point failure in May 1968 resulted in the addition of a "Fifth Amendment" to the British Building Regulations of 1970, which made it mandatory in Britain for buildings of five or more stories to be designed for the possibility of progressive collapse. This amendment applies to all structures of more than five stories and is not limited to designs using precast panels. It requires that every building be designed using either of the following alternatives:

1. The designer shall ensure that the notional removal of any of the structural components essential to the stability of the building does not produce the total collapse of the structure and that any resulting "local" damage or collapse be restricted to the stories above and below the one at which the notional removal of the component was made.

2. Structural members shall not collapse if subjected to the combined dead and imposed loads acting simultaneously with a pressure of 5-psi (34.48 kPa) in any direction and any extra loads transmitted from adjacent parts of the structure subjected to this 5-psi pressure.

Figure 1.12 The collapse of this Boston high-rise building during construction was a result of many construction inadequacies.

Figure 1.13 Premature removal of shores during construction led directly to the collapse of this 24-story flat-plate high-rise building in Virginia.

Figure 1.14 A natural-gas explosion triggered the progressive collapse of this precast-concrete building in London, demolishing an entire wing. Three people died in the collapse.

The 5-psi pressure was chosen because the Ronan Point structure failed at a calculated 5.5-psi (37.92 kPa) pressure. At the time the amendment was issued there was much criticism of the 5-psi basis, and claims of a 15 percent increase in construction costs were raised.

The New York City code, the Connecticut building code, the BOCA code and ANSI A-58.1 code are the only ones known to me where progressive collapse is addressed in one form or another.

The New York City code, as amended in 1989, states under "Rules and Regulations Relating to Resistance to Progressive Collapse under Extreme Local Loads," Par. 1, General Considerations:[15]

> Unless all members are structurally connected by joints capable of transferring 100% of the member's working capacity in tension, shear, or compression, as appropriate, without reliance on friction due to gravity loads, the layout and configuration of a building and the interaction between, or strength of, its members shall provide adequate protection against progressive collapse under abnormal load, where progressive collapse is interpreted as structural failure extending vertically over more than three stories, and horizontally over an area more than 1000 sq. ft. (93 m²) or 20 percent of the horizontal area of the building, whichever is less.

The BOCA (Building Officials and Code Administrators) 1981 code, Sec. 901.2, "Progressive Collapse," states:[16]

> Buildings and structural systems shall provide such structural integrity that the hazards associated with the progressive collapse, such as that due to local failure caused by severe overloads or abnormal loads not specifically covered herein, are reduced to a level consistent with good engineering practice.

The 1983 Connecticut building code, Sec. 701.2, includes an identical requirement.

ANSI (American National Standards Institute) A-58.1-1982 indirectly implies design for progressive collapse in Par. 1.3, General Structural Integrity:[17]

> Through accident or misuse, structures capable of supporting safely all conventional design loads may suffer local damage, that is, the loss of load resistance in an element or small portion of the structure. In recognition of this, buildings and structural systems shall possess general structural integrity, which is the quality of being able to sustain local damage with the structure as a whole remaining stable and not being damaged to an extent disproportionate to the original local damage. The most common method of achieving general structural integrity is through an arrangement of the structural elements that gives stability to the entire structural system, combined with the provision of sufficient continuity and energy absorbing capacity (ductility) in the components and connections of the structure to transfer loads from any locally damaged region to adjacent regions capable of resisting these loads without collapse.

The Ronan Point report of the Court of Inquiry stated:[18]

> It is the common aim of structural engineers so to design their structures
> that if one or two component parts or members fail due to any cause, the
> remaining structure shall be able to provide alternative paths to resist the
> loads previously borne by the failed parts. [*Author's comment:* It *should be* the
> common aim; unfortunately *it is not* practiced in actual design.]

The report goes on to say:

> The Comité Européen du Béton produced and published in March 1967 a
> comprehensive code covering the design and construction of systems build-
> ings. This Code draws attention to the danger of progressive collapse in the
> following words: One can hardly overemphasize the absolute necessity of
> effectively joining the various components of the structure together in order
> to obviate any possible tendency for it to behave like a "house of cards." . . .

And finally, after many years of failure to do so, the new 1989 ACI-318-
89 Building Code added Sec. 7.13, Commentary Requirements for *structural integrity* (emphasis mine):[19]

> In the detailing of reinforcement and connections, members of a structure
> shall be effectively tied together to improve integrity of the overall structure.
>   Experience has shown that the overall integrity of a structure can be sub-
> stantially enhanced by minor changes in detailing of reinforcement. It is the
> intent of this section of the Code to improve the redundancy and ductility in
> structures so that in the event of damage to a major supporting element or an
> abnormal loading event, the resulting damage may be confined to a relatively
> small area and the structure will have a better chance to maintain overall
> stability.
>   With damage to a support, top reinforcement which is continuous over the
> support, but not confined by stirrups, will tend to tear out of the concrete and
> will not provide the catenary action needed to bridge the damaged support. By
> making a portion of the bottom reinforcement continuous, catenary action
> can be provided.
>   Requiring continuous top and bottom reinforcement in perimeter or span-
> drel beams provides a continuous tie around the structure. . . .

I personally believe that these rather inexpensive additions to details in
ACI 318-89 will go a long way to reduce the devastating effects of progressive collapses.

It is obvious that the loss of life and damage resulting from disasters such
as the Bailey's Crossroads, Commonwealth Avenue, Cocoa Beach, Ronan
Point, and L'Ambiance Plaza would have been greatly reduced had these
provisions been adhered to.

The first major collapse of a lift-slab structure occurred during construction in Bridgeport, Connecticut, on April 23, 1987. This tragedy ended with

Figure 1.15 This progressive collapse of a lift slab occurred during construction, resulting in the death of 28 workers.

28 workers killed and enormous property loss and involved complicated litigation (Fig. 1.15).

The collapse of a hung ceiling at the Journal Square PATH train station in New Jersey came to national attention when two passengers in the station, standing several hundred feet away from the point of the original failure, were killed by the falling ceiling.

It is interesting to note that most codes are still completely silent on the provisions required to guard against this last type of collapse.

1.2.3 Nonperformance

Since all construction projects go through four basically predetermined phases, failure may be caused during any one of them. If one phase is faulty, no grade of excellence on the part of the other phases will prevent nonperformance or failure of the facility. The phases are:

1. Concept and feasibility
2. Design, details, and specifications — contract documents
3. Performance of the work, actual construction, control, guidance, and supervisory inspection
4. Owner and public use of the completed facility

Each of the four phases must include: knowledge (training and education), competence (experience), and care (control).

Some projects, as they are conceived, are basically unbuildable, and no satisfactory result can be obtained. There are designs and specifications which are produced with inherent flaws that lead to failure and even collapse. We also often see proper designs that are not executed with care, and then a local condition of instability leads to progressive failure and sometimes to total collapse. And there are completed structures that suffer from misuse, overloading, undermining, lack of maintenance, lack of protection against climatological and environmental deterioration, and unexpected structural distress.

During the first phase it is extremely important that the owner convey to the designers the proposed nature of the desired facility or structure.

It is equally important for the designers to know and understand the *exact* use of the structure and the loads to which it will be subjected. It is essential that designers insist on *getting from their clients written data* relating to use, load intensity and type, and local conditions such as maximum winds and flood and earthquake likelihood. If the client does not have these data, the architect and engineer must insist on a search.

Prior to start of design work, the architect and the engineer shall confirm to the client *in writing* their understanding as to the purpose and use of the facility and provide detailed descriptions of the dead and live loads that are intended to be used in the design.

It is also important that the owner be advised on the types of materials that are planned to be used and their limitations.

During the design and detailing phase, absentee direction or complete lack of control may result in incomplete or faulty drawings and specifications. Taking shortcuts in the design office, reuse of nonapplicable specifications, and casual perusal of shop drawings are sure ways to failure.

On the other hand, in the course of construction, contractors must learn that problems must be corrected, not buried, so that unexpected and unexplainable troubles do not surface in later work. Contractors should also not lose sight of the fact that they are particularly vulnerable in the area of temporary structures, i.e., formwork, scaffolds, shoring and reshoring, temporary cofferdams, bracing for excavations, and crane supports.

And last, the user of the completed facility must avoid abuse of the structure. It is no mystery why a structure designed for residential use collapses when it is later actually used for heavy commercial storage.

1.3 Causes of Failure

There are numerous ways in which causes of failure may be classified:

1. Unpredictable
2. Design
3. Detailing and drafting
4. Material
5. Workmanship
6. Inspection

1.3.1 Unpredictable

While causes 2 through 6 are controlled by *humans,* and hence are "predictable," cause 1, "unpredictable," often includes failures resulting from accidents such as explosions and vehicle or aircraft impacts. Sometimes this failure is referred to as an "act of God." Another example of this type of failure is a collapse resulting from a hurricane causing unforeseen and excessive loads. Even here we may have an element of human error, since sometimes the building code requirements are not sufficiently rigorous.

Explosions. Gas explosions are considered to be the most common cause of major accidental damage to a structure. Because gas burns relatively slowly compared to the detonation of high explosives, providing vents in the form of windows, doors, or other openings, or lightly attached curtain walls will help reduce the pressures generated by explosions. Lightweight cladding will be blown off the face of the building, thereby immediately dropping the developing pressures.

When explosions cause only local damage, the risk to the safety of the overall structure is low; when they damage main structural elements such as columns, bearing walls, or girders, however, the risk is high that the entire structure may subsequently collapse.

The famous Ronan Point collapse in East London, England, in 1968 drew attention to the potential hazards of gas explosions in high-rise buildings (Fig. 1.14). This disaster pointed for the first time to the danger of progressive collapse of prefabricated precast systems.

An accurate description of the explosion was given in this account by the Royal Armament Research and Development Establishment, which investigated the explosion:

> At about 5.50 a.m. on 16 of May 1968, Miss Hodge, the tenant of Flat 90 in Ronan Point, went into her kitchen to make an early cup of tea. When she struck a match to light the gas there was an explosion, and this led to the partial collapse of the building and the death of four people.
> . . . the living room and most of the bedroom of Flat 90 had completely disappeared. The kitchen, bathroom and hall were still there, but were damaged. . . .

The first order of business in the investigation was to establish the origin and geometric location of the center of the explosion. This was done by plotting the direction of the blast wave from the dished surfaces of the refrigerator door and the fuse box, bent pipes, and displaced walls. The plot pointed to the center of the kitchen as the origin of the blast. The type of explosive was then established by well-known forensic procedures that recognize the broad differences between two types of explosives: condensed (solids and liquids) and dispersed (gases and airborne dusts).

It was determined without question that a dispersed explosive was involved, since it would have required several pounds of condensed explosive charge to cause the same amount of damage. Also, Miss Hodge would have suffered severe injuries had solid or liquid explosive been the source. By a process of elimination (direct proof was not possible), it was determined that "town gas" was the culprit.

According to the report of the Court of Inquiry:

> So far as the explosion is concerned it will be seen hereafter that it occurred as the result of an unusual and unhappy combination of events unlikely to be repeated in the future, and for which no blame attaches to any of those concerned with the construction of Ronan Point.
> . . . The collapse exposed a weakness in the design. It is a weakness against which it never occurred to the designers of this building that they should guard. . . .

Structures are vulnerable to accidental overloading resulting from detonation of high explosives, but it is hardly realistic to design common, everyday structures for such an eventuality. Structures that may become military or terrorist targets could be designed as "hardened" structures, so that they can effectively resist explosion loading of various intensities.

Vehicle impact. While I am unaware of a vehicle impact having caused the collapse of a major high-rise structure, a great number of low-rise buildings have been damaged or even completely destroyed by aircraft or helicopter accidental impacts. The famous case of the twin-engine B-25 bomber that flew into the 79th floor of New York's Empire State Building in 1945 is an excellent example of localized damage without progressive collapse. Only walls and structures directly hit were damaged.

No doubt, if a large jet transport plane crashed into the side of a modern high-rise structure, built according to present-day standards, damage would be much more substantial, with the possibility of a total collapse.

Rough estimates for the likelihood that a high-rise building will be hit by an aircraft during a 60-year life are between 1 in a million and 1 in 10,000, depending on the size and location of the structure.

Because of the steady increase in size and weight of trucks on today's highways, the risk of substantial damage to structures from truck impacts has greatly increased.

Redundancy provisions which are useful in the event of gas explosions will also be beneficial in vehicle impact cases. However, while gas explosions may occur on any level of the structure, vehicle impact is limited to the lower floors.

1.3.2   Design

Proving a design error is usually the easiest task, since the design data and evidence are all there in the contract documents, i.e., the drawings and specifications. Thus, any error or omission by a designer, whether the architect or the engineer, can be readily established.

In the Cocoa Beach collapse (see Sec. 2.6.3), for example, it was clearly established that punching shear stresses were extremely high. The original structural calculation sheets did not include any punching shear computations, leading every investigator to conclude that a blatant and tragic omission by the design engineer was the most probable cause of the collapse.

Cases of gross design omissions are rare. Figures such as 20 to 50 percent of all failures attributed to design error have often been mentioned. Statistics regarding failures are often inaccurate, since many failures are concealed and not reported.

Often, shop drawings prepared and developed by the contractor are erroneously considered to be the work product of the designer. Sometimes, design drawings do, indeed, lack essential and critical details, which are left for development by the contractor. Contractors eager to quickly proceed with a project often fall into the trap of assuming the role of de facto designers. When a failure occurs in such cases, the finger-pointing starts.

The contractor's responsibility is to develop details in accordance with typical details prepared by the design team. The contractor is also required to engage professional help to design temporary structures required during the course of fulfillment of his contract.

It is *not* the contractor's responsibility, however, to design new connections and elements critical to the performance of the structure, or in any other way become involved with design of portions of the final structure.

One method of reducing the potential for failure is the introduction of deliberate redundancies in the design. By using this method, even a twofold increase in load would not fail a beam fixed at both ends. Thus, if reinforced concrete beams are reinforced top and bottom at their ends, or steel beams built with moment connections at their supports, their load resistances will increase many-fold. Similarly, when floor-framing systems with parallel beams are converted into a system of integrated cross-grids, the redundancies or the multiplicities are greatly increased. As a rule, the greater the multiple redundancies provided, the greater the margin of safety.

Superficially, it would appear that there is a simple solution to reduce the risk of design failure. The factors of safety merely have to be increased so that the risk becomes negligible. Unfortunately, this solution is not always feasible, as our society is by nature unwilling to pay an extravagant price to achieve this goal. We must not forget that risk cannot be entirely eliminated, but only reduced to an acceptable level.

1.3.3 Detailing and drafting

The proper detailing of engineering designs is an essential link in the planning and engineering process. Some of the most devastating collapses in history have been caused by defective connections or detailing. The collapse of the Hyatt Regency walkway in Kansas City was triggered by improper connection of the tie-rods at the double-channel crossbeams. The disastrous results of the progressive collapses of concrete high-rise structures, such as Virginia's Bailey's Crossroads and Boston's Commonwealth Avenue, could have been limited had there been continuous reinforcing rebars within the column periphery.

Well-designed and carefully executed structures are known to have failed as a result of sloppy drafting (see Sec. 2.11.2). As an example, a retaining wall at a Long Island, New York, automobile dealership, though founded on dense glacial gravel and backfilled with good coarse sand, cracked and then tipped during the backfilling operations. The 100-ft-long

(30.48-m) wall was intended to serve as the rear enclosure of a garage. Two-foot-wide (0.61-m) pilasters within the wall were to carry the roof structure. The structure was in a small community, and the specifications stated "Contractor shall obtain all permits, including the filing of plans with the proper authorities." The contractor did, indeed, file the architect's plans with the local authorities, and a permit to construct the wall was issued by the local building inspector. In the official records we found the notation: "Have checked the design of the retaining wall and find same OK and according to general engineering practice."

The steel supplier, the construction superintendent, and the reinforcement-placing foreman were interviewed and all expressed surprise [at the unbelievable error of using ¼-in-diameter (6-mm) rebars instead of 1¼-inch (32-mm) rods], but all disclaimed responsibility because the work had been strictly inspected and approved. We actually found a record of such inspection by both the local building inspector and by the architect's representative.

While drafting has always been within the scope and responsibility of the design professional, the task of assigning responsibility for detailing the structure and its connections has always been in dispute. In Europe, designers are charged with the task of detailing as an integral part of the design. The practice in the United States, on the other hand, is to assign this portion of the work to contractors, so that they are free to develop innovative techniques to reduce cost and speed construction.

However, the preparation of so-called shop drawings by the contractor is often performed by individuals who are not necessarily trained as engineers. Thus, important engineering decisions are sometimes made by nonprofessionals. In the Kansas City Hyatt Regency walkway collapse, the effect of the detailing change in the box girder–hanger rod connection was so far-reaching that it entirely changed the design intent. It is absolutely clear that *without the detailing change, the failure would not have occurred.* This change, in fact, caused the load in the hanger rod to double in magnitude. The dead load by itself was so close to the ultimate failure load that even a very small live load was sufficient to cause the failure. What made matters worse was the lack of redundancy in the system, causing a rapid progressive failure of adjacent hanger rods.

This collapse, which was referred to as "the most devastating structural collapse ever to take place in the United States," brought home the message that nonstandard details must be designed by the structural engineer and not by a detailer/drafter.

Also, because the architect or the engineer of record is charged with the responsibility of checking or reviewing shop drawings, some contractors take the position that these drawings are thus adopted by the professionals. Whenever errors in shop drawings have resulted in failures, the responsibility was in question.

1.3.4 Computers

The advent of computers and their almost universal use today has brought to the construction field new aids and higher accuracy and reliability. However, even with these advances, computers have spawned new types of failures. These are failures caused by overreliance on the almighty super-power of computers, which supposedly can do no wrong.

It is amazing to note that many engineers using computers for analysis forget that computers are not aware of one basic engineering fact which every novice recognizes: that cables cannot resist compression. A cable is simulated in practically all computer programs by a bar element which will resist tensile and compression loads. Thus, when an engineer computer-designed a stadium glass window frame using cables, the entire initial tension was overcome by lateral wind load, causing failure of the window.

Similarly, when investigating a failure of a precast wall panel in Maryland, an investigating engineer, in defense of his theory, "proved" that high erection stresses, as indicated by computer analysis, caused the precast panels to crack. Upon review of this computer analysis, I discovered that:

1. No equilibrium existed between the loads acting on the panel and the imaginary reactions.
2. The computer output deformations (which were never reviewed by this investigator) showed deformations in the order of *5 ft (1.52 m)*, which was obviously a total impossibility.

Thus, engineers must be warned that, while computers are excellent tools, basic checks of stability and equilibrium must be performed by manual computations to guard against errors.

1.3.5 Material

Human-made materials such as concrete, mortar, brick, steel, glass, and plastics may be defective as a result of the manufacturing process. Concrete, in particular, is sensitive to material failure because it contains a great number of ingredients.

Natural materials such as timber or stone are widely used in construction. Sometimes low-strength timber units that include knots and other imperfections, or cracked stone panels, are erroneously incorporated into a structure. These obviously should have been weeded out. In such instances, it was the selection process that failed.

Structural steel is assumed by many to be isotropic. Many catastrophic failures resulting from lamellar tearing have brought home the message that steel indeed is anisotropic; its properties are not identical in all directions. In these cases, the material manufacturer failed to inform the user of the risk in using thick steel plates and also failed to educate the user on how to minimize these risks.

Environmental conditions such as temperature or humidity are sometimes the culprit. Concrete can freeze at low temperatures and crack at high temperatures. Similarly, brick may be insufficiently baked at rather low temperatures and then fail either by repeated freezing-thawing cycles or by excessive expansion.

The recent development of high-strength structural steels [above 50-ksi (345-MPa) yield point] unfortunately brings with it increased brittleness. Also, the fact that both the modulus of elasticity and the fatigue performance of these high-strength steels have not improved poses new challenges to the steel industry.

Similarly, the ability to produce high-strength concrete [in the 18-ksi (124-MPa) range] requires the concrete industry to further study the behavior of this material in tension and evaluate its modulus of elasticity as well.

In general, engineers should be cautioned that high-strength materials often increase the risk of failure by excessive deformation rather than by cracking, fracture, or plastic collapse.

Material failure is either a failure of selection or a failure in the manufacture process. Materials themselves never fail. They follow the laws of nature and physics.

1.3.6 Workmanship

Workmanship is the manual aspect of a craft in which competence is achieved through apprenticeship and the continuing exercise of the craft. Since workmanship is a manual component it cannot be expressed as a written standard.

During the Middle Ages, the worker's skill at a craft was used to control the quality of the building process. The level of building competence was maintained and controlled through membership in the various craft guilds.

In recent years we find ourselves lacking a reliable measure of workmanship. Speaking about masonry workmanship, for instance, we use terms such as *excellent, good, average, poor,* or *sloppy,* which are based on subjective assessments rather than rigid standards. What may be good workmanship in one engineer's opinion may be poor in another's.

We are aware that workmanship is a critical component in the success or failure of certain construction elements. The quality of workmanship of mortar joints of an exterior wall can be the difference between a tight wall and a permeable one. Nevertheless, the acceptability of workmanship may be arbitrary, leading to frequent legal conflicts.

1.3.7 Inspection

Whether execution of a contract conforms with the requirements of the design and its intent may be determined by continual monitoring of the construction process.

There are various degrees of inspection, from *spot inspection, progress inspection, periodic inspection, inspection by a "clerk of the works,"* all the way to *full-time inspection.*

We often find in construction the misconception that good inspection by itself can cure all evils. On the other hand, everyone experienced in construction knows well that there is no way an inspector can fully guard against a greedy contractor who is intent on taking improper shortcuts by omissions or second-rate substitutions.

Inspection can best serve as a deterrent to flagrant contract and code violations. It may also well serve to prevent constant and continuous visible errors, often caused by ignorance rather than by ill intent.

The consequences of the lack of inspection are well demonstrated in the following story from a book by L. Sprague DeCamp.[20] In the year 148 A.D., Roman engineer Nonius Datus, a builder of aqueducts throughout the Roman Empire, returned to visit an aqueduct under construction at Saldae, Algeria. This is what he said:

> I found everybody sad and despondent. They had given up all hopes that the opposite sections of the tunnel would meet, because each section had already been excavated beyond the middle of the mountain. As always happens in these cases, the fault was attributed to me, the engineer, as though I had not taken all precautions to ensure the success of the work.
>
> What could I have done better? For, I began by surveying and taking the levels of the mountain. I drew plans and sections of the whole work, which plans I handed over to Petronius Celer, the Governor of Mauretania; and to take extra precaution, I summoned the contractor and his workmen and began the excavation in their presence with the help of two gangs of experienced veterans; namely, a detachment of marine infantry and a detachment of Alpine troops. What more could I have done?
>
> After four years' absence, expecting every day to hear the good tidings of the water at Saldae, I arrive; the contractor and his assistants had made blunder upon blunder. In each section of the tunnel they had diverged from the straight line, each towards the right, and had I waited a little longer before coming, Saldae would have possessed two tunnels instead of one!

In response to the early engineer's lament about what more he might have done, DeCamp says: "Stay on the job and don't wander off for four years at a stretch."

It is important to emphasize that it is imperative that the level of inspection be intense at the initiation of construction in its various stages, so as to weed out all errors, misunderstandings, and other defective work as early as possible, before costly damage is caused. Manhattan's North River caisson failure (see Chap. 5) is an example where inspection revealed deficiencies in the concrete. Unfortunately, these defects became known too late, after enormous damage had already resulted.

1.4   Modes of Failure

In a broad sense, a failure of a structural element occurs when it ceases to function satisfactorily. Failures may be classified as follows:

1. Elastic buckling
2. Fracture
3. Plastic deformation
4. Creep
5. Wear
6. Corrosion
7. Instability

The three most common forms in which structural failures occur are instability, fracture, and elastic buckling.

1.4.1   Instability

One of the basic laws of physics is the law of equilibrium. Schoolchildren know that for each action there is an equal and opposite reaction, a fact based on Newton's famous laws. There is also the familiar expression that "gravity always works." In short, when we have static equilibrium, we have no motion and no dynamics.

In engineering terms, we prefer to refer to the three equilibrium equations of:

$$\Sigma H = 0$$

$$\Sigma V = 0$$

$$\Sigma M = 0$$

where H are horizontal forces, V are vertical forces, and M are moments.

Whenever boundary conditions are not properly studied and understood, and all forces are not accounted for, we end up with failures. Once equilibrium of forces has been established, we concern ourselves with the strength of the individual members.

Overturning of retaining walls, toppling of high-rise structures, blowing off of roofs in high winds, and uplifting of marine structures during high tides are all examples of a lack of equilibrium.

Retaining walls will be stable if they have sufficient resistance to both rotational overturning and lateral sliding. This may be ensured by providing either the proper geometry (in particular, a wide base) or sufficient deadweight in the wall itself, which may be supplemented by the weight of soil for stability. Tiebacks may also be provided. Sliding may be resisted by frictional forces at the base.

Similarly, high-rise structures which are commonly subjected to high winds must have either sufficient dead load for resistance against overturning or, especially in narrow "sliver" buildings, resistances must be provided in the form of an extremely heavy base or reliable tie-downs to rock.

1.4.2   Fracture

Fracture may start at some point of localized stress concentration, unlike plastic deformation or creep, which involve moving appreciable volumes of material before noticeable structural damage is done. From a small point of initiation, a simple fracture can spread across a part to cause a total failure.

This mode of failure occurs in brittle materials under a steady load and in ductile materials under repeated stress. In the latter, the crack is a result of exhaustion of the ductility of the material — fatigue — and, once started, will spread even under relatively low stress.

Ductile and brittle fractures

Ductile mode.   In ductile failures, small cavities are initially formed by slip; then, as the loading continues, they are joined together and eventually grow to form a crack. This crack spreads with the aid of an intense shear distortion near the crack tip and moves in a zigzag pattern (generally perpendicular to the tensile force) toward the surface.

Brittle mode.   Brittle or cleavage fractures are also initiated by slip that forms small cavities and cracks. These cracks are spread by the load, initiating new cracks in adjacent grains along the cleavage plane. The cracks then join by tearing the grain boundaries. Each crystal tends to fracture on a single plane; however, this plane varies slightly from one crystal to the next in the aggregate. Surfaces of brittle fractures sometimes have distinctive appearances. From the origin of fracture, a characteristic chevron or herringbone pattern is formed which points toward the fracture origin.

See Figs. 1.16 and 1.17 for examples of ductile and brittle failures.

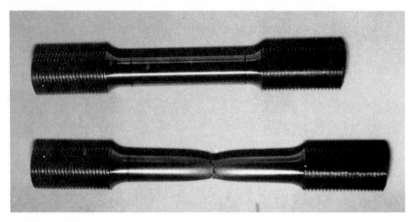

Figure 1.16 Ductile failure with its typical "necking."

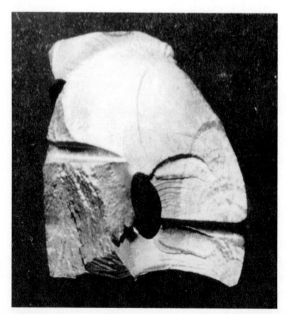

Figure 1.17 Close-up of brittle failure resulting from metal fatigue.

1.4.3 Elastic buckling

This mode of failure, or at least its initiation under stress, takes place within the elastic range of the material. For example, a long, slender column, when compressed lengthwise, will begin to buckle before the material is stressed beyond the elastic range. The critical load at which buckling begins depends on stiffness rather than strength of the material. Elastic buckling must be differentiated from the subsequent buckling that takes place in the plastic range of the material.

Because buckling (sometimes referred to as *elastic instability*) starts at loads much smaller than the compressive strength of the material, it may be evident quite early. Any initial bow within the column, or any other material imperfection, will seriously reduce buckling strength. The addition of lateral loads on a column or any other compression member will also reduce its buckling strength.

Pure compression failure is rare in materials such as concrete or steel. Because buckling is a function of the length-to-width ratio to a higher degree than other factors, it is clear that very long columns are susceptible to buckling failure. Braces in temporary constructions, such as cofferdams and trench excavations, are always the first to fail.

We must not forget that cross-sectional area and the modulus of elasticity are also important factors. While columns have failed in buckling because of material flaws, such failures are rare. More often, compression members fail in buckling because of inadequate cross-sectional area and excessive length rather than low modulus.

1.5 Types of Failures

Failures may be classified as construction, service, or maintenance failures.

1.5.1 Construction

Construction failures occur prior to and during construction. Prior to construction, errors occur in concept and in design. There are many factors leading to design failures. The most common in this group are:

1. Neglecting to account for temperature, shrinkage, and creep effects

2. Failure to ensure adequate stability

3. Failure to ensure adequate support and reaction

4. Neglecting to account for lateral loads

5. Neglecting deflection effects

Statistics show that failures during construction most often occur as a result of the following three causes:

- Formwork failures and collapses
- Inadequate temporary bracing
- Overloading and/or impact during construction

1.5.2 Service

Failures resulting from errors in service are most commonly caused by accidental overloading of floors and roofs and intentional abuse of the structure in many other ways.

Accidental overloading may be a result of very high wind loads (hurricanes), earthquakes in nonearthquake zones, floods, malfunction of safety or relief valves, and impact by airplanes, vehicles, or other equipment.

Abuse or misuse of the structure may be in the form of unusually heavy loads, impact loads, vibration loads, high temperatures, and induced deformations.

Many failures become known during the service life of the structure, but are actually caused during the design or construction period. Distress can be in the form of cracking, spalling, bowing, excessive deformation, or settlement.

1.5.3 Maintenance

Poor maintenance is a failure to properly care for a structure during its life. The two most commonly encountered causes of failure from lack of maintenance are deterioration and corrosion. These two are the sources of strength reduction leading to the eventual failure of many structures. Deterioration is the loss of integrity, strength, and aesthetics of materials. Corrosion is a particular type of degradation and is unique to metals. The effects of corrosion of metals are especially devastating, and billions of dollars are lost in the United States every year as a result of corrosion failures.

Repairs are an integral part of maintenance. It is important to determine the correct causes and mechanisms of failure; otherwise improper repairs may be made because true forces causing the failure have not been properly determined and eliminated. For example, a cracked concrete slab will recrack after epoxy repair of the original crack when the tensile forces that caused it have not been removed (Fig. 1.18).

In the event the forces acting on cracks cannot be eliminated, then use the principle of "Rather than fight them — join them!" Accept them as "active cracks" and repair them with flexible materials (such as sealants) which are capable of movement without failure (Fig. 1.19).

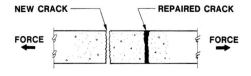

Figure 1.18 The forces causing a crack must be eliminated. Otherwise, the crack will reappear nearby.

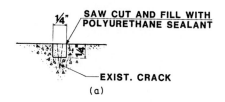

(a)

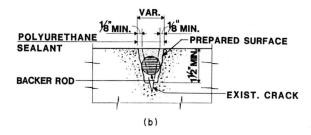

(b)

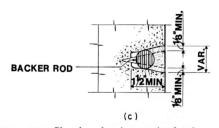

(c)

Figure 1.19 Sketches showing repair of active cracks of various widths. (a) ¹⁄₁₆- to ¼-in (1.6 to 6.4 mm) cracks. (b) Cracks over ¼-in (6.4 mm) on horizontal top surfaces. (c) ¼- to ¾-in (6.4 to 19.1 mm) cracks on vertical and horizontal surfaces. (*Note:* If crack extends through the entire depth of concrete, the other side shall be treated similarly.)

Where both stiff and flexible repair materials are used improperly in tandem (Fig. 1.20), they will also fail.

1.6   Risk

1.6.1   Insurance

There are enormous risks involved with construction activities. All parties to a construction project realize that construction is a risky business and constantly try to limit or mitigate their risks. Regardless of the steps taken

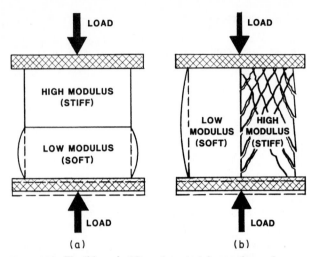

Figure 1.20 Flexible and stiff repair materials must be used properly by taking into account the direction of the forces in action. (*a*) Proper repair, (*b*) improper repair. Modulus = Modulus of Elasticity = Material Stiffness.

in this direction, they must — without exception — purchase construction insurance to cover their activities in the event of losses.

As problems arise, and often these will involve delay claims, or when a serious failure or collapse occurs, the finger-pointing starts. Everybody seems to sue everybody else, and third parties also get involved. To defend against such actions, attorneys are retained, and a circus starts. The cost of litigation skyrockets. Eventually there is a settlement, or the case may go to court, a process that often continues for years.

It was estimated recently that out of every dollar that the construction industry spends on insurance, 50 to 60 percent goes for litigation costs. Therefore only about 40 percent is actually left to pay for damages.

The idea of "wrap-up" insurance keeps coming up as a way to reduce the costs of litigation, so that more money will be left to pay damages in the event of a loss. It is the considered opinion of knowledgeable people in the insurance industry that this kind of insurance has many other advantages. However, this opinion is not shared by all.

The wrap-up concept was first tried on a number of projects in the early 1950s, but for many reasons did not survive. It was used at New York City's 1963–1964 World's Fair and Seattle's World's Fair in 1962, the Verrazzano Narrows Bridge (the nation's largest suspension bridge), New York's Lincoln Center, Alcoa's Century City in Los Angeles, New York Port Authority power projects, and many other substantial projects.

Generally, only worker's compensation insurance and public liability insurance were provided on these projects by the owner. In some cases, the program included other coverages.

Two forms of the program were advocated. Under one, the owner or general contractor obtains and pays for the policies, and one insurance company insures everybody involved with the project. The owner stipulates that all parties should exclude the cost of insurance from their bids or fees.

In the second method, the parties are given the name of one insurance agent from whom to purchase their insurance, so one insurer covers everybody on the project.

In recent years, the same idea, with variations, is being reapproached. This insurance, also known as *package* or *umbrella* insurance, basically involves the use of only one insurance company to provide all of the physical damage, worker's compensation, and liability insurance for every employer on a construction project. Furthermore, insurance is provided for one year after the project is completed to cover the guarantee period of the contract and the risks of design errors and defective workmanship. The program includes also the various bonds required by the construction contract.

The owner or the general contractor buys the insurance package. All the participants in the project receive the required insurance protection from the same company whether they like it or not. By contract they are also required to do so whether or not the insurance company providing the policy acts as the insurer for the various subcontractors. This contract also often includes one indemnity clause, which provides that the contractor will indemnify the owner for injuries to employees.

Those who advocate wrap-up insurance claim that it saves money. It prevents a pileup of premiums resulting from duplication of insurance and from "hold harmless" agreements. Because of centralized marketing, the owner reduces the internal costs of administering many different contractors' policies. It has been recommended that the program be used only on larger projects, say, above $50 million. However, it has been stressed even by those who advocate the program that the key to its success is the requirement that the construction contract and the insurance program be written together.

General-contractor organizations, subcontractors, labor organizations, and insurance brokers are opposed to this type of insurance. Labor argues that such insurance jeopardizes equitable settlement of workers' claims, while contractors are concerned about their relations with their insurance companies.

There were problems with unclear demarcation lines and gray areas between the wrap-up policy and the contractor's other insurance policies. Questions arose as to when coverage under the contractor's policies ceases and the wrap-up program takes over.

In many jurisdictions, contractors do not enjoy legal cutoff from liability after completing their construction operations. Contractors may find themselves forced to pay for a special "tail" policy after the end of the 1-year coverage of the wrap-up. There were also claims that many contractors do not, in fact, allow credit in their bids for the owner's wrap-up program; thus the cost advantage may not be fully realized.

It is my opinion that this type of insurance coverage should be carefully studied again, with modifications and improvements added so as to make it successful. The potential for success exists, but the implementation of the program must be revamped so it will become an effective tool for *all* the participants in a construction venture.

Improvement of the construction process itself will contribute to reduction in the number of conflicts as well as present insurance difficulties.

1.6.2   Liabilities

Black's Law Dictionary contains more than 50 definitions for *liability,* varying from *amenability or responsibility* to *condition of being actually or partially subject to an obligation.* Another dictionary defines the word as *the quality or state of being liable.* In construction life, *liability* became a term that implied responsibility of the various parties for the consequences of their acts.

In recent years, the liability situation has worsened to a point where exorbitant claims have forced some insurers to distance themselves from the construction field. At the same time, insurance premiums skyrocketed.

Over the years, there have been many efforts by architects and contractors to work out agreements on an equitable division of responsibility among the various elements of the industry. For years, the discussions have been fruitless, leaving the liability issue unresolved. Recently, in 1988, the ASCE issued a proposed manual under the title *Quality in the Constructed Project.*[21] This guideline attempts to clarify and define the roles and responsibilities of the various participants in a construction project. Unfortunately, all indications are that a very complicated and cumbersome document will emerge, which will do very little to reduce the level of construction failures. It is also expected that it will take years to finalize and shape this document.

The legal trend toward fixing liability on third parties started in the 1970s, involving cases of manufactured products. In earlier years, the purchaser of a defective or low-quality product was able to claim damages from the party who sold it. It was generally understood that the seller and the buyer had a legal relationship with one another, which manifested itself in a written or verbal agreement. This legal agreement was accepted as holding one party responsible to the other. The manufacturer or the designer of the product, on the other hand, was considered a third party and was actually excluded from the relationship.

As early as 1916 this concept was changed as it related to manufactured products, and it later was extended to include those who provided professional services. When later a young boy fell from a terrace of an apartment building, the lawyers acting on behalf of the boy brought action not only against the owner of the building but also against the design architect. The owner of the building had a rental agreement with the tenant, while the architect, who had designed the building some six years earlier, had no legal relationship with the tenant. And it was alleged by the boy's parents that the architect prepared an unsafe design by failing to show on his drawing a proper railing around the terrace which would have prevented the boy's fall. While the architect in this case was let out of the lawsuit because of a legal technicality, the court still held that architects and builders are liable "for their handiwork."

This lawsuit opened a new era of legal responsibility and liability which is still with us today. Thus came the demise of the legal term known as *privity of contract,* according to which architects and engineers had a legal relationship with only the owner who hired them to perform a service.

Another interesting case was reported by Victor O. Schinnerer, founder of a prominent insurance agency, in a talk he gave in September 1961 in New York. He told about an architectural firm in the South that designed and supervised the construction of a hospital. There was nothing wrong with the design, but there was something wrong with the plumbing subcontractor's shop drawings: he failed to install a pressure relief valve that was called for in the architect's plans. Before the architect was notified that the boiler had been installed, the subcontractor ran a test, the boiler exploded, and a worker was killed. His widow filed suit against many parties, including the architects and the consulting engineer.

She obtained a judgment of $58,700 from the architects. In the words of the lower court judge, the architects were supposed to "snoop, pry and prod," and, if they had done so in this case, they would have discovered the omission of the safety valve. The Circuit Court of Appeals not only affirmed the decision of the lower court, but increased the judgment to $83,000, stating:

> In view of the circumstances herein shown, we believe a duty existed on the part of the architect to use reasonable care toward the contractor and his employees whom the architect had every reason to anticipate would be involved in the construction of this particular project. An architect employed to prepare plans for and supervise construction of a building or a facility . . . must exercise reasonable diligence and care under the circumstances to protect against injury to those who may be reasonably foreseen to be imperiled by defective or improper construction or lack of adequate supervision.

The engineer in this case was released from the suit when it was shown that:

1. The shop drawings which the plumbing contractor used were not approved by the engineer, but only initialed by the architect.

2. The contract between the architect and the engineer provided a reduced fee for the engineer with the implied understanding that he would not have to supervise the construction.

An enlightening definition of negligence liability of architects and engineers was given by Professor George M. Bell:[22]

> An Architect-Engineer does not warrant the perfection of his plans nor the safety or durability of the structure any more than a physician or surgeon warrants a cure or a lawyer guarantees the winning of a case. All that is expected is the exercise of ordinary skill and care in the light of the current knowledge in these professions. When an Architect or Engineer possesses the requisite skill and knowledge in a reasonable manner, he has done what the law requires. He is held to that degree of care and skill and that judgment which is common to the profession.

We also often find cases where contractors follow to the letter plans prepared by architects and engineers on behalf of the owner. These contractors execute the project with good workmanship only to find later that the plans were defective.

Most courts hold that the contractors are not liable for damages caused by a defect in the plans and specifications furnished by the owner, if they perform the work without negligence.

It does make sense that responsibility should rest with whoever prepares the plan and presents it to the owner with the implied representation that it is adequate for the purpose. The precise laws and statutes vary from state to state, and advice of legal counsel is always recommended. The nature of the defect in the plans and whether the defect should have been obvious to any responsible contractor often become issues in litigation.

Many years ago Associate Justice Louis D. Brandeis, in a decision by the United States Supreme Court, said of such circumstances:

> If a contractor is bound to build according to plans and specifications prepared by the owner, the contractor will not be responsible for the consequences of defects in the plans and specifications. This responsibility of the owner is not overcome by the usual clauses requiring builders to visit the site, to check the plans and inform themselves of the requirements of the work.

1.7   Ten Basic Rules

Dr. Jacob Feld, my late partner, one of the first engineers to publicly lecture and write about construction failures, recommended the following rules. They represent a distillation of knowledge gained from years of investigation of structural stress and failure.

1. Gravity always works, so if you don't provide permanent support, something will fail.

2. A chain reaction will make a small fault into a large failure, unless you can afford a fail-safe design, where sufficient residual support is available when one component fails. In the competitive private construction industry, such design procedure is beyond consideration.

3. It only requires a small error or oversight — in design, in detail, in material strength, in assembly, or in protective measures — to cause a large failure.

4. Eternal vigilance is necessary to avoid small errors. If there are no capable crew or group leaders on the job and in the design office, then supervision must take over the chore of local control. Inspection service and construction management cannot be relied on as a secure substitute.

5. Just as a ship cannot be run by two captains, a construction job cannot be run by a committee. It must be run by one individual, with full authority to plan, direct, hire, and fire, and full responsibility for production and safety.

6. Craftsmanship is needed on the part of the designer, the vendor, and the construction teams.

7. An unbuildable design is not buildable, and some recent attempts at producing striking architecture are approaching the limit of safe buildability, even with our most sophisticated equipment and techniques.

8. There is no foolproof design, there is no foolproof construction method, without guidance and proper and careful control.

9. The best way to generate a failure on your job is to disregard the lessons to be learned from someone else's failures.

10. A little loving care can cure many ills. A little careful control of a job can avoid many accidents and failures.

References

1. *Engineering News Record,* June 4, 1981.
2. "Structural Failures in Public Facilities," Report by the House Committee on Science and Technology, Report 98-621, March 15, 1984.
3. "Building Structural Failures — Their Causes and Prevention," Santa Barbara Science Foundation, Santa Barbara, Calif., November 1983.
4. D. E. Allen, "Errors in Concrete Structures," *Canadian Journal of Civil Engineering,* vol. 6, no. 3, September 1979, pp. 465–467.
5. J. Fraczek, "ACI Survey of Concrete Structure Errors," *Concrete International,* vol. 2, no. 12, December 1979, pp. 14–20.

6. M. Yamamoto and A. H.-S. Ang, "Reliability Analysis of Braced Excavation," *Structural Research Series No. 497,* Dept.of Civil Engineering, University of Illinois, Urbana, Illinois, 1982.

7. M. Matousek, "Measures Against Errors in the Building Process," Institute of Structural Engineering, Swiss Federal Institute of Technology, Zurich, Switzerland, 1982.

8. M. Matousek, "A System for a Detailed Analysis of Structural Failures," *Structural Safety and Reliability,* Elsevier, Amsterdam, 1981, pp. 535–544.

9. Bruce Ellingwood, "Design and Construction Error Effects on Structural Reliability," *ASCE Journal of Structural Engineering,* vol. 113, no. 2, February 1987.

10. R. E. Melchers, M. J. Baker, and F. Moses, "Evaluation of Experience," *Proceedings,* IABSE Workshop on Quality Assurance Within the Building Process, Rigi, Switzerland, 1983, pp. 9–30.

11. R. Rackwitz and B. Hillemeier, "Planning for Quality," *Proceedings,* IABSE Workshop on Quality Assurance Within the Building Process, Rigi, Switzerland, 1983, pp. 31–52.

12. Ellingwood, op. cit.

13. A. C. Walker, "Study and Analysis of the First 120 Failure Cases. Structural Failures in Buildings," *Proceedings of Institution of Civil Engineers,* 1981.

14. "Two Testing Experts Dead in Dearborn Motel Collapse," *Detroit News,* March 8, 1963.

15. The Building Code of The City of New York, Department of General Services, Municipal Building, New York, September 1, 1988.

16. Building Officials and Code Administrators International, Inc. (BOCA), Tenth Edition, 1987.

17. "Minimum Design Loads for Buildings and Other Structures," ANSI Standard A-58.1, American National Standards Institute, New York.

18. "Report of the Inquiry into the Collapse of Flats at Ronan Point, Canning Town," HMSO, London, 1968.

19. Building Code Requirements for Reinforced Concrete and Commentary, ACI-318-89, American Concrete Institute, Detroit, Michigan, 1989.

20. L. Sprague DeCamp, *The Ancient Engineers,* as cited by DPIC "Communique," October 1979.

21. "Quality in the Constructed Project," Preliminary Edition, American Society of Civil Engineers, New York, 1988.

22. T. G. Roady and W. R. Anderson, *Professional Negligence of Architects and Engineers,* Vanderbilt University Press, Nashville, 1988.

2

Concrete

2.1 Concrete

Concrete is a very versatile construction material, with many advantages but with numerous disadvantages as well. It can be cast-in-place, with or without steel reinforcement, or precast with steel reinforcement, or prestressed for maximum strength.

2.1.1 Cast-in-place concrete

Cast-in-place concrete is the most commonly used form of the material. After the mixing process, the concrete, in its semifluid state, is poured into prepared forms most commonly constructed of wood or steel. While the "paste" is hardening, it cures to attain its design strength. The versatility of forms makes possible the creation of cast-in-place concrete in practically limitless shapes and surface finishes.

2.1.2 Precast concrete

Compared to cast-in-place concrete, which by definition is manufactured *in situ,* precast concrete is made away from the final intended location of the concrete element. Precasting is typically performed in a special, well-equipped casting yard, but sometimes is done at a designated location on the construction site itself.

2.2 Concrete Components

The material's distinguishing characteristic is the many components that are required for its production. As a result, there is an enormous number of variables that occur during its manufacture, and the risk of something going wrong is great. Furthermore, the probability of producing the identical material twice is practically nil.

The following are *just some* of the variables involved:

1. Cement
2. Sand
3. Broken stone
4. Water
5. Admixtures (e.g., water reducers, air-entraining agents)
6. Forms, shores, and reshores
7. Conveying
8. Casting
9. Vibrating
10. Weather protection
11. Curing
12. Finishing
13. Stripping

To this partial list must be added everything related to the reinforcing of concrete with steel elements. To name just a few:

1. Reinforcing steel bars or rebars
2. Wire reinforcing
3. Rebar detailing
4. Rebar strength
5. Rebar shape
6. Rebar length and diameter
7. Rebar placing and spacing
8. Rebar corrosion
9. Rebar coating
10. Rebar anchorage

Finally, in precast, prestressed concrete, there are these additional variables:

1. Prestressing wires
2. Posttensioning tendons

3. Connections

4. Strength of concrete at de-stressing

5. Stressing of wires

6. Camber

7. Transporting and erecting finished members

Thirty variables have been identified. Be assured that there are many more. A failure in any one can be sufficient to cause serious consequences, as the following case histories will attest.

2.3   Failure of Concrete

Concrete failure occurs in two principal areas: in the material itself and in the individual structural elements.

Material failure will often surface during the manufacturing process — during casting or during the hardening of the paste. During casting operations, when some of the basic requirements are ignored, failure ensues. When concrete is not vibrated properly, hollows and honeycombs result (Figs. 2.1 and 2.2).

Failure also occurs many years later during the life of the structure. Deterioration of concrete in use is often caused by freeze-thaw cycles. Whenever air-entraining agents have not been incorporated in the concrete mix, the concrete will deteriorate when exposed to the environment (Fig. 2.3).

Concrete parking decks are particularly subject to freeze-thaw damage when exposed to snow, rain, and low temperatures. The application of deicing salts is especially harmful because it accelerates the deterioration process (Fig. 2.4).

2.4   Defective Ingredients

2.4.1   Cement

An improper amount of cement in the mix has been and continues to be a very frequent problem in the concrete industry. Low cement/sand ratios result in low-strength concrete, since the strength is closely related to the amount of cement in the mix. On the other hand, high cement/sand ratios will quite often contribute to excessive shrinkage.

Cement is also sometimes accidentally mixed with impurities which will result in retardation of the concrete, causing delay in the hardening process coupled with loss of strength of the cured concrete.

Figure 2.1 Vibrating fresh concrete in the form is essential to successful cast-in-place concrete. It is obvious that this requirement was totally ignored in the residential housing wall shown in this photo.

2.4.2 Sand

Sand may contain impurities such as clays, salts, and vegetable matter. Improperly washed beach sand may include high levels of chlorides which are harmful to concrete, especially to metal embedments such as reinforcing bars, pipes, and connections.

When improperly proportioned sand is used (without uniform gradation of aggregate size), the resulting strength of the concrete may be lowered.

2.4.3 Water

Unless water is potable, its impurities will definitely affect the strength, color, and durability of the concrete. Needless to say, high water/cement ratios will produce low-quality concrete.

Figure 2.2 Failure to properly vibrate concrete during casting results in honeycombs in walls and buttresses. Reinforcing bars placed without sufficient cover and clear spacing make matters worse.

2.4.4 Admixtures

Admixtures, although they do have certain advantages, have also been the source of many problems in the manufacturing of concrete. One of the most frequently recurring problems is improper proportioning. Delayed setting of the concrete, accelerated setting of the concrete to the point that it cannot be finished in time, and excessive amounts of air-entraining agents are a few of the problems encountered.

2.5 Failed Elements

Concrete elements such as beams, girders, columns, footings, and piles may fail by fracture, excessive deformation, or long- or short-term deterioration as a result of exposure to the environment, water, chlorides, and other chemical agents.

Figure 2.3   The importance of air entrainment in concrete cannot be overestimated. As shown here, the coping on the right performed perfectly under severe exposure; the one on the left disintegrated.

Figure 2.4   After a man's foot fell through this waffle slab deck, the entire structure was sonically tested for defective concrete. The result is shown here.

Failure by fracture is caused by high stresses. When these stresses reach such a high level that they exceed the strength of the concrete, it will fail.

Concrete often fails in tension because its tensile strength is a small fraction of its compressive strength. The assumption that tensile strength is 10 to 12 percent of the compressive strength is an approximate one. A more accurate relationship is given by the well-known modulus of rupture formula:

$$f_t = 7.5 \sqrt{f'_c}$$

where f_t = 28-day concrete tensile strength and f'_c = 28-day concrete compressive strength.

In reinforced-concrete structures, the accepted assumption of the design concept is that the reinforcing steel rebars resist all the tensile forces. In reality, there is load sharing between the concrete and the steel rebars, and this complex "joint effort" continues until the concrete cracks and fails. After concrete failure, the general assumption is that the steel rebars resist the entire load.

A concrete failure due to deficiency in shear resistance is the most serious type of failure, for most shear failures are preceded by little, if any, advance warning.

The definition of what constitutes failure of concrete is also subject to various interpretations. The most common definition of concrete failure is the appearance of the first crack. This definition is sufficient for laboratory usage, but is inadequate for practical purposes.

As an example, I would like to refer to the punching strength of concrete slabs. Many laboratory tests have been conducted in the past in order to evaluate the failure load of slabs. In particular, these tests attempted to establish the effectiveness of reinforcing bars placed in and going through the direct periphery of columns. The "official" results of these tests, using the so-called first-crack definition, were that the rebars do not increase the punching shear capacity of the slab. Such tests were the basis for the elimination of the ACI code requirement to place a certain percent of the reinforcing steel within the column periphery.

In real life, however, rebars are very effective in preventing a total failure, and especially are highly efficient in preventing a progressive collapse (Fig. 2.5). These facts were confirmed by tests conducted at the University of Washington, Seattle. The latest revision of the ACI code, issued as ACI-318-89 with the structural integrity requirement, is a correct step in this direction (refer to Chap. 1 for additional discussion).

2.6 Slab Failures

The most common and, quite often, most devastating failures are punching shear failures. These are failures which are triggered by a localized failure around a single column at an upper level floor, resulting in a progressive failure going all the way down to the lowest level (Fig. 2.6).

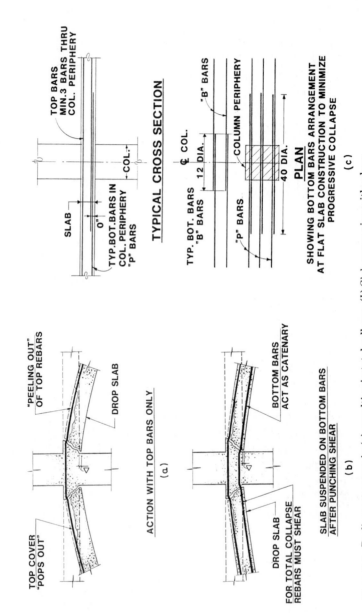

Figure 2.5 Peeling out occurs in (*a*), resulting in total collapse. (*b*) Slab suspension with rebars acting as catenary results where bottom rebars are provided. (*c*) Total flat-plate failure by punching shear is minimized by introduction of rebars going continuously through the column periphery.

Figure 2.6 This devastating domino-effect failure during construction resulted in heavy loss of life.

Such a domino-effect failure can be stopped by localizing it to a single level. Because punching shear is the weakest link in the chain, the use of *larger-size columns* coupled with *continuous reinforcing bars running through the column periphery at both the top and the bottom of the slab* will go a long way toward minimizing the possibility of progressive collapse. The continuous bottom reinforcing will develop a secondary load-carrying system, after the initial punching shear failure of the slab has taken place. Thus, these bottom bars, being well-anchored within the column, form a catenary net, holding the failed slab in place.

The most devastating and well-known collapses resulting from punching shear are those at Bailey's Crossroads, Virginia; Commonwealth Avenue, Boston; and Cocoa Beach, Florida.

2.6.1 Bailey's Crossroads collapse

This catastrophic collapse (Figs. 2.6 and 2.7) occurred at a project known as the Skyline Center under construction at Bailey's Crossroads, Alexandria, Virginia, on March 2, 1973. This was a residential complex of several 30-story (26 typical floors plus four lower floors) cast-in-place concrete structures. The facility also included posttensioned concrete garage parking decks.

The first two of three practically identical reinforced concrete towers had already been completed, and the third one was already at the 24th floor level when an entire wing of the building collapsed. The first assumption was that the collapse was related to a construction crane which toppled and came down with the concrete floor slabs. This assumption was later proved to be wrong.

The flat-plate design called for 8-in (203-mm) slabs spanning up to 27.75 ft (8.46 m) maximum. A 4-ft-thick (1.22-m) foundation mat was used to support the structure, and all indications were that the foundations had no effect whatsoever on the collapse.

There were, however, three columns with relatively high punching shear stress — high stresses, indeed, but still within the stress allowed by the code and by acceptable engineering practices. It was at one of these three columns that the collapse was determined to have started. A detailed finite-element analysis of the stress conditions at the columns was performed by us and confirmed the high but still acceptable punching shear stresses at columns 67, 68, and 84 (Fig. 2.8).

The construction had proceeded at a rate of one floor per week (7 to 8 days). Two climbing cranes were used to haul the concrete up to the placing level. The structural drawings called for reshores at the third floor below the fresh concrete — one shore midway between columns in both directions and one shore at the midpoint of each bay. Photographs taken just before the collapse show little or no reshoring below the levels of the fresh concrete. Records of concrete casting operations also show that the pace of

SHORES

NO SHORES
OR RESHORES

Figure 2.7   Photo shows formwork shores at top level and only one level of reshores on floor below (arrow). Premature removal of shores led to the catastrophic collapse of one of two towers of the Skyline Center project.

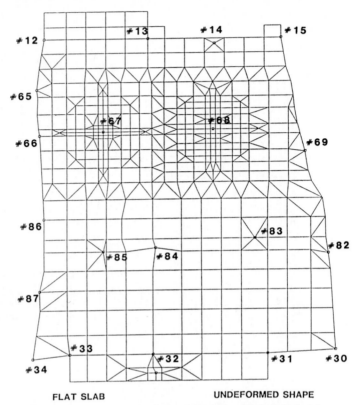

Figure 2.8 Finite-element analysis of the design of the failed slab confirmed high but still acceptable punching shear stresses at columns 67, 68, and 84.

construction accelerated 5 days prior to the collapse; i.e., the combined age of pairs of floors supporting one level of fresh concrete dropped dramatically after February 26.

On the day of the collapse, concrete was being placed on the 24th-floor forms, with placement starting early in the morning and ending at noon. After lunch, removal of the formwork between the 22d and 23d floors started. Shortly afterward, a big sag became apparent in both the 23d and 24th floors. The sag was followed by a total collapse, when floor after floor fell successively, one on top of the other.

In the garage section, the loads and impact of the collapse of the tower spread through the entire structure by successive shear failures, leaving many of the columns free-standing.

Fourteen workers lost their lives and more than 30 were injured as a result of the collapse.

A review of the activities during the day of the collapse revealed that the removal of shoring did not follow the removal of the formwork at the 22d and 23d levels. The reshores between the 21st and 22d levels were reported to have fallen as the 22d floor was being unloaded by the removal of the shores between the 22d and 23d floors.

Concrete cores extracted from the upper slabs indicated concrete strength of between 700 and 1400 psi (5 to 10 MPa), well below expected strength.

We have concluded, as have all other investigators, that the collapse was triggered by punching shear failure at the 23d floor slab because of the premature removal of formwork between the 22d and 23d floors. There were no preconstruction plans of concrete casting, formwork plans, removal of formwork schedules, and reshoring program.

The Bailey's Crossroads collapse has raised the issue of the extent of the engineer of record's responsibility for the success and safety of formwork design and inspection. While the engineer's contract specifically stated that he had no responsibility for field inspection, nevertheless a jury found him negligent. This was so because the pertinent code required that "a competent architect or engineer" must provide supervision "where requested by the building official."

It is my opinion that the engineer of record should perform inspection services in the field or ascertain that such services are being provided by other professionals. Total responsibility for safety of construction operations and methods and means of construction must remain with the contractor who controls these operations.

Lessons

1. The contractor should prepare formwork drawings showing details of formwork, shores, and reshores.

2. The contractor should prepare a detailed concrete testing program, to include cylinder testing, before stripping forms.

3. The engineer of record should ascertain that the contractor has all the pertinent design data (such as live loads, superimposed dead loads, and any other information which is unique to the project).

4. Inspectors and other quality control agencies should verify that items 1 and 2 above are being adhered to.

5. Uncontrolled acceleration of formwork removal may lead to serious consequences.

6. Top and bottom rebars running continuously within the column periphery must be incorporated in the design.

2.6.2 Commonwealth Avenue collapse

On January 25, 1970, a 17-story concrete high-rise cast-in-place building collapsed under construction in Boston, Massachusetts. Four workers lost their lives and 20 were injured as a result of this collapse. An intensive investigation was launched by the various parties to establish the cause of the failure. The mayor of Boston appointed a separate independent investigation committee, which concluded after an in-depth study that "practically every element of the construction process failed in this case." This committee identified the following as "A Catalog of Irregularities and Deficiencies":

1. Lack of proper building permit
2. Insufficient concrete strength
3. Insufficient length of rebars
4. Lack of proper field inspection
5. Various structural design deficiencies
6. Improper formwork
7. Premature removal of formwork
8. Inadequate placement of rebars
9. Lack of construction control

There were also other irregularities listed.

One of the life-saving features of the collapse was the fact that there was early warning of the imminent failure in the form of a two-stage collapse; i.e., after the initial failure, there was a "hesitation" period which enabled some of the workers to evacuate the building. Actually, after the failure at the roof level, a level which fell onto the floor below, approximately 20 minutes elapsed. Then the collapse continued with the rest of the floor slabs failing in succession, stacking one on top of the other on the ground floor level. Only part of the tower remained standing (Fig. 2.9).

Our analysis, which was confirmed by many of the investigators, determined that the slabs failed in punching shear as a result of a combination of premature removal of forms coupled with insufficient concrete strength at the time of the collapse. The maturity of the concrete was affected by the low temperatures prevailing in January.

The records of the concrete coring program showed strengths as low as 700 psi (5 MPa) at the time of the collapse.

The following account tells the story of the tragedy as seen by eyewitnesses.

Phase one: Shear failure in the main roof at column E5 (Fig. 2.10)

Phase two: Collapse of the roof slab. Concrete placement in the mechanical room floor slab, walls, wall beams, and brackets commenced January 25, at about 10 a.m. A coffee break occurred at about 3 p.m. after the concreting had extended about 3 ft (0.91 m) east of line 5. Messrs. N. and O. remained

Figure 2.9 Commonwealth Avenue structure after collapse.

on the pouring level. Just after the coffee break ended at about 3:25 p.m., they felt a drop in the mechanical room floor of about 1 in (25 mm).

Mr. P., the labor foreman, was standing on the south side of the mechanical room floor when he felt the same settlement.

Mr. F., the carpenter foreman, was on the south side of the roof slab. He had also noticed a settlement of the roof slab. After discarding the thought that the formwork for the mechanical floor had failed, he immediately went to the 16th floor by way of a ladder in the east stair opening. After pausing at the stair opening, he proceeded directly to column E5 on the 16th floor. As he stated it: "I can't believe my eyes. I see this slab coming down around the column." He estimated a drop of 5 to 6 in (127 to 152 mm) at this time and described a crack in the bottom of the slab extending from column E5 toward column D8. Mr. F. immediately shouted a warning to the workmen on the 16th floor that the roof was going to collapse, ran back

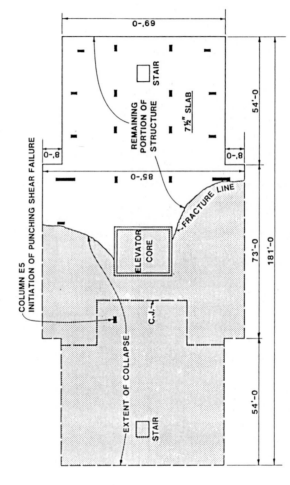

Figure 2.10 Plan of the roof slab showing the collapsed area (shaded) and the location of column E5 where punching shear initiated the progressive collapse.

to the east stair opening, climbed up the ladder, shouted a warning to the men on the top, then climbed down to the 15th floor where he and several others remained on an east balcony until the collapse of the roof occurred.

Mr. N. had called the attention of his foreman, Mr. M., who went up onto the mechanical room floor to observe the sag. The sag must have increased rapidly because Mr. M. got trapped there and later had to be rescued by the crane.

Mr. C. did not notice the sag of the roof slab. He heard Mr. F.'s warning and a second one by Mr. M. After he saw everybody running, he went to the 15th floor via the ladder in the east stair opening. As he stated it: "We stood there a couple of minutes, and then the roof let go, and I saw everything starting to shake, so we jumped from the scaffold onto the balcony. That's when the whole roof collapsed, and then it stopped." Mr. C. then went across the 15th floor to the west stairwell and descended to the ground.

In his testimony, Mr. F. mentioned hearing cracking noises about 1 p.m. while working on column C7. He likened the noises to the sound of ice cracking. As he stated it: "I do a lot of ice-fishing, and we heard the noise, and my boss said to me, 'Holy mackerel, what kind of noise is that?' Approximately 15 to 20 minutes after that, the same noise." Although Mr. F. dismissed these sounds at the time, it is quite possible they were emanating from the roof slab under the *west side* of the mechanical room floor.

Mr. F. was working on the northeast corner of the 16th floor. By the time he reached the stairwell, the sag had increased to 4 ft (1.22 m). After the east half of the roof slab had collapsed, he saw it sloping down to the 16th floor and smelled gas. Fearful of an explosion, he backed down, crossed the 15th floor, and ran down the west stairs.

Mr. D., the steel foreman, was in the office when the crane oiler told him that something had gone wrong. As he rode up, the roof slab collapsed. He saw the compressor falling off the roof along with other debris and dove out at the 8th floor.

While talking with Mr. D., Mr. P. could see that, slowly but surely, the whole bay E5-E6-F6-F5 was sinking in. He could hear creaking and groaning and saw the reinforcing steel popping up out of the mechanical floor slab. Mr. D. returned to the office, but Messrs. M. and D. remained by the materials hoist. Within a minute or so the general collapse occurred.

Phase three: The general collapse. The general collapse is clearly and succinctly described in a written statement by Mr. D. E., a resident of 1959 Commonwealth Ave., which was adjacent to the new structures:

> I stepped outside and noticed the top floors, one or two, had collapsed, and sent debris flying out and down. At the same time, I raised the camera and began to take pictures. I took three while the structure was falling. They were taken in the minimum time it takes to trigger the shutter and advance the film. The collapse progressed in what is best described as domino fashion—

that is, the weight of what had already collapsed seemed to blend into where a floor was, then this floor would fall, still somewhat defined by the balcony rail. Then as that floor reached the next, the debris from the floor above become indefinable. The building did not seem to fall at once, but in a floor by floor progression, nor did it appear to fall away from the core that remained after the collapse, but as each floor fell, it fell completely, in that nothing remained around the central core.

Extent of the collapse.   About 120,000 ft^2 (11,000 m^2) of floor slab collapsed into the basements and lay in layers like pancakes. Consider another dimension of the collapse: about 8000 tons of construction materials fell to the ground.

The remaining structure had to be demolished because it created a safety hazard and was impractical to repair because of the extensive damage.

The similarity between the last two cases discussed is obviously apparent. The *lessons* listed for Bailey's Crossroads are applicable here as well.

It is important to cite from the Mayor's Investigation Committee report:[1]

> *It must be recognized that no set of laws, regulations, or procedures can, by themselves, prevent building failures. As a case in point, the failure at 2000 Commonwealth Avenue WOULD NOT HAVE OCCURRED had the existing laws, regulations and accepted procedures been understood and mutually respected by the several parties involved. We believe that this fact contains the most important lesson to be learned from the tragedy.*
>
> The loss of life in this tragedy is particularly distressing. We wish to express a sense of dismay that the project could have proceeded through so many successive phases without anyone recognizing and making appropriate remonstration against the irregularities and deficiencies which took place. The public interest cannot tolerate a repetition of this series of events.

2.6.3 Cocoa Beach collapse

In spite of the fact that this failure involved a relatively low-rise structure, the consequences of the collapse were catastrophic. The five-story cast-in-place concrete structure collapsed in Cocoa Beach, Florida, on the afternoon of March 27, 1981, leaving 11 workers dead, 23 injured, and a totally collapsed, unsalvageable structure (Fig. 2.11).

The 242-ft-long (74-m) structure was 58 ft (18 m) wide and consisted of 8-in-thick (203-mm) slabs constructed with flying forms, made up of wood decks on aluminum trusses. The typical column size was 10 × 18 in (254 × 457 mm). The almost completed structure was already at the fifth floor (Fig. 2.12) on the morning of March 26. Workers were finishing the concrete at this floor later in the afternoon when they heard a loud crack which sounded like wood splitting. Eyewitnesses reported that the center of the fifth floor went down first, with the balance of the floors following. Some witnesses described the fourth floor as coming down first while the fifth

Figure 2.11 Collapse of a five-story, cast-in-place concrete apartment house in Florida killed 11 and injured 23 workers. Photo taken immediately *after* the collapse.

Figure 2.12 The Cocoa Beach structure was nearly completed before collapse.

was still hanging. Then the fifth floor collapsed and carried the fourth with it onto the third floor.

While most investigators agreed that there was a startling design flaw in the original plans, our own analysis and review concluded that there were many construction inadequacies as well, all of which contributed to the final collapse.

One look at the design documents is sufficient to flag the glaring discrepancy between a slab thickness of 8 in (203 mm) and the long span of 27 ft 8 in (8.43 m). The significance of the use of a small column (10 × 18 in; 254 × 457 mm) for a span of over 27 ft (8.23 m) should have been apparent and should have served as a warning sign.

However, all these flags apparently were not sufficient to alert the constructors of this project to the imminent dangers.

Another fact that was very disturbing was the indication that the design engineer was made aware of severe cracking of the floors and big deflections of the slabs several days prior to the actual collapse. The cracks were described as "spider-web-type cracks" (Fig. 2.13). There were reports of deflections as large as 1¾ in (44 mm) several days prior to the collapse. The structural engineer was requested to recheck his design, which he did, reporting back that it was "OK."

One glaring omission, which became apparent after a review of the design calculations provided by the structural engineer, was the fact that

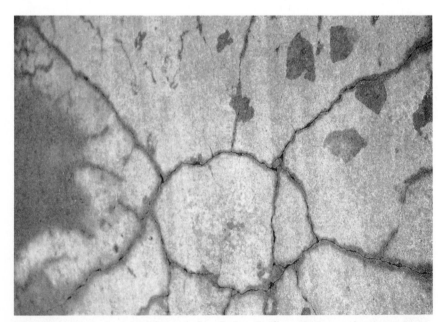

Figure 2.13 Cracks in floor slabs were described as "spider-web-type, and were similar to those shown here."

Figure 2.14 With debris removed, the remains of the columns still stand.

punching shear calculation was not performed at all. Another startling omission was the lack of a minimum slab thickness calculation. This simple check shows that an 11-in (279-mm) slab thickness will be required with the given spans and column size. It is quite apparent to the experienced design engineer that increasing the size of the columns would have been a simpler solution. This would also have resulted in a more economical design, besides providing the needed space for casting concrete between the vertical column bars.

I visited the site of the collapse some time after the debris had been removed. The remains of the columns were still standing (Fig. 2.14), with piles of slabs stacked one on top of the other. Measurements of effective depth of the reinforcing bars confirmed the low placement of the rebars. A close review of the rebar shop drawings showed that the detailers intended to use one of the layers of the top column rebars as support bars, which resulted in an effective depth as low as 5.3 in (135 mm). This depth is approximately 1 in (25.4 mm) less than was expected by the design engineer using a ¾-in (19-mm) concrete cover. The individual "high chairs" used were 4¼ in (108 mm) high (Fig. 2.15).

One of the questions on the investigators' minds, and for that matter on everybody else's, was: Why didn't the structure collapse earlier; i.e., why had it survived that long?

The answer to this question was given by the review of the construction procedures, especially the shoring and reshoring methods. An examination of Fig. 2.16 will show that the dead loads of the structure were initially carried by the shores and the reshores, which transferred the loads to the ground. Only after the removal of the reshores below the first floor level was the total weight of the structure actually carried by the concrete slabs, through their punching shear capacity at the columns.

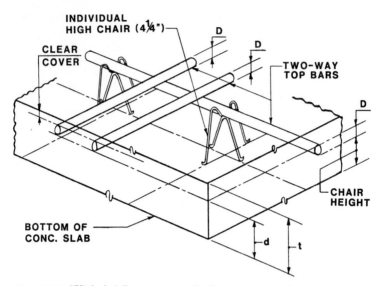

Figure 2.15 "High chair" support setup for the main top reinforcing bars resulted in low effective depth and reduced punching shear resistance. d = average effective depth, t = slab thickness, D = bar diameter. (Based on NBS reports.)

While all tests and data showed that the concrete strength was proper, improper control of concrete strength determinations prior to removal of forms was revealed. It was noted that *laboratory*-cured cylinders were used in order to determine actual *field* strength of the slabs prior to stripping,

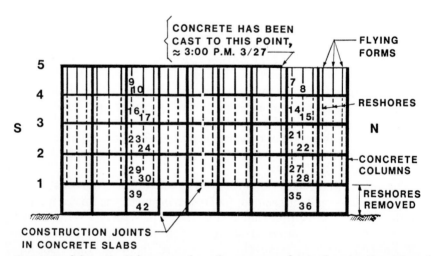

Figure 2.16 Schematic of job progress shows the sequence and state of construction at time of collapse at 3 p.m., March 27. Numbers within schematic indicate age in days at time of collapse. (Based on NBS reports.)

when proper procedures and common sense dictate testing of *field*-cured cylinders.

The conclusion of our investigation was that the cause of the failure was:

1. A design error in using an 8-in (203-mm) slab lacking the required punching shear capacity.

2. A construction error using "high chairs" of the improper height, resulting in reduced effective depth of the rebars. The low effective depth considerably reduced the structural strength of the slabs.

3. Absolutely no rebars going through the column periphery to avoid progressive collapse.

4. Once the fifth floor failed, it triggered a progressive collapse that impacted the lower levels.

5. Lack of construction control for properly designed formwork and shoring and reshoring plans and sequences. There appeared to be an insufficient number of shores, but the actual number could not be confirmed.

6. Inadequate reshoring. Although the analysis by the NBS, which conducted a full and in-depth investigation of this collapse, noted that "full reshoring was assumed in the three stories below the flying forms,"[2] this assumption was not consistent with some of the data we reviewed.

7. There were sufficient warnings in the form of excessive slab deflections and spider cracks at the slabs in the vicinity of the columns. These warnings were not heeded. The structure was doomed at this point, with a remote possibility for repair. [*Author's note:* It is my opinion that the workers could have been saved had proper attention been given to the warnings.]

It is most interesting that common to all three cases described above were:

1. Punching shear failure

2. Improper procedures for form removal and reshoring

3. Cold temperatures in winter (January to March period)

Lessons

1. *A punching shear strength check is critical to the success of a flat slab,* since punching shear is the most common failure mode of concrete slabs.

2. Minimum depth of a flat slab must be checked to assure proper strength and acceptable deflections.

3. Reinforcing bars, both at the top and the bottom of the slab, should be

placed directly within the column periphery to avoid progressive collapse. This can easily be accomplished routinely in all flat-slab jobs at no additional cost at all.

4. Proper construction control must be provided in the field, including design of formwork by professionals. This must include shoring and reshoring plans, procedures, and schedules, with data on minimum allowable stripping strength of the concrete.

5. When there are failure-warning signs of any type on a construction site, work must stop. All aspects of the project must be carefully evaluated by experienced professional help. Immediate evacuation of the structure must be considered.

6. Special care must be taken during cold weather to evaluate the *actual in-place* strength of the concrete. It is also a fact that the level of construction carelessness increases in the winter months.

2.6.4 L'Ambiance Plaza collapse in Bridgeport, Connecticut

On April 23, 1987, during construction of a 16-story apartment building in downtown Bridgeport, after several floor slabs had been lifted and temporarily parked on steel block plates at their final position, the entire structure suddenly collapsed (Fig. 2.17). At that time, most of the frame of the building had been completed.

This tragic collapse, which ended with 28 workers killed and many more injured, was one of the worst construction accidents of the century. With few exceptions, all those working beneath the confines of the lifted slabs were crushed by tons of falling concrete slabs. The entire collapse took only about 5 seconds. The speed with which this collapse progressed may be appreciated when considering that it takes only about 2.5 seconds for a free fall from a height of 80 ft (24.38 m), the distance from the top slab to the ground.

The structure, consisting of two connected rectangular towers (east and west) approximately 112 × 62 ft (34 × 19 m) each (Fig. 2.18), with 7-in (178-mm) thick posttensioned floor slabs, had been designed to be constructed by the lift-slab method. This system utilizes individual posttensioned concrete floor slabs cast on the ground-floor slab, one on top of the other. The cured slabs are then lifted into position by large hydraulic jacks. The slabs are temporarily hung on steel jack rods until they are lifted into final position. For this project, two types of jacks were used: a standard jack of 150 tons nominal capacity and a "superjack" with 300 tons nominal capacity. The lifting system is shown in Fig. 2.19.

The load was transferred from the slab to the steel column by steel shearhead collar assemblies (Fig. 2.20), creating a hole in the slab through

Figure 2.17 This concrete lift-slab apartment building collapsed during construction in Bridgeport, Connecticut. The exact cause of the collapse has not been determined.

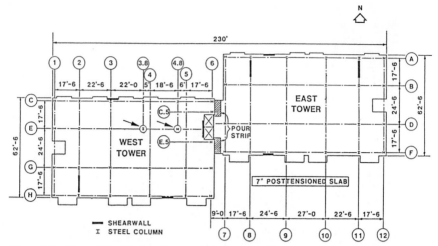

Figure 2.18 General layout plan showing the two rectangular structures connected by narrow pour strips. The entire collapse took less than 10 seconds.

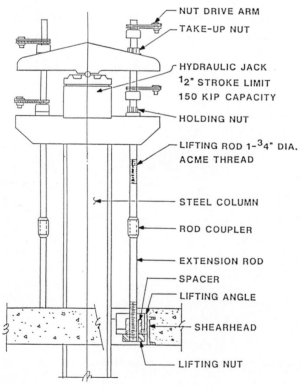

Figure 2.19 The lifting assembly is shown with the jack mounted on top of the steel column. The slab is lifted on two 1¾-in steel rods. (Based on NBS reports.)

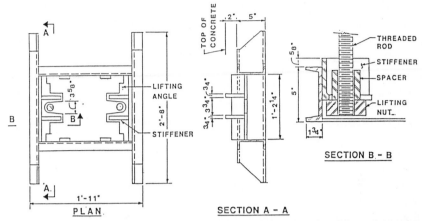

Figure 2.20   The steel shearhead is shown with the two lifting rods. The slipping of the lifting nut from under the bearing angle is alleged by NBS to have initiated the failure. Dimensions shown are typical.

which the slab was lifted. The slabs were rested temporarily on welded column plates/blocks (Fig. 2.21) and then welded permanently through steel adjustment wedges which controlled the final floor slab elevation.

Eyewitness reports stated that the collapse started with one distinctive loud noise, then progressed horizontally to the adjacent tower and then

Figure 2.21   Punching shear failure of the concrete slab at the steel shearhead is shown. The shearhead is shown supported on steel wedges and bearing blocks.

continued vertically. The recorded measured wind speed at a nearby airport at the time of the collapse was approximately 23 mi/h (37 km/h). It was reported that the collapse started at a high elevation of the west tower. One witness said that he "heard a loud noise . . . like steel breaking under pressure." He then noticed that the floor slab directly over his head "was cracking just like ice breaking." (See similar collapse description of the Commonwealth Avenue structure, Sec. 2.6.2.)

It was generally agreed by most investigators that the collapse started at levels 9, 10, or 11 of the west tower near column E4.8 or E3.8 (see arrows in Fig. 2.18). There was never any consensus as to whether the initiation was the failure of the lifting system or failure of the concrete slabs. Typical punching shear failure may be seen in Fig. 2.21. Steel shearhead collars separated from the columns and stacked loosely one on top of the other (Fig. 2.22). Some inherent floor slab design weaknesses were noted in the arrangement of columns adjacent to the floor slab openings and the elevator shaft.

Other mechanisms of failure such as the following were also considered:

1. Foundation failure

2. Lateral instability of the entire system

3. Weld failure

4. Instability of one or more individual columns

5. Failure resulting from lateral soil pressure acting on the foundation walls

6. Prestress slab failure

What had to be established was which failure was the trigger, and which ones were the consequences of the initial failure. At the collapse site we found many examples of welding failures, punching shear failures, and column instabilities.

An investigation by the National Bureau of Standards came to the conclusion in a September 1987 report[3] that:

> The most probable cause of the collapse was the failure of the lifting system in the west tower during placement of a package of three upper level floor slabs. The failure most probably began below the most heavily loaded jack (column E4.8) or an adjacent jack (column E3.8).

The report also pointed out:

> . . . excessive deformations occurred in the lifting angle of the shearhead collar assembly at the location of the initial failure. This was followed by one of the jack rods in the lifting assembly slipping off the lifting angle in the shearhead supporting the package of three slabs. This failure mechanism was duplicated in laboratory experiments.

Figure 2.22 The remains of the concrete slabs embedded at the shearhead collars are shown stacked one on top of the other after the progressive collapse.

I should mention here that one of the inadequacies of the testing of the shearhead collar by NBS was that it did not include the effects of the restraint against steel collar rotation, provided by the confining concrete slab.

During our preliminary investigation, we examined failed sections, surveyed the debris, and analyzed the entire structure and its lifting system. We were able to make the following observations and preliminary conclusions:

1. There were several inadequacies of the original design and/or construction. Good practice and Chapter 18 of the ACI 318-83 code require that "A minimum of two tendons shall be provided in each direction through the critical shear section over columns." This was not followed in the L'Ambiance Plaza structure.

2. The construction procedure which was employed dictated the casting of the shear walls at a later date, leaving long spans temporarily unsupported, resulting in excessive deformations.

3. Since it takes a failure at only one column to generate a progressive collapse, the individual failure of many elements of the supporting systems at columns E4.8 and E3.8 cannot be ruled out. It is interesting to note that the NBS tests did not include the confinement of the concrete directly next to these columns.

4. The failed concrete at the two suspected columns was neither recovered and positively identified, nor was it tested to determine its strength.

5. The NBS testing was based on many support assumptions (eccentricities and shearhead-to-slab connection), which are not substantiated by facts.

6. The "ice crushing" noise heard by eyewitnesses is a definite proof of concrete failures.

7. The fact that it took less than 10 seconds for the propagation of the collapse is a clear indication of a built-in weakness in the structure itself rather than the method of its installation.

Our own investigation, like those of others, was not completed because of an unusually prompt legal settlement. Since the various investigations were aborted, all the facts from which we can derive lessons for the future may never come to light. The legal settlement should be welcomed, however, because it produced funds to compensate the families of the deceased and injured.

Following the collapse, the question of the adequacy and safety of the lift-slab method was raised. This was the first serious failure of a lift-slab structure, a system that had been in use for over 40 years. Hundreds of structures have been completed successfully by this method of construction. We should not lose sight of the fact that there have been a great number of progressive collapses of cast-in-place concrete structures. No doubt, the lift-slab industry needs to reaffirm the safety of its system. The same provisions which apply to all concrete structures (such as built-in redundancies) to avert progressive collapses are applicable to lift slabs. Section 7.13 of the new ACI 318 code requirements (see Sec. 1.2.2 in this book) should also benefit the lift-slab industry.

Lessons

1. Step-by-step procedures for lift-slab construction should be developed and controlled by an independent licensed professional engineer.

2. Temporary lateral bracing must be provided at all construction stages.

3. Redundancies in concrete punching shear and connections must be provided in the structure.

2.6.5   Openings in flat slabs adjacent to columns
— The Jackson, Michigan, collapse

The placement of openings immediately adjacent to columns reduces the punching shear strength of slabs. Sometimes, such openings will reduce this strength dangerously. Openings for plumbing slots or heating and ventilating ducts should be placed away from the columns as far as possible. This advice was not followed in a construction site in Jackson, Michigan.

At first glance, this tragic failure resembled a formwork collapse. However, our investigation determined that it resulted from very high punching shear stresses in the flat slab near the columns.

The reinforced-concrete flat-plate floors (no column capitals or drop panels) were 10 in (254 mm) thick, supported on square columns spaced 24 ft (7.32 m) on centers in both directions. Columns rested on concrete-filled pipe piles driven to bedrock. Typical columns were 25 in (635 mm) square at the basement level, decreasing to 20 in (508 mm) square between the second and fourth floors.

First and second floors were several weeks old at the time of the accident, and forms and shores had been removed. The third-floor concrete was at least 20 days old; forms had been removed. The slab was reshored to the second floor, and was carrying the formwork for the fourth floor. Concrete had been placed in the fourth-floor forms only a short while when most of the east wing, an area about 72 × 144 ft (22 × 44 m), dropped all the way to the cellar. The other three wings were little damaged except where they adjoined the collapsed section.

Significantly, almost all columns remained standing full height after the collapse (Fig. 2.23). Top-story column forms remained in place and very

Figure 2.23 Almost all columns remained standing full height after the collapse, indicating a punching shear failure.*

little reinforcement projected from the free standing columns at any floor level. Plans indicated 10 × 14 in (254 × 356 mm) duct openings in the slab along two adjacent faces of some interior columns, which, of course, prevented slab steel from running through the columns. The design called for a complex reinforcement assembly around each interior column within the slab thickness, but how these assemblies could be placed within the zone of high shear and still permit the duct openings was not clear. Figures 2.24 and 2.25 show the openings adjacent to the columns and the dropped slab after failure. Often, flat plates are designed with too many slots adjacent to a column, leaving insufficient space for rebars.

The importance of the rebars going continuously through the column periphery in order to avoid a progressive collapse was discussed earlier in Secs. 1.2 and 2.5.

Lesson.   Avoid locating slots directly adjacent to columns supporting flat plates.

2.6.6 The case of the missing capitals

This case involved a collapse resulting from a punching shear failure caused by a construction error. A three-year-old concrete plaza deck, serving as a roof of a garage, collapsed in midtown Manhattan in December 1962, without warning, crashing down on parked cars.

The roof consisted of a 16-in-deep (406-mm) waffle slab with 3 ft

---

* Dov Kaminetzky, "Failures During and After Construction," *Concrete Construction,* Volume 26, No. 8, August 1981, p. 641.

Figure 2.24 Slab openings (arrow) adjacent to columns considerably reduced punching shear strength.

(0.92 m) of earth cover. About half of a symmetrical entrance plaza, an area of 45 × 50 feet (14 × 15 m), failed. The other half remained in place, apparently in sound condition. Failure was a clean punching shear, with little effect beyond the shear cut (Fig. 2.26).

In spite of the builder's long experience with this type of construction, the 12-in-deep (305-mm) square (nonconical) concrete caps that should have extended 10 in (254 mm) beyond the column faces had been omitted at all of the columns in the failure area (Fig. 2.27). They had also been omitted at all of the nine columns in the symmetrical area that did not fail.

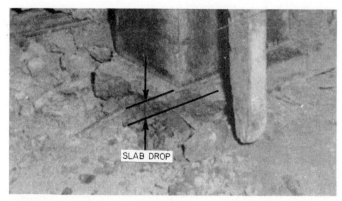

Figure 2.25 Top view of slab at column shows dropped slab from punching shear failure.

Figure 2.26 Collapse of concrete plaza deck resulting from a punching shear failure was traced to a construction error.

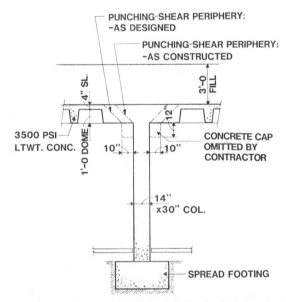

Figure 2.27 Concrete caps at the top of each column were omitted by the contractor because he mistook lines represented by the concrete caps for outline of the spread footings.

The only difference in conditions between these two areas was a stopped-up drain in the failed area. This resulted in a frozen earth cover on the deck that failed; the other half of the deck was well-drained. Unknown to anyone, the factor of safety of this structure was a dangerously low 1.05.

With water unable to drain from the soil because of a plugged-up drain, the weight on the garage roof increased, precipitating the collapse.

But the key reason for the failure was that the contractor failed to construct the called-for concrete cap at the top of each column. In looking at the plan view of the engineering drawings, the contractor mistook the lines representing the concrete caps for the outline of the spread footings. Had there been better field inspection of this job, or had the drawings been clearer, this failure would have been prevented.

The failed slab was reconstructed with column capitals and new columns. The other half of the deck was considerably strengthened by new girders, capitals, and column jackets.

Lessons
1. Essential elements of a structural system such as concrete capitals must be clearly indicated on the design drawings and so identified.
2. Inspection in the field should concentrate on the critical elements of the structure, to verify conformance with the design.

2.6.7 Torsional cracking of slab edges and reentrant corners

Unstiffened edge slabs of flat-plate floors require top reinforcement at the exterior edges to prevent torsional cracking at the stiff connection to the column. Columns should be so located as to avoid reentrant corners at exterior edges of flat plates that have shallow beams or no beams at all. The load transfer will induce high torsional stress in the slab edge or the face of the shallow spandrel and split it near the column.

Examples of this type of failure are Miami, Florida, apartments; Williamsburg, Virginia, homes; and an East side New York high-rise building.

2.6.8 Miami apartments — Corner crack failure

Concrete flat slabs of a 13-story Miami high-rise cracked near the corner column and also near the reentrant corner (Fig. 2.28). Our 1973 investigation revealed excessive shear stresses due to combined vertical shear and torsion. In addition to the design deficiency, top reinforcing bars were placed lower than required. It was also found, by reviewing construction records, that the shores were removed prematurely. Figs. 2.29 and 2.30 show the typical diagonal shear cracks.

Because defects were noted during construction (the eighth floor had

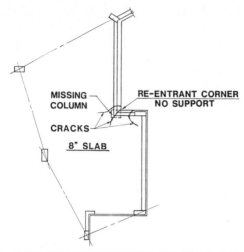

CAUSE	OVERSTRESS
DESIGN DEFICIENCY	30%
INSUFFICIENT HEIGHT OF STEEL	50%
PREMATURE REMOVAL OF SHORING	60%
	140%

Figure 2.28   Plan showing cracking and distress at reentrant corner where column was omitted by design engineer.

Figure 2.29   Diagonal shear crack at exterior of corner of Miami highrise follows classic 45° angle.

Figure 2.30   Top view of slab cracking of Miami high-rise.

already been cast), an additional masonry pier was added at the reentrant corner.

The new pier accomplished two purposes:

1. Reduced the load on the corner column, thereby reducing the vertical shear stresses.

2. Eliminated the eccentric load at the corner, causing reduction of the torsional edge stresses.

Lessons

1. Avoid unsupported reentrant corners where possible.

2. Reduce loads on corner columns and/or increase the size of corner columns to reduce torsional shear stresses.

3. Carefully inspect effective depth of top rebars over columns.

4. Avoid premature removal of forms, shores, and reshores.

2.6.9   Williamsburg homes — Slab failures

A startling similarity to the previous case occurred in a Williamsburg, Virginia, construction project in April 1985. Observe the close resemblance of the diagonal shear crack in both cases (Fig. 2.31 compared with Figs. 2.29 and 2.30).

In the same project, the designer introduced metal stud walls as bearing walls for support of the flat slabs at the periphery (Fig. 2.32). The assumption of an elastic distribution of load was erroneous, because concrete

Figure 2.31 These photographs show almost identical failures to those in Figs. 2.29 and 2.30, although taken 10 years later at another project, in Virginia. They were a result of punching shear failure.

columns cast monolithically with the slab will be stiffer than the metal stud walls; thus part of the load shifted to the concrete columns. This unfortunately caused very high punching shear stresses in the slab, resulting in cracking. Correction was by introduction of a concrete jacket cast around the top of the existing column. Figure 2.33 shows details of the jacket, while Fig. 2.34 shows the completed jacket.

Figure 2.32 The designer introduced metal stud walls for support of the floor slabs at the periphery. The walls did not function that way. As all loads were transferred by shear through the rigid concrete slab, the overloaded slab failed, as shown in Fig. 2.31.

A steel corner column also was added for additional support of the slab to correct the situation.

Lessons
1. Avoid hybrid support systems. Otherwise, take full account of differential stiffnesses and shifting loads, increasing the shear resistance of the

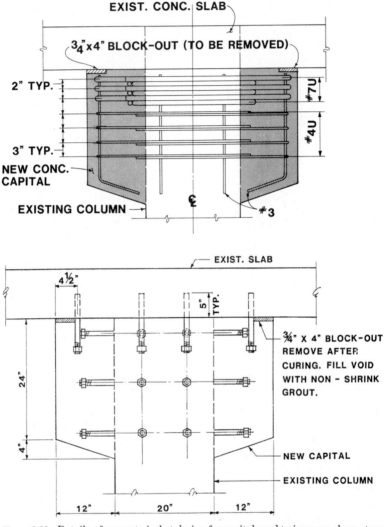

Figure 2.33 Details of concrete jacket design for capital used to increase shear strength at existing columns.

slabs accordingly. This may be accomplished by the use of larger columns or by the introduction of capitals or drops.

2. Take into account the torsional stresses at corner columns, where, in addition to the vertical shears, *the bending moments in both directions must be accounted for.*

2.6.10   East side high-rise slab cracking

The concrete frame of this 38-story reinforced lightweight concrete high-rise in New York City had been practically completed when cracks were

Figure 2.34 Concrete jacket was cast around the top of each column to reduce punching shear stresses in the slabs and eliminate cracking.

noted around all exterior columns at the circumference of the building and at the reentrant corners. An interesting observation was that only the upper floors exhibited this type of distress. The lower levels, which had columns of larger dimensions, as expected, in order to support the higher loads, did not crack.

Columns varied in size from 16 in (406 mm) above to 36 in (914 mm) at lower levels. While the columns themselves were adequate to support the design loads, they did not provide sufficient shear contact surfaces.

Initially, there were allegations that the cracks were caused by an accident in which bundles of rebars fell and hit the slab with impact, by construction loads or low-strength concrete.

An analysis of the 9-in (229-mm) slab (without spandrel beams) readily showed that the shear stresses at the edges were excessive. The vulnerable reentrant corner (Fig. 2.35) was not reinforced with diagonal bars, and the fact that it was not supported at the corner induced many tensile stresses which exceeded the diagonal tension capacity of the concrete. The cause of the cracking was determined to be high torsional shear stresses combined with low concrete strength.

One floor slab with a cracked area was subjected to a full-scale *in situ* load test. The slab failed after sustaining the required load for 22 hours (just short of the 24-hour ACI 318 requirement). The punching shear fracture is depicted in Fig. 2.36. It was then decided to repair the cracks

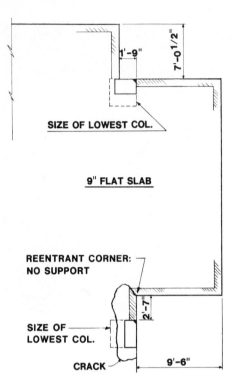

1'-9"

7'-0 1/2"

SIZE OF LOWEST COL.

9" FLAT SLAB

REENTRANT CORNER:
NO SUPPORT

2'-7"

SIZE OF
LOWEST COL.

CRACK

9'-6"

Figure 2.35 Failure at reentrant corners and punching shear fractures around columns of the upper 18 floors of the 38-story high-rise led to severe slab cracking.

using pressure-injected low-viscosity epoxy and, in addition, to provide support by positive means.

The repair solution of adding new piers, used at the Miami apartments, was out of the question in this case, since the reentrant corner was located in the middle of a kitchen. We therefore decided to utilize structural steel brackets which were bolted to the face of the column to support the corner and to resist the twisting action (Fig. 2.37). Anchor bolts were set into 10-in-deep (254-mm) holes, and each bolt was tested for pullout resistance to twice the design requirement. The proof tests were all satisfactory.

In addition, to reduce the shear stresses at the exterior columns, the size of the exterior columns above the 20th floor was increased by casting new concrete bonded to the existing column with stainless steel pins and epoxy bonding agent. The concrete was cast from the top through holes in the slab (Fig. 2.38).

Lessons

1. Where possible, avoid unsupported reentrant corners at the exterior of unstiffened slabs. Otherwise, provide sufficient diagonal bars to reinforce the corners.

Figure 2.36 Punching shear failure at the exterior columns shows typical 45° shear plane, after 22 hours of a full-scale load test. Slab dropped on shores.

2. Keep the shear stresses generated by torsion at exterior columns of flat-plate slabs as low as practically possible by increasing the size of these columns.

2.6.11 The case of the soil-overloaded garage roof deck

This multilevel parking deck collapsed immediately when the soil fill was placed on its roof (Fig. 2.39). The cause of failure became evident when the height of the soil fill was measured and found to be excessive. It was a clear case of lack of communication among the architect, engineer, and contractor. Design changes shown on field sketches failed to arrive at the field. The

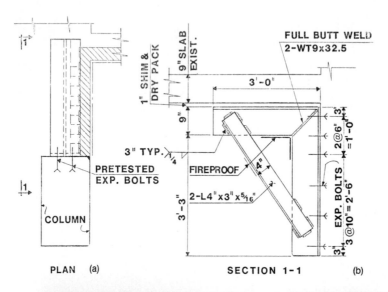

PLAN   (a)   SECTION 1-1   (b)

Figure 2.37 Structural-steel brackets bolted to face of concrete columns were used to support reentrant corners and resist twisting action. Actual bracket is shown in photo above (see arrow). Bracket details are described in (a) and (b).

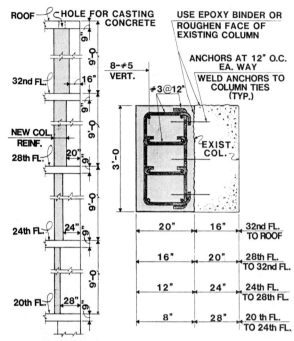

GENERAL SECTION   DETAIL 'A'

Figure 2.38 Size of existing concrete columns was increased to a total of 36 in from 20th floor to the roof, to increase shear periphery and resistance to punching shear.

Figure 2.39 This Washington, D.C. waffle flat-slab deck failed under excessive soil load.

Figure 2.40 The failed section was adjacent to an expansion joint. This progressive collapse crushed the roof deck plus the three parking levels below. Note the slabs sheared off completely at the columns, leaving the vertical bars exposed (arrows).

roof deck as constructed was not adequate to support the heavy soil loads (Fig. 2.40) and collapsed, progressively demolishing the lower levels.

Lessons
1. Avoid small-scale field sketches. All design changes should be made on the original full-scale drawings with official submittals to the entire construction team.
2. For a reminder of the effects of lack of communication, see Fig. 1.11.

2.7 Expansion Joints

Expansion joints are often confused with control joints, contraction joints, and construction joints, which are, in fact, completely different types of joints. *Construction joints* are "convenience joints" used during construction to separate two sections of concrete during casting. *Control joints* can mean practically anything. Sometimes the term refers to a groove provided in a concrete slab or wall with the expectation of dictating to the concrete where a future crack should develop. Sometimes the term refers to joints used to control shrinkage or cracking of wall sections.

On the other hand, *expansion* or *contraction joints* are always associated with movement of two adjacent sections, the direction of the movement being the only difference between the two. The proper terminology here is actually *movement joints.*

Expansion joints or, to be precise, movement joints, have continually been associated with problems. The big question is usually: Should an expansion joint be introduced? The answer is simple: Unless you are sure that you can design and construct a proper expansion joint, you might just as well eliminate it.

Quite often, engineers or architects fail to recognize that expansion joints move not only across their length but also in two other directions. In fact, expansion joints move in all directions (x, y, and z) and also rotate in three planes. Unless the design accommodates all these movements and rotations, the expansion joint is doomed to fail.

2.7.1 Arlington Plaza collapse*

The collapse of the plaza deck of a parking facility located near Washington, D.C., was actually triggered by the failure of the main expansion joint, separating two sections of the garage.

A section of the slab supporting a statue on the plaza level collapsed and fell onto and demolished the slab directly below (Fig. 2.41).

---

* Dov Kaminetzky, "Structural Failures and How to Avoid Them," *Civil Engineering,* ASCE, August 1976, pp. 60–63.

Figure 2.41 Photograph of upper plaza level of collapsed area adjacent to expansion joints shows demolished sculpture.

Our investigation which followed revealed deficiencies in placement of reinforcing bars (Fig. 2.42) and also reduced thickness of the support seat (Fig. 2.43). The expansion joint did not operate as planned and unexpected stresses developed, cracking the concrete at the expansion joint toes along typical diagonal shear patterns (Fig. 2.43). The large horizontal movements could not be accommodated by the building paper which was placed over the seat. Moreover, the rotation at the joint generated additional point pressures at the edge of the concrete bearing seat, causing it to crack.

One of the best ways to design a proper expansion joint is to introduce neoprene pads to provide for both rotation and vertical movement, while the large horizontal movement may be accommodated by the use of sliding teflon pads (Fig. 2.44). This method was used at the failed Arlington Plaza.

Figure 2.42 Close-up of failed joint shows bent low dowels and steel angles.

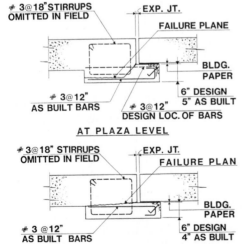

3@18" STIRRUPS
OMITTED IN FIELD

EXP. JT.

FAILURE PLANE

BLDG.
PAPER

3@12"
AS BUILT BARS

3@12"
DESIGN LOC. OF BARS

6" DESIGN
5" AS BUILT

AT PLAZA LEVEL

3@18" STIRRUPS
OMITTED IN FIELD

EXP. JT.

FAILURE PLAN

BLDG.
PAPER

3 @12"
AS BUILT BARS

6" DESIGN
4" AS BUILT

AT LOWER LEVEL

Figure 2.43 Failure pattern shows shear of bracket support at expansion joint. Note misplaced rebars, omitted stirrups, inadequate concrete shelf support.

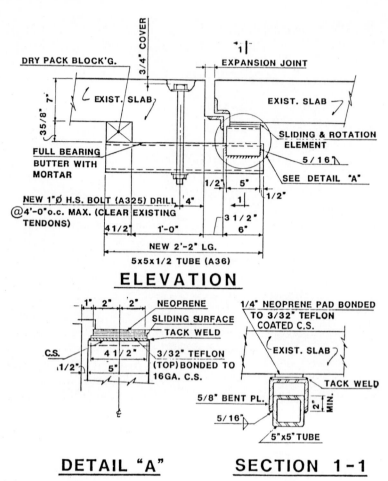

Figure 2.44 Method of repair at failed slab and reinforcement of the slabs in balance of the structure, a preventive measure. Proper sliding and rotating joints were also provided.

Structural-steel tubes were also placed below the slab at the joints and were supported by steel bolts inserted through the slab at 4-ft (1.22-m) intervals.

Lessons

1. Movement joints or so-called expansion joints must be designed and constructed to properly accommodate all movements (three directions) and all rotations (three planes).

2. Serious consideration should be given to the elimination of expansion joints in borderline cases, where the length of the two sections is not excessive. In such cases, the resulting stresses must be provided for by proper reinforcing bars.

3. The use of building papers or any other materials which are likely to bind and transfer loads by friction should be avoided at expansion joints.

2.8 Brackets and Daps (Notches)

The introduction of brackets and daps (notches) is a common technique used to save headroom in tight situations. They have become a favorite of the precast-concrete industry. Unfortunately, while use of such brackets and daps by designers and constructors is well-intentioned, the results are often unsatisfactory — and sometimes quite disastrous.

The source of the difficulties is that when elements of concrete are joined together, movement must always be expected. Where free movements are permitted, buildup of stresses is avoided. Where movements and rotations are resisted by friction, stresses result and distress is certain to follow.

2.8.1 Prestressed girders cracking — Seattle industrial plant

A large industrial structure had 40 × 44 ft (12 × 13 m) bays of precast, prestressed beams supported on precast girders with brackets at one-third points. Shortly after construction, these notched beams and the brackets on the girders cracked along the line of the expansion joint (Fig. 2.45).

The beam bearing locked on the asbestos packing layer, restricting movements. Analysis showed tension stresses at the notch corner which were aggravated by horizontal forces due to volumetric changes. Cracks also developed at the brackets, which were not designed for the additional stresses generated.

The forces acting at the interaction between the girder and the beam are shown in Fig. 2.46. Note the tensile stresses caused by the lower strands acting alone at cross section 1-1.

Because the cracking was intensive and occurred throughout the entire structure, repairs were immediately ordered. The repair method utilized hanging saddles that transferred the bearing reaction from the brackets to the top of the girders (Fig. 2.47).

Lessons

1. Brackets and daps (notches) should be avoided where possible.

2. Where they are used, they should be constructed to permit free movement. Avoid building papers to reduce friction forces. Notches should be heavily reinforced, possibly by posttensioning along the diagonal.

3. Avoid bearing at the edge of the brackets and corbels, and use elastomeric pads (Fig. 2.48).

4. Provide steel-reinforced edges at ends of corbels (Fig. 2.49).

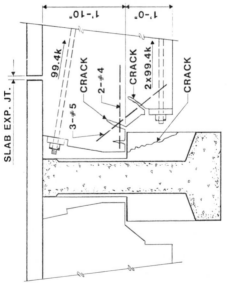

**ROOF SECTION SHOWING CRACKS
IN PRESTRESSED BEAMS**

Figure 2.45 (a) Cracks in prestressed beam and support bracket resulted from inadequately reinforced corners and locked expansion joints. (b) Actual cracks in beams.

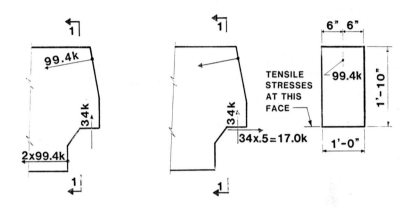

<u>FULL DEAD LOAD</u>   <u>CONTRACTION</u>   <u>SECTION 1-1</u>

Figure 2.46 Load pattern shows cause for overstressing at corners.

2.9 Cracks in Concrete

Cracking is one of the most common forms of failure of concrete. All cracks in concrete are undesirable, in spite of the fact that some hairline cracks in areas not exposed to the environment are not necessarily injurious.

Committee 201 of the American Concrete Institute defines a crack as "an incomplete separation into one or more parts with or without space between." Cracks are classified by direction, width, and depth. They may be longitudinal, transverse, vertical, diagonal, or random. They may be full-depth, partial-depth, or surface cracks.

Generally, cracks may be divided into the following four categories:

1. *Pattern cracking.* Fine openings on the concrete surface in the form of a particular pattern. They often may be the result of a decrease in volume of the material near its exposed surface, or increase in the volume of the material below the surface, or a combination of both.

2. *Checking.* Shallow cracks which develop at close spacing but at irregular intervals. These cracks develop on the surface of the concrete.

3. *Hairline cracking.* Small cracks usually less than 0.005 in (0.127 mm) wide. These may appear at random or in a set pattern and are usually at the exposed surface of the concrete.

4. *D cracking.* The progressive formation of a series of fine cracks on the surface of the concrete, at close intervals, parallel with the slab edges

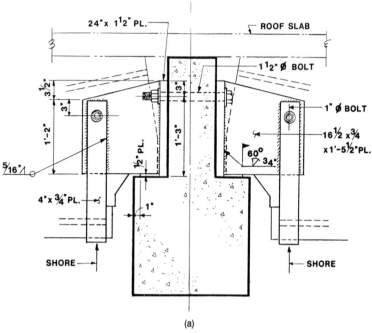

(a)

(b)

Figure 2.47 (a) Proposed repair of cracked prestressed girders by installation of steel hanging saddles. (b) Actual repair.

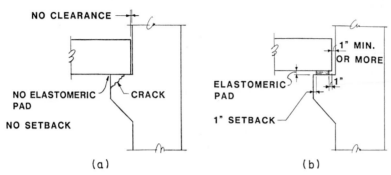

Figure 2.48 Typical end-bearing details of (a) improper design causing cracks compared to (b) correctly specified bearing pads to eliminate cracking.

and joints, and sometimes following and curving around the corners of the slabs.

2.10 Formation of Cracks

Cracks may be further classified, depending on the time of their formation, i.e., those which form prior to hardening of the concrete, and those which form after the completion of the setting and hardening process. Jean-Pierre Lévy described these cracks in 1954[4] and noted the now-famous *crack circles* (Fig. 2.50):

1. *Cracks which form prior to setting.* These may be caused by subsidence of the subgrade (such as crushed stone or sand as shown in Fig. 2.50a). The subgrade may settle because of inadequate compaction of the underlying soils. Temporary loading, movement of the forms under pres-

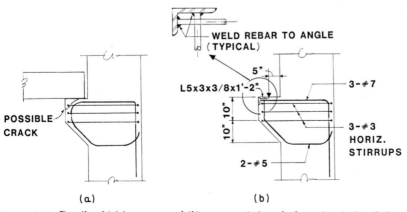

Figure 2.49 Details of (a) incorrect and (b) correct reinforced edges of typical corbels.

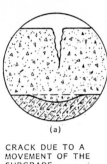

(a)

CRACK DUE TO A
MOVEMENT OF THE
SUBGRADE.

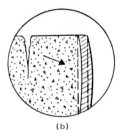

(b)

CRACK DUE TO A
MOVEMENT OF THE SIDE
FORM.

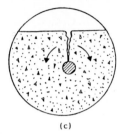

(c)

SHRINKAGE AT RIGHT
ANGLES TO A REBAR.

(d)

SHRINKAGE PERPENDICULAR
TO THE AGGREGATE
SURFACE.

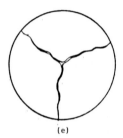

(e)

THREE-BRANCH CRACK
CAUSED BY DRYING
SHRINKAGE.

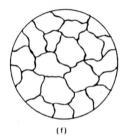

(f)

NETWORK OF FINE CRACKS
DUE TO CARBONATION OF
THE CEMENT.

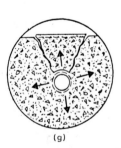

(g)

CRACKS DUE TO EXPANSION
OF A CORRODED REBAR.

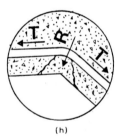

(h)

CRACKS ARE CAUSED BY
THE STRESSES TO WHICH
A BENT REINFORCEMENT
ROD IS SUBJECTED.

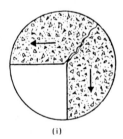

(i)

TENSILE STRESS
CONCENTRATION AT A
REENTRANT CORNER.

Figure 2.50 Famous crack circles.[4]

sure of fresh concrete (Fig. 2.50b), swelling of forms, weak forms (common occurrence with metal deck forms), settlement of mud sills, and deflection of structural supports (such as steel beams) are other factors leading to cracking.

Shrinkage during the hardening process also takes many different forms. We have cases of shrinkage at right angles to the aggregates (Fig. 2.50d) or the reinforcing bars (Fig. 2.50c) and of plastic shrinkage and drying shrinkage (Fig. 2.50e) during setting. Casting of concrete against porous or dry supports should be avoided, since it can cause too-rapid drying, which is generally followed by cracking.

2. *Cracks which form after hardening.* Cracks form in hardened concrete whenever the tensile strength of the concrete is exceeded. Tensile stresses by themselves are generated by various mechanisms. They may be the result of a directly applied tensile force, such as dead, live, wind, impact, and vibration loads. They are sometimes caused by external restraint, such as when the end of a concrete member is prevented from moving freely. When the structure is restrained at its ends or along its length, it may crack with or without outside tensile forces. Temperature variations or shrinkage result in dimensional reduction, which together with restraint will cause tensile stresses, sometimes resulting in cracking.

Improper detailing of reinforcing bars may introduce tensile stresses at corners (Fig. 2.50h). Corrosion of rebars also can result in cracking (Fig. 2.50g). See Chap. 6, "Corrosion."

While cracking in concrete has been intensively studied for more than 60 years, there are many details relating to the formation of such cracks and their initiation, propagation, and continuity which are not completely clear even today. The cracking process in concrete and mortar is a highly complicated mechanism and does not follow the simple fracture-mechanics rules applicable to metals and other ductile materials.

2.10.1 Shrinkage

Shrinkage, sometimes referred to as *drying shrinkage,* is affected by many factors, all of which can be controlled to reduce the amount of shrinkage prior to and after hardening of the concrete. These factors are:

1. Excessive water content

2. Aggregate type, composition, surface, and grading

3. Rapid flow of water to the surface

4. Excessively high cement content

5. Moisture conditions during curing

6. Water losses by rapid evaporation, bleeding, or chemical reaction

7. Carbonation of the cement (Fig. 2.50*f*)

8. Conditions, rates, and temperature of placing

The control of these factors, especially during the period between the first few days of the initial set and final hardening, is critical, since this is when most of the shrinkage takes place. Simply put, concrete shrinkage is a reduction of volume caused by a loss of water. Shrinkage by itself is harmless. The problem is that most concrete elements in the real world are restrained. This restraint of free movement is the culprit.

Some of the most useful tips to control shrinkage are:

1. Avoid mixes which are overrich with cement.

2. Avoid aggregates containing clay or dirt.

3. Use a low water/cement ratio (while reducing shrinkage, such ratios also yield excellent concrete for protection of rebars).

4. Start curing immediately after completion of troweling.

While the importance of reduction of shrinkage by control of the ingredients and manufacture of the concrete is self-evident, the designers of the structure are in a position to further reduce shrinkage effects by:

1. Introducing movement joints (expansion joints), control joints, and other isolation joints at reasonable intervals.

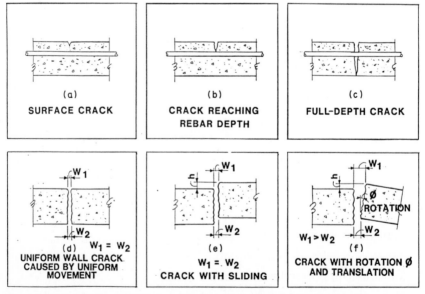

Figure 2.51 Crack squares further explain formation of crack types in concrete.

2. Minimizing sharp reentrant corners (Fig. 2.50*i*) and cutouts and providing reinforcing at these locations when they are unavoidable.

3. Providing sliding supports where possible to reduce the restraint.

4. Equalizing the friction along the concrete, especially at slabs on grade.

5. Using prestress where it is found to be cost-effective.

In conclusion, the designers and constructors of concrete structures should strive to *reduce shrinkage and, at the same time, increase freedom of movement.*

It should be kept in mind that cracks cannot be totally eliminated; however, they can and should be reduced to a minimum and controlled.

To better understand the behavior of cracks, close attention should be paid to the details and formation of the cracks, since cracks are not "all the same." The *crack squares* shown in Fig. 2.51 further explain the various formations of cracks.

Cracks are often classified in accordance with their width. I made one such attempt;[5] a summary of this classification is shown in Table 2.1.

2.11   Errors in Placing Reinforcing Bars

It takes a gross error in sizing reinforcing bars to cause a collapse or catastrophe. The more common occurrence is that of mislocated rebars. The errors of totally missing bars, placing top bars at the bottom of the slab, or placing bars on "high chairs" (special rebar support accessories) of improper height (or placing them with no "high chairs" at all) are the most damaging ones.

2.11.1   Midwestern university — Student union building

The waffle-type slab of the new roof of a college student union building developed serious cracks at the intersection of a number of transverse and longitudinal joists (Fig. 2.52).

The key reason for the failure is shown in Fig. 2.53. This illustration shows transverse sections through the roof. The top diagram shows that the design called for lapping the steel reinforcing bars. The bottom diagram shows the waffle slab as actually constructed. Instead of overlapping the rebars, the builders left a gap, causing the cracking shown in the photograph. Nondestructive magnetic examination in the field and later probes into the joists (Fig. 2.54) confirmed the existence of a gap in the bottom bars at the point of the crack.

Our investigation revealed additional deficiencies. Concrete cores showed low compressive strength even though the job cylinders had been

TABLE 2.1 Classification of Visible Damage

Degree of damage	Description of typical damage*	Approximate crack width, in (mm)†
	Hairline cracks of less than about 0.005 in (0.13 mm) are classified as negligible.‡	
1. Very light	Isolated light fracture in building. Cracks in exterior walls¶ visible on close inspection.	1/64 to 1/32 (0.40 to 0.80)
2. Light	Light fracture visible on the inside on floors/partitions and on the exterior of buildings. Doors and windows may stick slightly.	1/32 to 1/8 (0.80 to 3.2)
3. Moderate	Moderate cracks are visible on the inside and on the exterior of buildings. Doors and windows stick. Utility pipes and glass may fracture. Water and air penetration through exterior.	1/8 to 1/2 (3.2 to 12.7)
4. Extensive	Severe cracks are visible on the inside and on the exterior of buildings. Windows and door frames are skewed and "locked-in"; floors slope noticeably. Walls lean or bulge noticeably; some loss of bearing in beams. Utility pipes and glass broken, and water and air penetration through exterior.	1/2 to 1 (12.7 to 25.4)
5. Very extensive	Very extensive cracks are visible on the inside and on the exterior of buildings. Broken pipes and glass. Full loss of beam bearing; walls lean or bulge dangerously. Structure requires shoring. Danger of instability.	1 or wider (25.4 or wider), but depends on number and location or cracks

* In evaluating the degree of damage, consideration must be given to its location in the building and structure (e.g., points of maximum stress).

† Crack width is only one aspect of damage and should not be used alone as a direct measure of damage.

‡ Refers to existing "old" cracks. New cracks must be monitored for possible increase in width.

¶ A criterion related to visible cracking is useful since tensile cracking is so often associated with settlement or movement damage. Therefore, it may be assumed that the start of visible cracking in a given material is associated with its limit of tensile strain.

SOURCE: Dov Kaminetzky, "Rehabilitation and Renovation of Concrete Buildings," *Proceedings of Workshop*, National Science Foundation and American Society of Civil Engineers, New York, N.Y., February 14–15, 1985, pp. 68–78.

acceptable. Truss bars were also found to be erroneously substituted with straight bars.

The stress analysis we immediately performed showed that the stresses generated by dead loads at the actual location of cracking (point of counterflexure) were relatively low. Volumetric changes such as shrinkage and temperature expansion and contraction were found to be the actual initiators of the cracks. The cracking did not cause the roof to collapse. But as a precautionary measure, the roof, constructed of a two-way grid of concrete

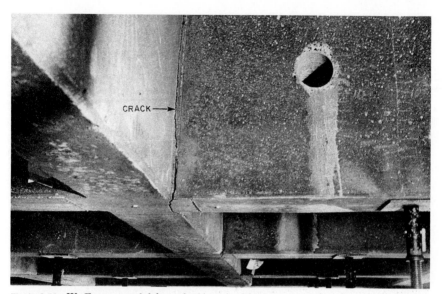

Figure 2.52 Waffle-type roof slab cracks (arrow) at the intersection of transverse and longitudinal joists were a result of gaps in rebars.

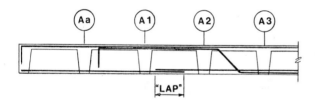

REINFORCEMENT DETAILS SHOWING LAP AS INDICATED ON DESIGN- AND SHOP DRAWINGS

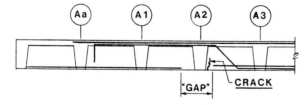

REINFORCEMENT DETAILS AS FOUND ON THE SITE

Figure 2.53 Transverse sections through roof show reinforcement details, as designed and as built. Instead of correct *lap* (overlapping) of rebars, *gap* was left.

Figure 2.54 Close-up of probe shows *gap* rather than *lap* specified in design.

joists, was strengthened by installing a network of steel beams below the joists.

Lessons

1. Long concrete structures must be analyzed for temperature and shrinkage stresses.

2. Reinforcing bars should be provided to resist these stresses. The bars must be continuous (or properly lapped) at points of stress.

2.11.2   Long Island automobile dealership's retaining wall

An unbelievable collapse really happened on Long Island, N.Y. (Fig. 2.55). Some engineers think that it belongs in Ripley's "Believe It or Not." The 20-ft-high (6.1-m) retaining wall (Fig. 2.56) behind an automobile dealership in Manhasset, Long Island, collapsed immediately on placement of the soil backfill. Amazingly, we found that the main reinforcing bars, which were correctly designed to be 1¼ in (32 mm) in diameter (no. 11), were erroneously substituted with ¼-in-diameter (6.35-mm) bars (no. 2 bars). The error was confirmed by examination of the broken concrete sections in the debris.

The cause of the error was poor drafting; a dimension line covered the 1 of the 1¼ inch dimension (circled area in Fig. 2.56). The incredible fact was

Figure 2.55 Failed and cracked retaining wall.

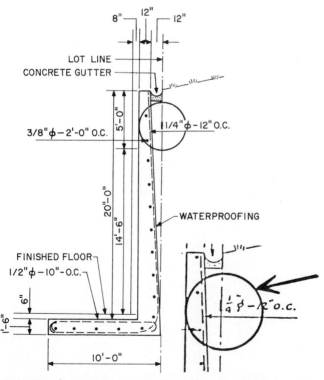

Figure 2.56 Cross section of retaining wall which collapsed immediately after placing of soil backfill because ¼-in rather than 1¼-in-diameter rebars were used. Error occurred because correct rebar diameter was covered by a dimension line. See circled area enlargement. Copy of actual drawing is shown.

that this error escaped all the standard construction control systems such as checking of drafting, review of shop drawings, and field inspection.

Lessons

1. Sloppy drafting may lead to serious consequences.

2. Utmost attention must be given to details *at all stages of construction.*

2.12   Errors in Detailing

Reinforcing bars must be properly anchored in the compression zone. Where rebars are not detailed and anchored properly at obtuse angles, tensile stresses and cracks will develop (Fig. 2.50*h*). As an example, I recall several transfer girders supporting a high-rise condominium in Florida, where the rebars were not thus detailed. As a result, heavy cracks (Fig. 2.57) developed, threatening the safety of the entire structure. Steel brackets and columns had to be added to secure the building.

The error of short rebar length may be a very devastating one. In these instances the cracks always develop at the end of the rebars. These are serious structural cracks which cause strength reduction, often failures, and sometimes even result in total collapse.

2.12.1   Downstate parking garage

A seven-story parking facility was designed and constructed as cast-in-place long-span slabs. Shortly after removal of the formwork shores, con-

Figure 2.57   These enormous concrete transfer girders cracked because of an error in detailing the rebars.

Figure 2.58 Timber shores (arrows) were used to temporarily support cracked slabs of a seven-level parking garage.

tinuous cracks appeared at exactly the same location on each level. For safety reasons the structure had to be temporarily supported with timber shores (Fig. 2.58).

The structure consisted of 18-in (457-mm) waffle slabs spanning 51 ft (15.5 m), with 11-ft (3.40-m) overhanging cantilevers on one side (Fig. 2.59).

After the cracks were discovered, our analysis of the drawings revealed that the rebars as detailed were too short. Their length was determined by using the cantilever length to establish the tie-back development length. Unfortunately, the controlling length in this case was the middle span, which was considerably longer than the cantilever length.

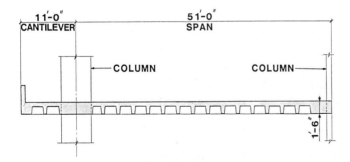

ELEVATION

Figure 2.59 Cross section of parking garage shows relationship between cantilever slab and main span.

The resulting cracks developed at the end of the rebars (Fig. 2.60). Prestressing was considered for repairs but was found to be too costly. The eventual repair method utilized an overlay slab which was bonded to the existing slab with epoxy bonding compound and with mechanical pins shot into the slab (Fig. 2.60).

In order to confirm the effectiveness of the system after repair, a full-scale in-place load test was successfully performed, in total compliance with ACI-318 code requirements.

Lessons

1. Reinforcing bars must be extended *beyond* the point of maximum stress.

2. Tie-back development length of top rebars in cantilever slabs is *often* controlled by length of cantilever, *but not always*. Their length sometimes is determined by the length of *a long adjacent span*.

3. It is recommended, where possible, to load-test a proposed repair technique. It is especially important where the technique does not have a "track record" and/or where the repair will be applied to many repetitive locations.

2.12.2   New York viaduct

The serious consequences of short bars became apparent immediately upon completion of concrete girders on a highway viaduct. When the viaduct was opened to traffic, several of the girders cracked, and had to be shored immediately. Our review concluded that the location of the cracks exactly coincided with the ends of the top rebars (Fig. 2.61). These bars were not long enough to develop the required tensile stresses. (The detailer assumed that a length of 40 bar diameters for anchorage was all that was needed.) Accordingly, all girders which had already been cast were shored temporarily and posttensioned in order to reduce the tensile stresses at the top (Fig. 2.62).

Lesson.   Top reinforcing bars at cantilevered structures must be extended *beyond the point of inflection, but in no case should this length be less than the length of the cantilever itself.*

2.13   Shrinkage and Temperature Effects — Nassau County State College

On a freezing winter day I visited this building at a Nassau County, New York, state college and viewed the cracks which developed at a main circular concrete column (Figs. 2.63 and 2.64). The pattern was generally a tree of cracks in the vertical direction, primarily between two horizontal girders framing into the distressed column.

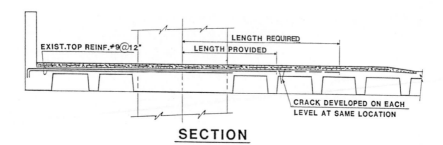

SECTION

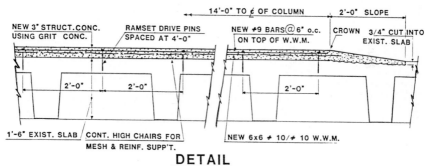

DETAIL

Figure 2.60 Detail shows crack locations and overlay method of repair.

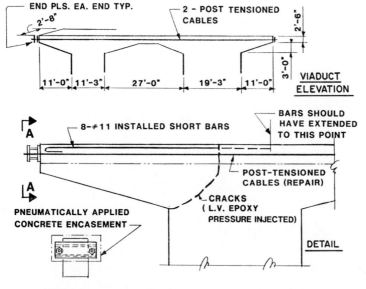

Figure 2.61 Epoxy repair of cracks and posttensioning were used to restore integrity of highway girders. Cracks were caused by rebar detailing error when bars were too short.

Figure 2.62 Viaduct after repair.

Figure 2.63 General view of cracked column at girder intersections.

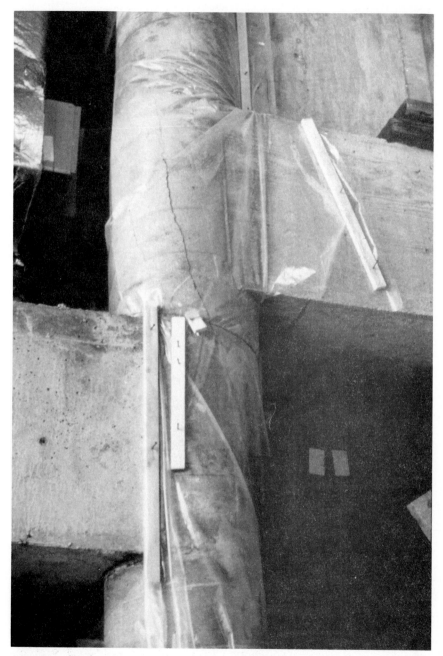

Figure 2.64 Close-up shows pattern of column cracks.

Our review of the geometry of the structure showed that a very long line of concrete girders terminated at the cracked column. An analysis of the stresses generated by a temperature drop combined with shrinkage of the concrete resulted in very large deformations which could not be accommodated by the rigid end column. The release of the stresses was effected, unfortunately, by the cracking of the column.

A field survey by pachometer indicated that the column lacked horizontal ties in the entire height between the two girders (Fig. 2.65). It was, therefore, not surprising that there was no tensile resistance at this particular location.

The proposed repair in this case involved inserting a sliding bearing at the column, thereby allowing future movement without distress (Fig. 2.66).

Lessons
1. Provide properly designed expansion joints in long runs of girders. This will reduce the deformations caused by shrinkage and temperature changes.

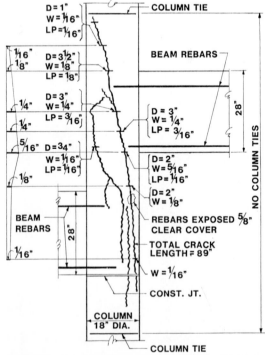

Figure 2.65 Detail diagram of cracks (crack intensity) indicates short horizontal rebars and missing column ties, where D = depth of crack, W = width of crack, LP = left side push out.

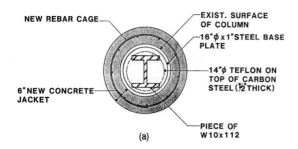

NEW REBAR CAGE

EXIST. SURFACE OF COLUMN

16"φ x 1"STEEL BASE PLATE

14"φ TEFLON ON TOP OF CARBON STEEL (½"THICK)

6" NEW CONCRETE JACKET

(a)

PIECE OF W10x112

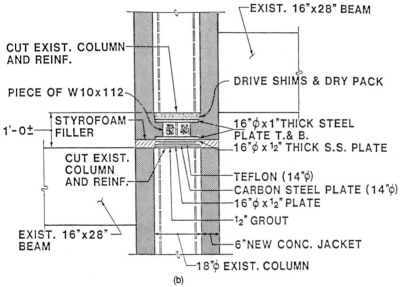

EXIST. 16"x28" BEAM

CUT EXIST. COLUMN AND REINF.

DRIVE SHIMS & DRY PACK

PIECE OF W10x112

1'-0±   STYROFOAM FILLER

16"φ x 1"THICK STEEL PLATE T.& B.

16"φ x ½" THICK S.S. PLATE

CUT EXIST. COLUMN AND REINF.

TEFLON (14"φ)

CARBON STEEL PLATE (14"φ)

16"φ x ½" PLATE

½" GROUT

EXIST. 16"x28" BEAM

6"NEW CONC. JACKET

18"φ EXIST. COLUMN

(b)

Figure 2.66 Proposed details of column repair, shown in plan (*a*) and elevation (*b*), include insertion of a sliding bearing for movement between girder intersections.

2. Provide resistance to large deformations at end columns, especially where the columns are relatively stiff.

3. Columns should include horizontal tie hoops at deep girders to resist localized tensions (Fig. 2.67).

4. Horizontal beam rebars should extend full length and be anchored in the column (Fig. 2.67).

2.14   Construction Joints — Construction Sequence

Construction joints are considered by many designers to be "convenience joints" placed by the contractor for the contractor's own convenience. This understanding is correct most of the time but not all the time.

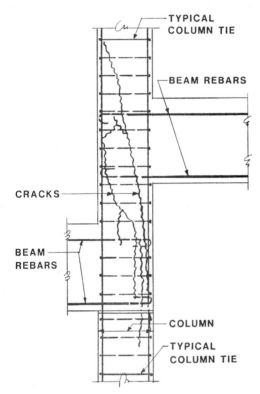

TYPICAL
COLUMN TIE

BEAM REBARS

CRACKS

BEAM
REBARS

COLUMN

TYPICAL
COLUMN TIE

Figure 2.67   Proper details of rebars and ties at offset intersection. Required additional bars and ties shown dashed.

Construction joints are often used to minimize shrinkage effects in slabs and walls. Engineers should insert in the project specifications limitations on the maximum size of slab or wall that may be cast in one continuous operation, such as ". . . maximum area of slab cast will be 800 ft² (74 m²) . . . " or "maximum length of wall cast in one continuous operation shall not exceed 80 ft (25 m) . . . . "

There are, however, construction joints which may dramatically affect the strength of a structure. Therefore, good construction practice will dictate that construction joints must be perpendicular to the member, normally vertical in slabs and girders. They should be free of dust or other foreign materials which may reduce the bond at the interface, and, most important, they *must be located at points of low shear stresses.*

The introduction of construction joints may affect the strength of the structure. Therefore, the engineer should be emphatic and forbid their placement without prior approval.

The collapse at a Denver highway interchange, described below, was caused by the introduction of a critical construction joint, which unfortunately was constructed with the engineer's knowledge — with grave consequences.

2.14.1 Colorado interchange collapse

A highway interchange was constructed using cast-in-place piers supporting AASHTO-type prestressed girders (Fig. 2.68). The original design intended that the tops of the piers would serve as the top deck for carrying traffic, i.e., the tops of the piers were at the same level as the top of the slab surface. This procedure would have ensured a monolithic pier for its full height.

For his own convenience, the contractor proposed a change. He asked to cast the pier in two stages. The first stage would cast the pier to the bottom of the slab. The second stage would consist of casting the slabs over the girders and over the pier in one operation (Fig. 2.69). This procedure had, in fact, the effect of introducing a construction joint in the pier, a joint that determined the ultimate fate of the pier. The new construction joint introduced by the contractor was accepted by the engineer.

The construction proceeded as follows:

1. The piers were cast to the bottom of the slab.

2. The precast girders were placed on top of the piers.

3. The pier's formwork and supporting shores were removed.

4. As the concrete slab was being cast, the entire structure collapsed (Fig. 2.70).

The impact of the introduction of the construction joint was to eliminate the effectiveness of the piers' top bars, since they were not cast with the piers. This procedure required the piers to support high loads, which obviously was not possible.

It is interesting that such an obvious error was missed by practically everybody. After the collapse, it was learned that one of the carpenters on the job had questioned the procedure but had been ignored.

Lessons

1. Construction which does not follow standard methods requires *preconstruction step-by-step written procedures.* These should be outlined by the engineer or proposed by the contractor and carefully reviewed by the engineer.

2. Horizontal construction joints within the depth of girders, piers, slabs, and beams must be avoided. They may be introduced when absolutely unavoidable, and then only after carefully engineered procedures, uti-

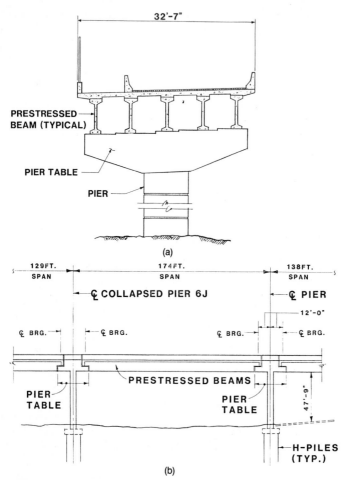

Figure 2.68 Typical cross section (*a*) and general elevation (*b*) of precast viaduct beams supported on pier tables at each end.

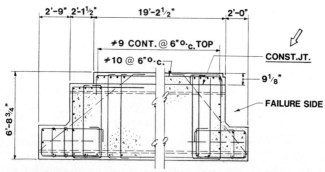

Figure 2.69 Pier table change shows construction joint introduced at top of pier table that led to collapse.

REMAINING WEST PORTION
OF PIER TABLE 6J

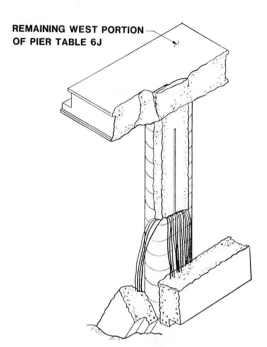

Figure 2.70   Diagram of failure of pier 6J with collapsed portion of pier table.

lizing measures such as adequate temporary shores and rebar dowels, have been carried out. These must ensure the adequacy of the structure to support all loads at all construction stages.

2.15   Precast Concrete

Precast-concrete systems have the tremendous advantage of being produced under good quality control conditions in the casting yard. This operation eliminates uncontrollable field conditions. It also brings with it considerable cost savings when the design involves a great number of repetitive units. Very rarely are there problems of concrete strength, porous concrete, and other related difficulties, which occur with cast-in-place concrete. The most common problems associated with precast units are *the connections* by which the units are attached. When precast units are also prestressed, which is common, other problems unique to prestressing are encountered.

The first instances of cracking and distress of precast and/or prestressed concrete structures attributable to end restraint appeared in the United States some 30 years ago. These same faults occur again and again in many forms and shapes, but with one consistent result: serious damage to concrete structures, often causing millions of dollars in financial losses. Several of these instances have resulted in total collapse where the seriousness of the initial distress was not recognized in time. Here, the cracking developed further and increased to such magnitude as to cause total loss of the shear resistance at the supporting ends.

Why has this constant recurrence gone unheeded despite warnings? The answer lies partly in legal restrictions imposed by some clients who wish to bury the facts in cases involving embarrassing circumstances such as loss of taxpayers' money. More important is insufficient publicity to make the construction industry aware of the serious dangers inherent in providing end restraint to precast elements. The damage is magnified when this restraint is often coupled with the introduction of *notches and brackets.* The following cases will illustrate these points of view.

2.15.1   Public school — Cracking of double tees

A school in New York state was constructed in the late 1970s of precast double-tee panels supported on a structural steel frame. The tees were bearing on either the top or the bottom flanges of the structural-steel girders. Alternate stems of the tees were welded at their ends to the supporting girders. For economic reasons, the construction stopped and the partially completed structure was exposed to the environment for more than 4 years. As a result, many of the welded stems cracked, predominantly those stems which had reduced sections resulting from bottom notches or top flange blocking (Fig. 2.71).

Here again, the welding at the ends restrained the panels from free movement and rotation, resulting in serious damage that had to be corrected by pressure-injected epoxy supplemented by steel shear plates. The restraining welds, obviously, had to be removed to prevent recurrence.

2.15.2   Antioch High School, California — Roof collapse

Cracks developed in the precast stems of the roof structure of a school built in California in the late 1950s:[6]

> . . . an extremely hazardous roof failure that occurred March 25, 1980, at Antioch High School . . . Due to the alert action of several students and school administration, no one was injured. . . . A section of the roof of the auditorium consisting of approximately 18 prestress concrete double tee joists, approximately 38 feet [12 m] by 70 feet [22 m] in area, fell from the center of the roof to the floor below [Fig. 2.72] . . . . The collapsed section of the roof framing consisted of factory fabricated pretensioned prestressed lightweight concrete double tee joists [Fig. 2.73] spanning 40 feet [12 m] between prestressed concrete girders . . . .

Figure 2.71 Cracked double-tee panel (arrow) in a New York school, a result of stems with welded ends that precluded free movement.

The joists were 14 in (356 mm) deep with the ends dapped to middepth, i.e., 7 in (178 mm). The length of the seat was 3¾ in (95 mm). Each stem of each double tee was welded at both ends at the steel-to-steel seat level. Hardly any supplementary reinforcing was provided in the dapped ends.

Figure 2.72 School auditorium shows collapsed roof debris resting on chairs.

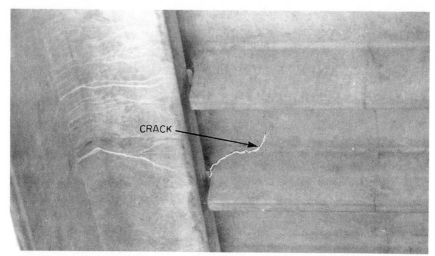

Figure 2.73 View from bottom up shows cracked double-tee joists (arrow) at reentrant corners.

As a result of this failure, a survey of public school structures which were constructed using precast-, prestressed-concrete construction reported:

> Out of some 30,000 projects approved by the office of the State Architect between 1955 and 1980, some 200 were located which used precast pretensioned prestressed concrete. Of this number 24 projects were found where buildings required minor repairs. Eight projects involved buildings which were temporarily closed or provided with temporary shoring until major repairs can be made.

The survey concluded by describing the problem this way:

> A potentially hazardous condition may exist in certain buildings which were constructed using precast pretensioned prestressed concrete framing members. This condition may exist where inadequate provision was made to allow for the effects of long-term shortening which occurs in such members. The result of this hazardous condition can be structural failure and collapse.

Lessons

1. Ends of precast beams, joists, slabs, and other elements should *not be welded.*

2. Provide sliding pads at bearing ends of precast elements.

3. Avoid excessive web cuts.

4. Where web cuts cannot be eliminated, the webs must be properly reinforced with additional rebars or strands.

2.15.3 West-Side apartments

A high-rise apartment house complex in the New York area was constructed in the late 1970s of hollow-core precast, prestressed slabs bearing on precast concrete walls and beams. At the typical floors, spalls at the bearing surfaces appeared as edge loading occurred (Fig. 2.74). At the lower garage level, precast girders cracked as a result of restraint provided by end welding plates. Tensile stresses developed at the nonconfined edges. Edge spalling occurred here, too.

On the roof, precast parapet sections were welded at their ends. This restrained the sections, limiting their movement and generating tensile stresses as a result of contractions due to shrinkage and temperature changes. These tensile stresses exceeded the maximum tensile strength of the panels, which cracked at their ends (Fig. 2.75). One of these parapet units was so weakened that a heavy wind blew it totally off the roof (Fig. 2.76).

Lesson. Design end details with provisions for movement to avoid locked-in stresses and resulting cracks.

2.15.4 Upstate mental health hospital

A hospital structure constructed in upper New York state in the early 1960s had precast panels, both columns and beams, as one of the main

Figure 2.74 Spalling at top of face of girder resulted from edge bearing (without bearing pads) of precast panels.

Figure 2.75 Precast concrete parapet sections cracked because of restraint by welding at both ends.

structural elements. Beams notched at both ends (Fig. 2.77a) were bearing on concrete brackets cast together with the precast columns (Fig. 2.77b).

Shortly after construction, serious cracks developed in both the notched beams and the column brackets. The cracks were so intensive that we had a full-scale notched beam load-tested at Columbia University's Carlton Lab-

Figure 2.76 View of top level and roof shows missing section after parapet and roof slab had been blown off by wind.

BLOCKED
END NOTCHES

(a)

(b)

Figure 2.77 Precast beams with large notches at both ends: (*a*) in casting yard and (*b*) in place.

oratory.* The beam failed at a substantially lower load than would be expected from the design (Fig. 2.78).

---

* Dov Kaminetzky, "Verification of Structural Adequacy," American Concrete Institute, Detroit, Mich., 1985; "Rehabilitation, Renovation and Preservation of Concrete and Masonry Structures," SP-85-6.

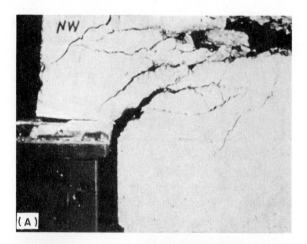

Figure 2.78 Testing sequence shows (*a*) severe cracking and (*b*) complete failure.

A rebar section removed from the failed tested beam was then put in a tension machine; it failed at 40-ksi (276-MPa) yield rather than the expected 60-ksi (414-MPa) yield specified on the structural drawings. A further study revealed that the project specification under the section "Precast Concrete" called for the rebars to be 40-ksi (276-MPa) yield grade.

Unfortunately, the fabricator of the precast panels (structural beams and columns) completely ignored the structural drawings and blindly followed the specifications. What is more amazing is the fact that many supervisors and inspectors were present at the casting yard during casting operations. A special construction management firm was also engaged on this project. The owner had representatives on site, and similarly the architect had periodic inspection responsibilities.

The problem of conflicting contract documents is a common occurrence; variations between the project specifications and drawings are a frequent phenomenon. Some specifications will include one of several general statements such as:

"In the event of conflict, the detailed specifications will control over the general specifications."
" . . . the drawings will control over the specifications."
" . . . the specifications will control over the drawings."
" . . . the more stringent requirement will control."

It must be perfectly clear to contractors who are confronted with conflicting requirements in the contract documents that they have no authority to make a determination as to which requirement governs, even if the specifications contain control statements such as the four listed above. Contractors *must* consult the architect or the engineer, and request a *written* direction as to which of the conflicting directions to follow. The control statement will *weigh heavily* only as relating to payment for additional work and not as an excuse in the event of a failure or collapse.

In the upstate hospital case, it became apparent that the specifications for the structural precast panels were adapted by the architect from an architectural precast wall panel specification which required only 40-ksi (276-MPa) reinforcing bars. How the use of lower-grade rebars escaped so many of the usual procedural checks is still a mystery. The low-strength (40-ksi) rebars went unnoticed by the design team (architect and engineer), field inspectors, owner's representatives, shop drawing review process, the contractor, and precasting yard personnel.

The full-scale, in-place test at Columbia University also revealed that the bottom reinforcing bars did not extend sufficiently beyond the reentrant corner of the notch, leaving the corner exposed to tensile stresses

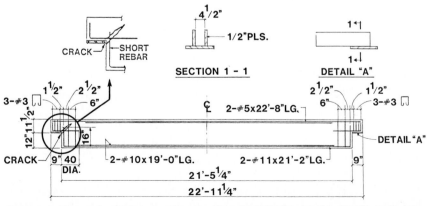

Figure 2.79 Precast beam elevation shows crack and reinforcing details, with discontinuous rebars at notches.

Figure 2.80 Girder repairs. Big steel plates were added.

(Fig. 2.79). It was also found that the steel bearing plates were not vertical and the concrete did not bond properly to the plates because of insufficient consolidation (vibration) of the concrete.

The structure was eventually repaired (Fig. 2.80) by using steel cradles (similar to those used at the Seattle industrial plant, Sec. 2.8.1) and new steel column supports (Fig. 2.81), at very high costs.

Lessons

1. Proper coordination between specifications and drawings is essential.

2. Contractors *must request written directions as to how to proceed,* in the event they become aware of conflicts in the contract documents.

3. Avoid notches and daps where possible.

4. Properly reinforce around notches when these become unavoidable.

2.15.5   Old Country Road garage

An enormous parking facility was constructed by using a precast-, pretensioned-slab system with individual concrete elements (Dox planks). Several years after it opened to the public, extensive spalls developed at the surfaces of the girders (Fig. 2.82). This created a serious hazard of falling pieces of concrete. The intensity of the distress increased because of the sloped surfaces of the girders (Fig. 2.83).

The investigation showed that proper provisions for movement and rotation were not incorporated in the construction of this facility. The loss

Figure 2.81   New steel columns were added for girder support.

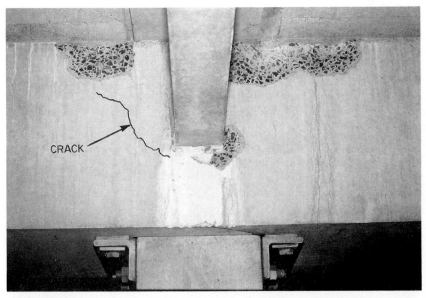

Figure 2.82   Spalling at bearing of precast beams and slabs. Note cracks that signal imminent further spalling.

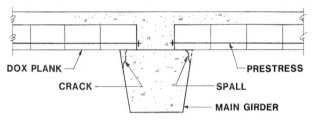

DOX PLANK

CRACK

PRESTRESS

SPALL

MAIN GIRDER

Figure 2.83 Cross section of girder shows sloping faces and cracks and spalls at bearing locations.

of prestress over time, coupled with thermal movements, was determined to be the major factor creating the damage.

Lessons

1. Provide for movements and rotation at supports.

2. Avoid inward-sloping surfaces of supporting girders.

2.15.6   Upstate shopping center and parking decks

A large piece of concrete with sharp edges (Fig. 2.84) falling inside a department store obviously gets attention. The structure of the entire upstate New York complex in which this event occurred was of precast, prestressed

Figure 2.84 When this sharp piece of concrete fell to the floor, quick corrective action was taken.

(a)

(b)

Figure 2.85 Cracked (a) and spalled (b) flanges of precast-concrete beams.

girders and slabs. The parking deck section was framed with basically the same components.

The girders were AASHTO-type precast beams, supporting double-tee slabs at the edge of their top flanges. At the failure area, several of these flanges broke, leaving a number of the quad tees without support. Some flanges were badly cracked (Fig. 2.85a), so loose spalls had to be removed (Fig. 2.85b).

Our investigation here showed to our amazement that the edge thickness

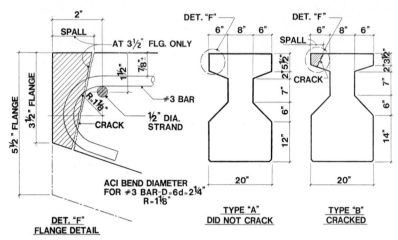

Figure 2.86 Details of two types of girders. Type A with 5½-in flange *did not* crack; type B with 3½-in flange *did* crack.

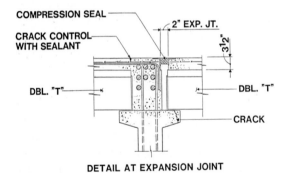

Figure 2.87 Damage was particularly severe at expansion joints, where movement was greatest.

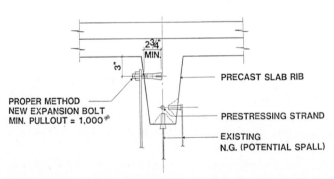

Figure 2.88 Other defects included failed anchors for support of utilities.

of the flanges was a mere 3½ in (89 mm) (Fig. 2.86). In other areas of the complex, the flanges were thicker (5½ in; 140 mm), with practically no damage at all. We also found that bearing pads were deficient and did not allow the quad-tee slabs to move and rotate freely. Damage was especially heavy at the expansion joints, where movements were at the maximum (Fig. 2.87).

There was also a list of other defects, such as anchors shot at the bottom of the tees to support pipes and other utilities (Fig. 2.88), cracked beams, and girders bearing improperly on columns.

The repair method utilized support brackets attached to the existing girders to reconstitute adequate support (Fig. 2.89). In addition, safety debris netting was installed for protection against possible future spalling (Fig. 2.90).

Lessons

1. Do not bear precast elements on relatively thin flanges.

2. Provide adequate bearing pads to allow for movement and rotation at the supports.

3. Hanging of pipes from the webs of precast concrete tees should be by well-embedded anchors installed at the *sides* of the tees, acting in shear rather in tension (Fig. 2.88).

2.15.7   Collapse of hockey rink roof — Brockport State College

An ice skating rink was under construction in Brockport, New York when an entire section of the roof collapsed (Fig. 2.91). The roof was constructed of thirteen 147-ft-span (45-m) AASHTO-type precast girders which were pretensioned at the casting yard and posttensioned in place after erection. The girders were cast in three segments, each approximately 50 ft (15 m) long, and were erected on cast-in-place concrete edge beams and on two high temporary timber towers at the interior. Six-inch-thick (152-mm) precast hollow slab sections (planking) spanning 14 ft (4.3 m) were placed on the girders. The slabs were made to act compositely with the girders by inserting steel dowels in their cores and filling them with concrete. A concrete section 12 in (305 mm) wide was cast in place to fill the space between the ends of the slabs.

After erection, the tendons inside each of the girder segments were connected together by means of special couplers. These couplers, located at the one-third point of the span, were then encased in 2-ft-long (0.61-m) concrete sections, and the entire assembly (147 ft long; 45 m) was then posttensioned and lifted off the temporary supports. A layout of the roof framing and the typical cross sections are shown in Figs. 2.92 to 2.95.

The first and second girders were fully posttensioned by noon of July 15, 1971, and work started on the third girder. It was during the tensioning of

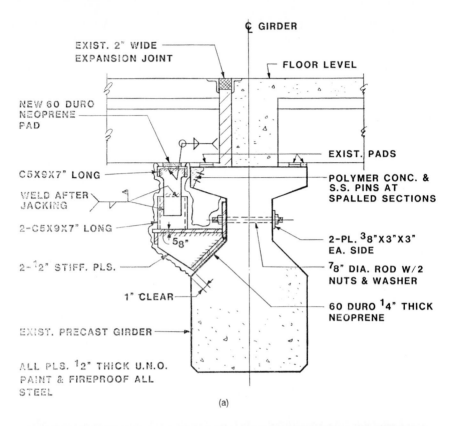

¢ GIRDER

EXIST. 2" WIDE
EXPANSION JOINT

FLOOR LEVEL

NEW 60 DURO
NEOPRENE
PAD

EXIST. PADS

C5X9X7" LONG

POLYMER CONC. &
S.S. PINS AT
SPALLED SECTIONS

WELD AFTER
JACKING

2-C5X9X7" LONG

$5_8"$

2-PL. $^3_8"$X3"X3"
EA. SIDE

$^7_8"$ DIA. ROD W/2
NUTS & WASHER

2-$^1_2"$ STIFF. PLS.

1" CLEAR

60 DURO $^1_4"$ THICK
NEOPRENE

EXIST. PRECAST GIRDER

ALL PLS. $^1_2"$ THICK U.N.O.
PAINT & FIREPROOF ALL
STEEL

(a)

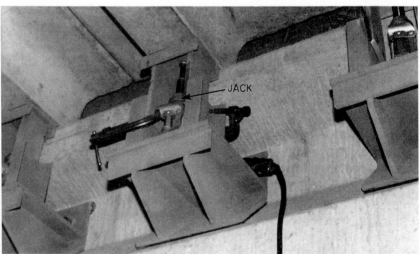

JACK

(b)

Figure 2.89 Repairs included quad-tee support brackets attached to existing girders, shown in sketch (a) and photograph (b). Existing precast slabs were jacked up prior to welding of seats.

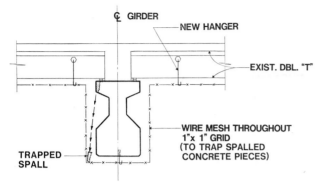

Figure 2.90 Safety netting was installed at girders for protection against possible future spalling.

this third girder, which received five-sixths of its full design posttension force, that the entire roof lifted and then collapsed, causing injuries to several workers.

On the following morning, a large crack was observed near the bottom, east of the coupler location. Crackling sounds were heard and a little later another large crack developed to the west of the same coupler location. A few minutes later, the eastern section of the girder collapsed, carrying with it a portion of the roof planking (Fig. 2.91). Minutes before the collapse, cracks were also noticed at the bottom flange at the coupler of the first girder.

Figure 2.91 Section of ice skating rink roof collapsed during posttensioning of 147-ft (45-m)-long precast-concrete girder, injuring several workers.

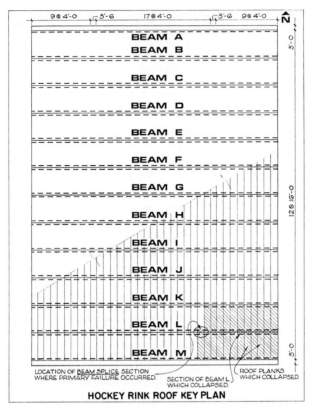

Figure 2.92 Key plan of skating rink roof shows location of beam splice where primary failure occurred.

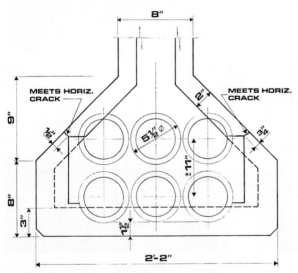

Figure 2.93 Typical cross section of roof girder that failed.

Figure 2.94 Actual coupler sleeves shown after collapse.

The investigation involved several engineering teams. It was agreed by all that concrete strength was not at issue. There were other major factors which were identified; however, there was disagreement as to the weight allocated to each one. These factors were:

1. At the location of the couplers, voids 6 in (152 mm) in diameter were formed, considerably reducing the compression area of the bottom flange.

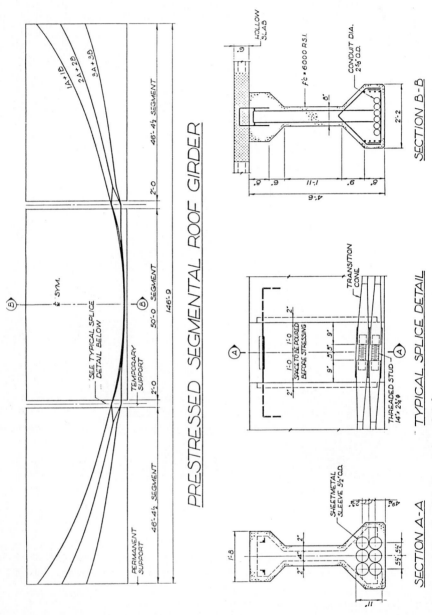

PRESTRESSED SEGMENTAL ROOF GIRDER

SECTION B-B

TYPICAL SPLICE DETAIL

SECTION A-A

Figure 2.95 Details of skating rink roof girder.

2. At the ends of the precast segments, cracks were found emanating from the cable holes to the exterior surface of the girder.

3. The cavities at the couplers which were formed by sheet-metal sleeves were found to be full of concrete, which locked in the cables, restricting them from moving.

4. The cores in the precast slabs were found to be only partially full of concrete in some locations, rendering composite action questionable.

5. There were allegations that some workers were jacking the tensioned girder on the tower, causing forces which negated the girder deadweight and uplifted the girder.

Our own analysis gave the most weight to factors 1 and 3. Factors 2 and 4 were considered to be of secondary importance. Factor 5, which was definitely an important one, could not be proved.

The design of the girders was deficient, since the stresses at the bottom flange were seriously exceeded. The construction was also flawed because it permitted concrete to enter the sheet-metal forms and lock the cables. Without free movement there could be no deformation and elongation of the tendons. In turn, without tendon elongation there could be no stress and load transfer to the tendons.

Field control also failed here, since a routine check of elongation of the tendons would have revealed that the tendons were locked in.

The *Journal of the Prestressed Concrete Institute* has published a *Recommended Practice for Segmental Construction*. This standard contains the following warning:

6.7 — Couplers

Couplers should be designed to develop the full ultimate strength of the tendons they connect. Adjacent to the coupler, tendon should be straight for a minimum length of 12 times the diameter of the coupler. *Adequate provisions should be made to assure that couplers can move during prestressing. The void areas around a coupler should be deducted from the gross section areas when computing stresses at the time of prestressing* (emphasis mine).

The two sentences emphasized above accurately sum up the causes for the collapse. Interestingly, the current ACI-318 code has this added provision:[7] "18.19.2 — Couplers shall be placed in areas approved by the Engineer and enclosed in housing long enough to permit necessary movements."

As in many other failures involving precast elements, the question of responsibility for the design was raised here, too. The engineer of record took the position that *the precast design was not his* but was the product of the precast contractor, who, in fact, prepared drawings and calculations for the system.

The precaster, on the other hand, claimed that he only prepared shop drawings and backup calculations for the engineer, who took the actual responsibility for the design. The question of responsibility, as in so many instances, was not resolved here. This case was settled by mutual agreement and not by a court decision.

In a somewhat similar case which occurred in Omaha in 1957, a 100-ft (30.5-m) prestressed roof girder heaved upward, cracked, and failed. It was one of five such girders, and had *7 of 14 prestressed rods spliced at one location.* There was too much prestress at that cross section, with 3-in-diameter (76-mm) sheaths at the splices compared to the 1⅜-in (35-mm) conduits encasing the rods elsewhere! Unfortunately the ice skating rink designers were not aware of this failure, which had occurred 14 years earlier.

Lessons

1. Avoid oversize couplers in segmental precast construction. When this is not possible, take into account in design the loss of section of the compression flange.

2. Ascertain that tendon paths are always clear, so that tendons may be properly stretched to obtain the correct prestress force.

3. Check actual tendon elongations constantly to verify the magnitude of the prestress force actually induced.

4. Avoid blind work. Filling slab core is one such operation, and there is no way to confirm that cores are indeed filled with concrete.

5. Never jack or shim a prestressed element within its span. By doing so, the downward action of the weight of the concrete is negated, and will create uplift and sometimes even a total blowup (explosion).

2.15.8 Rockland College*

Approximately 6 years after the completion of a college classroom facility in Suffern, New York, cracks were noticed at the stems of the exposed concrete joists. The damage was heavier at the spandrel beams. The school management was concerned, and on the advice of their consulting engineer ordered the erection of shores along the entire exterior periphery of the building (Fig. 2.96).

The four-story structure was constructed with a cast-in-place frame supporting precast, prestressed double tees, which formed the floors and the roof. It was essentially a standard construction, with these exceptions:

1. The lateral stability of the structure was based on the rigidity provided by the precast tees, rather than by cast-in-place crossbeams serving as

---

* Dov Kaminetzky, "The Three Orphans," *Concrete International: Design and Construction,* American Concrete Institute, Detroit, Mich., June 1988, pp. 47–50.

Figure 2.96 Typical cracks at stems of exposed concrete joists in 6-year-old, four-story college building. Temporary shores were erected along the entire exterior periphery of the structure until repairs were completed.

braces. In order to achieve this stability, the webs of the tees were required to be, and were, indeed, welded to the spandrel beams (Fig. 2.97). Only alternate webs were welded, which the designer (erroneously) assumed would permit movement of the tees.

2. The structural frame (reinforced concrete) along the periphery was completely exposed to the environment (Fig. 2.97).

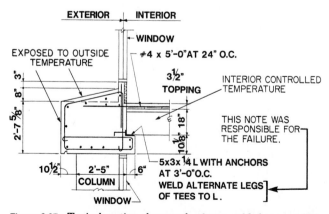

Figure 2.97 Typical section shows web of tees welded to spandrel beams. This section shows that the reinforced-concrete structural spandrel beam was exposed to the environment.

To establish the exact cause of the distress it was necessary, as usual, to first determine the exact pattern of the cracking and then to correlate this pattern with a consistent mechanism causing the distress.

After the condition survey that we performed, it became evident that the cracks were consistently adjacent to the welded web plates of the tees, and always along the exterior periphery. It was also evident that only the welded stems cracked while the unwelded stems were unaffected (Fig. 2.98).

The most interesting and unusual observation was that the intensity of the cracking and the width of the cracks increased along the exterior, starting from the centers of the wings (Fig. 2.99). This is a deformation pattern which is consistent with linear expansion.

Obviously, the fact that the spandrels were exposed to high-temperature effects, while the precast tees were kept at lower temperatures by the air conditioning, did not escape our notice.

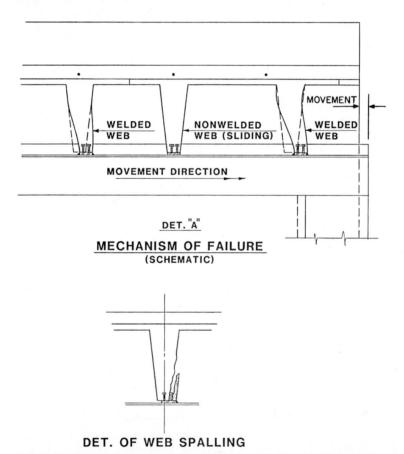

Figure 2.98 Schematic representation of failure mechanism (detail A of Fig. 2.99), with detail of web spalling.

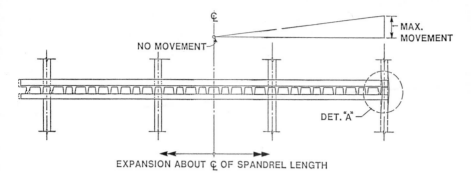

ELEVATION OF PRECAST TEES ON SPANDREL

Figure 2.99   The intensity of cracking and width of cracks increased from center of spandrel beams to wings, a deformation pattern consistent with linear expansion.

In an exploratory program launched by the university, it was found that reinforcing dowels called for in the design and shop drawings could not be found by pachometer readings. Even destructive probes into the top of the slab failed to find them. A controversy arose regarding the importance of these dowels to the failure mode. One opinion held that these dowels were crucial, and their omission was the major contributor to the cracking.

The other school of thought (to which we belonged) insisted that regardless of the existence or omission of the dowels, the stems would have cracked anyway, since the rigidity of the welds was the overriding factor involved. Even if the dowels had been installed, they would have been ineffective; they would have bent slightly and transferred the entire stress to the rigid welded plates (Fig. 2.100).

The tees were repaired by the addition of vertical side plates connected with through-bolts across the stems. The holes drilled through the stems were carefully core-drilled after determining the exact location of the strands (Figs. 2.101 and 2.102).

Lessons

1. Ends of precast elements *must not be welded,* since this restriction will cause buildup of stresses with resultant cracking. Welding details, when welding is absolutely unavoidable, should allow for movement and rotation at the support. This may be accomplished by the use of flexible steel-clip angle connectors.

2. Design engineers should carefully consider the implications of connecting different building components subjected to high differential temperatures. Exposed concrete spandrel beams are one such example.

3. Precast elements preferably should be designed by the engineer of record or a specialist engineer retained by the engineer of record. At the least, the engineer of record should submit to the contractor all the data

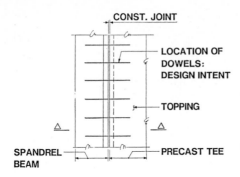

CONST. JOINT

LOCATION OF
DOWELS:
DESIGN INTENT

TOPPING

SPANDREL
BEAM

PRECAST TEE

PLAN

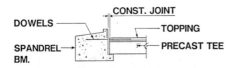

CONST. JOINT

DOWELS

TOPPING

SPANDREL
BM.

PRECAST TEE

SECTION
(a)

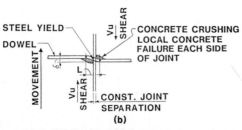

SPANDREL BEAM

Vu

SPANDREL BEAM MOVEMENT

DOWELS

C.J.

PRECAST
TEES

DET. "B"

Vu

SHEAR FORCE AT JUNCTION OF
TOPPING AND SPANDREL BEAM

STEEL YIELD

DOWEL

MOVEMENT

Vu SHEAR

CONCRETE CRUSHING
LOCAL CONCRETE
FAILURE EACH SIDE
OF JOINT

L

Vu SHEAR

CONST. JOINT
SEPARATION
(b)

DETAIL "B"

Figure 2.100 (a) Horizontal bar dowels were designed to connect the precast tees to the spandrels. (b) Even if the dowels had been installed, they would not have prevented the cracking of the stems and would have yielded and failed. δ = movement. L = deformed zone.

Figure 2.101 Vertical steel side plates connected with through-bolts across the stems were used for repair.

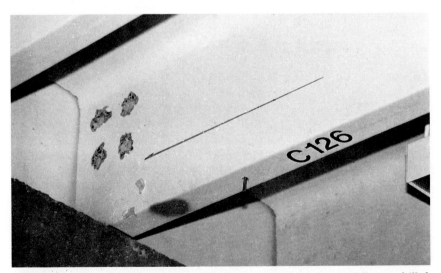

Figure 2.102 Holes drilled through the stems for support brackets were carefully core-drilled after exact location of steel strands was determined.

required to design the precast elements. These should include data such as live, wind, impact, and vibration loads. Where temperature effects are critical, these, too, must be submitted to the designer of the precast.

4. The fact still remains that an engineer of record who abdicates responsibility and leaves portions of the design to be performed by the contractor actually engages in self-deception. When failure occurs, the engineer of record is ultimately dragged into the case, when the finger-pointing starts.

2.15.9 Westbury shopping center parking garage

The design of a vast parking facility called for precast-concrete columns shaped in a T (hammerhead) form with precast fill-in girders. The deck itself consisted of hollow-core precast slabs bearing on the girders and side fascias. The precast girders were *not* prestressed, but were reinforced with mild-steel rebars.

After a few years of service the owners became aware of severe diagonal cracking in the girders combined with a heavy penetration of water through the decks. During a survey of the bottom of the deck, horizontal cracks in the bottom of the slabs were visible. There were also clear signs that the hollow slabs contained water in the cores. Rust marks were evident (Fig. 2.103), and laboratory tests indicated that the water had been trapped there for a long time. Accordingly, we probed the slabs, and streams of water poured from them (Fig. 2.104).

RUST MARKS

Figure 2.103 Rust marks and laboratory tests indicated water had been trapped for a long time in the hollow-core precast slabs.

Figure 2.104 Probing of the slabs produced streams of water.

Probes into the precast girders adjacent to the expansion joints (where cracking was heaviest) indicated that the joints were locked and therefore could not operate properly.

The diagonal tension cracks were so wide that one of the brackets on the columns completely failed and broke and had to be temporarily shored (Fig. 2.105).

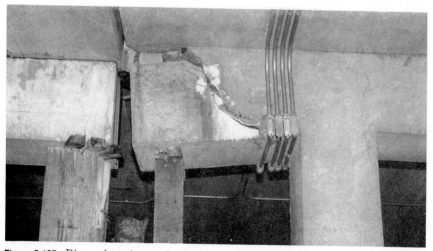

Figure 2.105 Diagonal tension cracks were so severe that one of the brackets on a column failed and had to be temporarily shored.

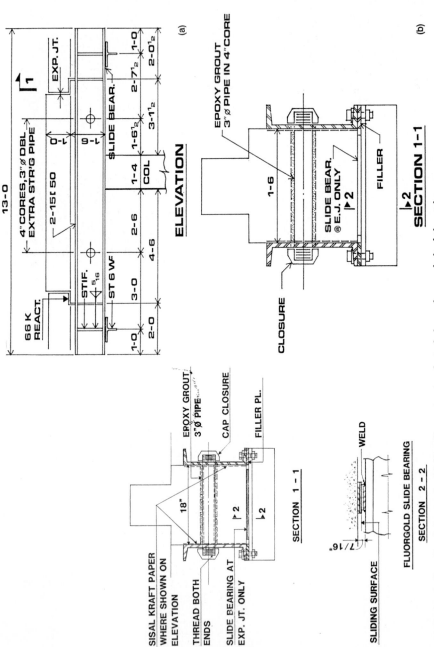

Figure 2.106 Details of the repair. (*a*) Two large structural-steel channels were bolted through large holes drilled in the girders. (*b*) At the ends of the channels, provisions were made to allow movement.

Figure 2.107 Completed repair.

We were faced with two tasks: (1) providing support for the cracked girders; (2) making provisions for movement at the expansion joints. To satisfy these requirements, we devised a repair method utilizing two large structural-steel channels bolted through large holes drilled in the girders (Fig. 2.106a). At the ends of these channels, special provisions were made for sliding elements (Fig. 2.106b), thus removing the undesirable restraint or lock. The completed repair is shown in Fig. 2.107.

Additional exploratory probes into the slabs indicated, fortunately, that the rust of the mesh within the slab was only surface corrosion. Therefore it was sufficient to install a new membrane covering to exclude any additional water.

Lessons

1. Expansion joints must provide freedom of movement. Lacking this freedom, these joints will serve only to attract and build up stresses, with resulting distress.

2. Parking decks must be protected from water penetration by membranes covered with concrete overlays. At a minimum, they should be coated with effective sealers.

2.15.10 Maryland naval facility — Failure of precast concrete wall panels

Precast wall panels enclosing a naval aircraft facility cracked during construction, bringing the project to a halt. Because of indecision on the part of the management of the project, no solutions to the problems were initiated, only finger-pointing.

The prefabricated wall panels were designed to enclose a one-story structural frame building. The specified concrete strength was 5000 psi (35 MPa) at 28 days. The panels were supported and attached to the frame by steel angles at both the floor and roof levels.

An unusual feature of the panels was their shape. The panels consisted of two thin (4- and 5-in; 102 and 127-mm) vertical sections joined by an 8-in (203-mm) horizontal portion to form a Z-like section (Figs. 2.108 and 2.109). The panels were reinforced with a single layer of welded wire mesh.

The cracks appeared consistently slightly above the 8-in (203-mm) center section and ran horizontally across the panels. The cracked panels (Fig. 2.109) had to be temporarily shored (Fig. 2.110).

As is typical, the contractor blamed the failure on errors in design, while the owner and the designers of the project alleged that the fault was with improper construction of the panels and also with the "rough handling" of the panels by the erection crews. It was also the contention of the owner that the cracks developed during lifting of the panels from the casting forms and in transporting them to the site.

Actually, one look at these panels will suffice to establish their uniqueness. The 4-in (102-mm) thickness specified for the upper portion was less

Figure 2.108 Prefabricated concrete wall panels to enclose a one-story structural frame building were of unusual shape.

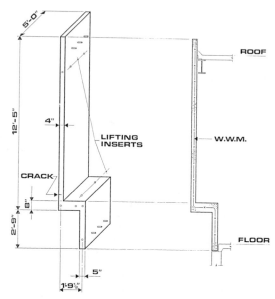

Figure 2.109 Cracks appeared consistently above 8-in-thick center section of panels, running horizontally. WWM = welded wire mesh.

Figure 2.110 Temporary shoring was used to support cracked wall panels.

than the minimum required for a 15-ft (4.5-m) panel by ACI code (*L*/30). The irregularity of the shape created many stress raisers. The single layer of reinforcing mesh was not sufficient to resist high tensile stresses.

Our review indicated that the connections were not developed at all by the design engineer. Connections shown by the contractor on the shop drawings and reviewed by the engineer failed to prevent buildup of tensile stresses in the concrete. The top connections were either locked or partially locked against vertical movements.

All concrete tests, either of cubes taken from fresh concrete during fabrication or of cores taken from the hardened concrete, showed compressive strength conforming to ACI code requirements.

In order to better evaluate the stresses developed under dead, live, and wind loads in combination with shrinkage and temperature effects, a computer model was created. In accordance with finite-element procedures, the panel was divided into many small triangular elements (Figs. 2.111 and 2.112). The analyses proved that the panels which were restricted from moving freely were subjected to very high tensile stresses during a rise in temperature. These stresses, when combined with stresses caused by the

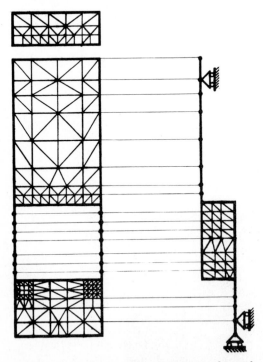

Figure 2.111 Finite-element analysis was used to evaluate panel stresses under applied loads in combination with shrinkage and temperature effects.

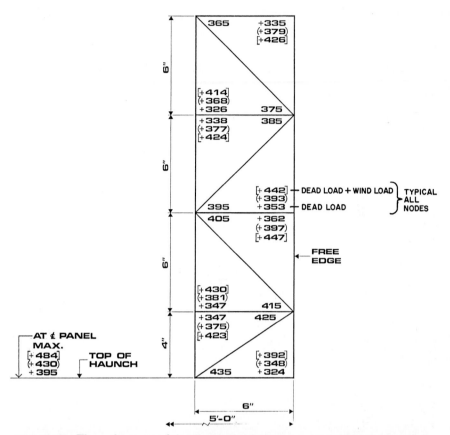

Figure 2.112 The analyses proved that the panels which were restricted from moving freely were subjected to high tensile stresses (expressed here in psi). +No. = dead load. (+No.) = dead load + 6.4 psi wind load. [+No.] = dead load + 6.4 psi wind load adjusted for corner weight and depth.

weight of the panel itself, exceeded the tensile strength of the concrete in the panels. The location of these high stresses correlated closely with the actual location of the cracks.

Lessons

1. For precast panels, minimum L/T ratios, where L is the length and T is the thickness, must be adhered to.

2. Connections for precast wall panels must allow movement so as to prevent buildup of stresses. It is recommended that such typical connections be developed by the engineer of record, who understands the design parameters, and shown on the design documents.

3. Precast wall panels exposed to exterior temperatures must be designed with full consideration of temperature variations during construction and during service life; variations between inside and outside temperatures must also be taken into consideration to avoid curling or even cracking of the panels.

2.16   Posttensioning

Compared to cast-in-place concrete structures, where the structure is actually passive and only reacts to loads, prestressed structures are subjected to loads controlled by the constructor. In postlensioning, in particular, these loads are often enormous (as high as several hundred tons). Therefore, this procedure brings with it very serious consequences when improperly used.

A case in point is the following structure, which failed as a result of improper prestressing procedures.

2.16.1   Washington, D.C. — Foundation mat

A 3-ft-thick (0.92-m) foundation mat was cast with draped tendons (Fig. 2.113) in accordance with the design requirements, and shortly afterward was posttensioned. Unfortunately, as soon as the high prestress loads were applied, the slab lifted upward and cracked (Fig. 2.114).

Figure 2.113 Three-foot-thick concrete foundation mat for five-story building prior to casting of concrete.

CRACK UPLIFTED

Figure 2.114 When design prestress loads were applied, slab lifted upward and cracked.

The error was discovered immediately because it was so obvious. It was amazing that the incorrect procedure eluded everybody involved with the project. The fact that the designer of the mat was a prestigious engineer only proves that no one is immune to mistakes.

As can be seen in Fig. 2.115, there could be no equilibrium of forces in the foundation mat without the presence of the building columns. Realistically, the design should have planned for the prestressing to be executed in stages, with the prestress forces increasing as more and more weight of the upper levels of the structure was imposed. This procedure was not followed, and the failure ensued.

The repair was carried out using rock anchors to provide the necessary vertical reactions to the prestressing forces (Fig. 2.116). The mat was repaired, and the prestressing was successful the second time around.

Lessons
1. Posttension sequence must be carefully reviewed on a step-by-step basis.
2. Total equilibrium of forces must be assured.

2.17   Anchorage Failure in Concrete

The most common forms of failure of anchorages in concrete are (1) pull-out failures, (2) edge-distance failures, and (3) embedment corrosion.

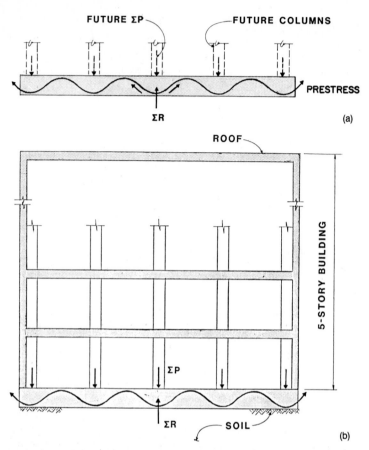

Figure 2.115 No equilibrium is obvious in (*a*). Error that led to failure was based on need for foundation mat to be fully loaded with building column loads of the final structure (*b*) at the time of prestressing.

2.18 Pull-Out Failures

Anchors fail almost always because of inadequate anchorage *in the concrete rather than fracture of the metal itself.*

Anchors are used primarily to connect precast concrete elements to each other, but also are quite often used to connect various trade items to concrete. These usually include items such as curtain walls, steel shelf angles, ceilings, pipes, mechanical equipment, and platforms. Temporary scaffolding and forms are also often anchored into concrete structures.

The pull-out capacity of an anchor depends on three basic factors:

1. The strength of the concrete
2. The depth of embedment of the anchor

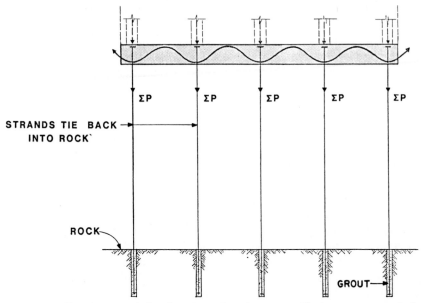

Figure 2.116   Repair was carried out by using rock anchors to provide the necessary vertical reactions to the prestressing forces.

3. The edge distance of the anchor

While failures involving low-strength concrete are common, of course, often the problem is caused by subjecting fresh concrete to pull-out loads before the concrete has matured sufficiently.

Inadequate embedment depth is most critical and cannot be corrected merely by increasing the strength of the concrete.

The bond between the anchor and the concrete is another important factor. Expansion bolts are subjected to loss of bond as a result of alternating loads such as winds and vibrating mechanical equipment.

A fact that is *often overlooked* is that edge distance has a tremendous effect on anchor capacity.

The following failures demonstrate the importance of anchorage, especially the first case with its high loss of life.

2.18.1   West Virginia cooling tower scaffold collapse

One of the costliest collapses of recent years, involving the death of 51 construction workers, was caused by the failure of a temporary structure. The patented jumpform scaffolds carrying workers had successfully been used for the construction of 36 similar natural-draft hyperbolic cooling towers without any problems. (Figure 2.117 shows an identical tower successfully completed.)

The scaffolds at the cooling tower at Willow Island, West Virginia, were anchored to the concrete walls by means of bolts embedded in the concrete. As described earlier, the strength of embedded anchors depends to a great extent on the strength of the concrete — *not at 28 days but at the time of application of the load.* The collapse (Figs. 2.118 and 2.119) occurred in April 1978, while low temperatures were still prevalent. The steel I beams supporting the scaffolds were anchored into a 5-ft-high (1.5-m) concrete section which was cast 24 hours earlier. As the anchors pulled out of the concrete, the workers suspended on the scaffold plunged 170 ft (52 m) to their deaths.

The investigation carried out by the NBS concluded that the concrete had not sufficiently matured to achieve the strength required to resist the pull-out loads developed by the scaffold. Since a carefully controlled in-place pull-out testing program was not used as part of the construction procedures, the low strength of the fresh concrete was not detected.

The major conclusion of the NBS report was that "the most probable cause of the collapse was the imposition of construction loads on the shell before the concrete of lift 28 had gained adequate strength to support these loads."[9]

Figure 2.117 Completed hyperbolic concrete cooling tower used jumpform scaffolds during construction.

Figure 2.118   Partially erected cooling tower at Willow Island, West Virginia, after collapse of scaffolding in which 51 workers plunged to their deaths.

Figure 2.119   Interior view of cooling tower after scaffolding failure.

Lessons

1. Ascertain the actual strength of concrete by in-place testing prior to subjecting the concrete to pull-out loads and prior to removal of forms and shoring.

2. A *written* detailed plan spelling out the step-by-step procedure for testing concrete by destructive and nondestructive methods must be instituted to ensure the safety of temporary structures such as scaffolds, forms, and shoring.

3. Adequate factors of safety must be used.

2.19   Edge Distance of Concrete Anchors

Another important factor affecting the anchorage capacity of inserts embedded within concrete elements is the edge distance of the anchor (Fig. 2.120). This distance not only affects the shear capacity of the anchor, but also, where it is small, reduces the pull-out load capacity.

The case of the failure of the double tees at Rockland County College (Sec. 2.15.8) demonstrates the weakness of stems to resist lateral loads caused by the welding of embedded plates. The following failure is similar and necessitated the expensive replacement of many precast wall panels.

2.19.1   West Virginia hospital — Precast wall panel failure

When several precast panels forming part of the fascia of a hospital cracked, an investigation was launched to find out why. Soon it became evident that stress relief in the form of "soft joints" had not been provided during the construction of the building.

An examination of several panels which cracked and were removed from the wall revealed that the embedments were installed very close to the edge of the panels and failed like the embedments shown in Fig. 2.121.

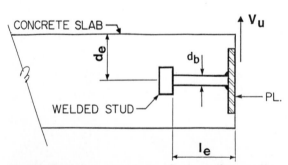

Figure 2.120   An important factor affecting anchorage of inserts into concrete is edge distance of the anchor. $\phi\, V_u = 3250\, (d_e - 1)\, \lambda\, (f'c/5000)^{1/2}$, where $\phi = 0.85$, $d_e =$ distance from stud to free edge, $\lambda =$ factor.

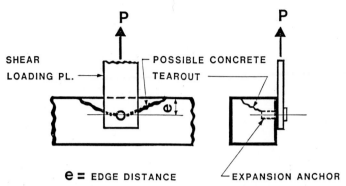

Figure 2.121 Failure mechanism of an anchor embedded too close to the edge of the panel.

Tables of load capacity as a function of edge distance have been developed by the Prestressed Concrete Institute,[9] and Professor Fred Roll of the University of Pennsylvania has originated charts that provide similar data. From these references sources (Fig. 2.122 and Fig. 2.123), it became

Maximum design shear strength, ϕV_u limited by concrete strength, kips													
		Normal weight concrete ($\lambda = 1.0$)						Sand-lightweight concrete ($\lambda = 0.85$)					
	Edge distance	Diameter, d_b in.						Diameter, d_b in.					
f_c'	d_e	¼	⅜	½	⅝	¾	⅞	¼	⅜	½	⅝	¾	⅞
4000 psi	2	1.4	1.4	1.4	1.4	1.4	1.4	1.1	1.1	1.1	1.1	1.1	1.1
	3	2.1	3.0	3.0	3.0	3.0	3.0	1.8	2.6	2.6	2.6	2.6	2.6
	4	2.1	4.7	5.4	5.4	5.4	5.4	1.8	4.0	4.6	4.6	4.6	4.6
	5	2.1	4.7	8.4	8.4	8.4	8.4	1.8	4.0	7.2	7.2	7.2	7.2
	6	2.1	4.7	8.4	12.2	12.2	12.2	1.8	4.0	7.2	10.3	10.3	10.3
	7	2.1	4.7	8.4	13.2	16.5	16.5	1.8	4.0	7.2	11.2	14.1	14.1
	8	2.1	4.7	8.4	13.2	19.0	21.6	1.8	4.0	7.2	11.2	16.1	18.4
	9 or more	2.1	4.7	8.4	13.2	19.0	25.8	1.8	4.0	7.2	11.2	16.1	22.0
Maximum design shear strength V_s ofstuds limited by steel strength (kips)		Diameter	¼	⅜	½	⅝	¾	⅞					
		V_s	2.2	5.0	8.8	13.8	19.9	27.1					

Figure 2.122 Shear strength of welded beaded studs and bolts.

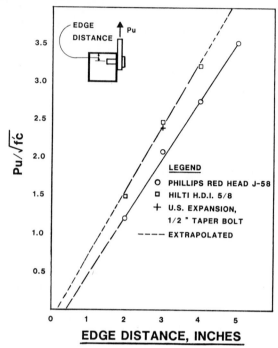

Figure 2.123 Chart of pull-out capacity versus edge distance. (see Prof. Fred Roll).

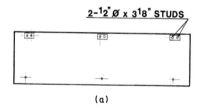

(a)

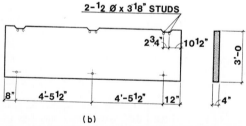

(b)

Figure 2.124 Diagrams showing that the precast panel failure was a result of anchors being placed too close to panel edges. (*a*) Panel prior to erection. (*b*) Panel after failure.

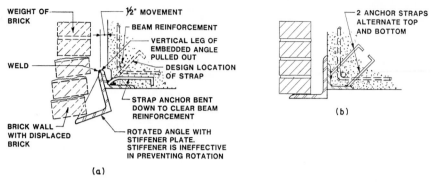

Figure 2.125 Anchor straps were improperly bent in the field to clear rebar placement. The brick shelf angle rotated and failed (*a*). Proper design is shown in (*b*).

apparent that the load capacities of the anchors, based on their proximity to panel edges, were sharply reduced, enough to render them ineffective in supporting the wall panels, leading to their failure (Fig. 2.124).

Lesson.   Embedded anchors must be placed away from the concrete edges. Failure to do so results in a great reduction in anchor capacity, leading to concrete failure in shear.

2.19.2   Alabama courthouse shelf anchorage and New York City high-rise curtain wall clips

The similarity between two failures which occurred hundreds of miles away from one another (in Alabama and New York City) is that both were caused by errors in placement of anchors. In both cases, anchor straps were bent to make room for the main reinforcing bars required for the structure (Figs. 2.125 and 2.126). Thus, the anchors, whose function appeared to be secondary in importance, were abused because of the lack of coordination in placement between the rebars and embedments by other trades.

In the courthouse case, the shelf angle welded to the failed anchor dropped sufficiently to crack the brick. The bent straps were visible to the eye after a thin concrete cover was removed (Fig. 2.127).

In the New York high-rise structure, several of the embedments cracked, which alarmed everybody connected with the project. Load tests by lateral hydraulic jacks followed. These tests proved the insufficiency of the anchors with the bent studs (Fig. 2.128). An angle which failed after the load test is shown in Fig. 2.129. Additional expansion bolts had to be added to supplement the strength of the connection to resist the weight of the curtain wall combined with wind loads (Fig. 2.130).

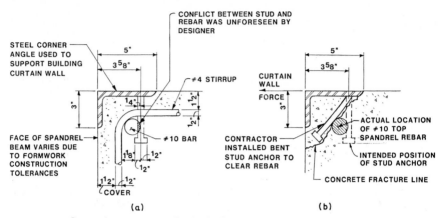

Figure 2.126 Straps for support angles for high-rise curtain wall were bent to make room for beam rebars. Concrete fracture resulted. Design intent is shown in (a) and as-build condition in (b).

Lessons

1. Two items of work cannot occupy the same space. Specifically, metal embedments and anchors must be so located as to allow the placement of reinforcing bars without conflict.

2. Where conflict does exist, the various trades must advise the engineer, who then will take the necessary steps to correct the conflict. No single contractor is permitted to move, shove, or bend embedments placed earlier by other contractors.

Figure 2.127 Exposed bent straps are visible (arrow).

Figure 2.128 Anchor strap embedment that cracked in New York high-rise building after an in-place load test.

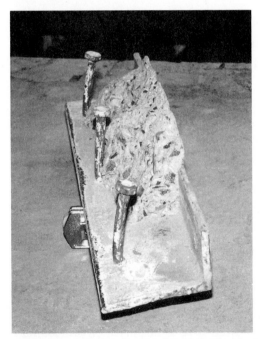

Figure 2.129 Angle bracket that failed after load test, showing bent studs.

Figure 2.130   For repair, plates were welded to existing angle brackets to allow use of supplemental anchors.

2.20   Concrete Formwork Failures

Many investigations into the causes of formwork failures bring to our attention the same repetitive factors:

1. Formwork is a temporary structure, with a relatively short life span.

2. Most formwork failures occur during the winter months.

3. Most formwork design and construction is not performed by professionals trained in this field.

In essence, formwork is a temporary container built for the purpose of holding the fresh, semiliquid concrete until it hardens. This container is often placed at high elevations and must be shored and braced properly. This is accomplished by an array of shores, diagonal braces, lacings, and mudsills (Fig. 2.131).

The key word is *temporary*. Because we all know that formwork is only a means to obtain the final concrete structure, the tendency is to erect and dismantle it in a hurry — forgetting that any failure of the formwork will certainly doom the structure. It is well known that the overwhelming number of concrete construction failures are related to formwork and shoring. Even sagging, settlement, or bowing of forms will often require their costly removal and the repeating of the casting operation.

While formwork is indeed a temporary structure, it is subjected to the same loads as the final permanent structure. It must support dead and live loads and, at the same time, resist wind loads. It must bear on firm foundations, otherwise it will settle and fail. Formwork is also required to resist lateral loads of many kinds, and has to be braced against buckling. So it

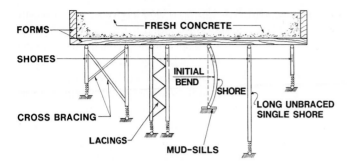

MECHANISM OF FORMWORK FAILURE

Figure 2.131 Formwork is made up of an array of shores, crossbracing, mudsills, and lacings.

should be obvious that we are dealing here with a substantial structure which should be designed, detailed, and constructed carefully and thoroughly. Unfortunately, often this is not done. The fact that the cost of formwork is generally over 50 percent of the cost of the structural concrete frame (Fig. 2.132) is often overlooked.

The interaction of the formwork with the concrete structure is even more complex. The whole operation of shoring and reshoring, the stresses induced in the structure by the jacking operations, and the stresses developed in the relatively young slabs and beams upon stripping, should all be coordinated and engineered properly. Otherwise, hairline or wider cracks will occur as a result of early stripping and improper shoring procedures.

The actual live loads which forms are subjected to include many types of temporary construction loads such as workers, equipment, and runways.

Concrete structural frame cost, $/ft^2				
	Total cost		Frame cost	
Formwork.	$5.25	39%	$5.25	51%
Equipment and labor for installation and removal				
Concrete.	$2.85	21%	$2.85	27%
Material and labor for placement and finishing				
Reinforcement.	$2.25	17%	$2.25	22%
Materials, accessories, and labor for installation				
General conditions.	$2.00	15%		
Bonds, insurance, utilities, administration, and material handling (cranes)				
Subtotal: Direct cost.	$12.35	92%		
Interest.	$1.11	8%		
Total frame cost ($/ft^2).	$13.46	100%	$10.35	100%

Figure 2.132 The cost of formwork is more than 50 percent of the cost of the concrete frame itself, yet it is often not designed by qualified engineers (1985 prices).

The effects of impact forces from dropping of loads, power buggies, and unusual construction vibrations must be included in the design live load [which is recommended as 50 lb/ft^2 (2.4 kPa) by the ACI-347 standard[10]].

Although forms are most vulnerable to wind load when the forms are empty, as the forms are being filled with concrete in its semiliquid state, reversal of loads often occurs. This, in some instances, will result in *uplift* forces on the forms which therefore *must be tied down for safety.*

The stresses generated during form removal, or stripping, may lead to failure or even to total collapse. Stripping is especially critical in formwork of large structures such as bridges, where *balanced* stripping is essential to the *stability of the formwork and the new concrete just cast.*

Premature stripping of forms may also cause damage to "young" concrete. Concrete is elastic even at early age, so it possesses these properties:

1. Compression strength

2. Tensile strength

3. Modulus of elasticity

Prior to stripping, good engineering practice today calls for testing only the first property. Even this procedure is often ignored. We *should* also test the last two properties, because:

1. Tensile strength is a measure of cracking resistance.

2. Modulus of elasticity is a measure of deflection.

Upon stripping, if these latter properties are low, we should expect to find cracks and deflections — and unfortunately we often do. Where a quality control program is in effect, low tensile strength or modulus will thus be detected *before* stripping, and cracking and excessive deflections will be avoided.

2.20.1 Lateral stability

Formwork is usually a crude structure supported by an array of posts which are not rigidly connected to the form at the top or to the floor at the bottom. In addition, the weight of the fresh concrete is almost entirely at the top, which affects the stability of the system.

This vulnerability requires that lateral resistance be introduced into the formwork to prevent the structure from becoming unstable. This resistance is most effectively provided by diagonal crossbracings which are placed at the main formwork members.

Considerable resistance against lateral sway may also be provided by casting the columns at least a day ahead of the floor slab. The overall cost of placing the concrete may be somewhat increased by following this procedure, but this small expenditure is relatively inexpensive insurance for safety and stability. At the same time, the columns are permitted to absorb their initial shrinkage, a procedure often specified but rarely followed.

2.20.2 Progressive collapse

Formwork structures are very susceptible to progressive collapse. Because the factor of safety of shores against buckling is relatively low, regardless of whether they are made of timber or of steel, a failure of a single shore may trigger a monumental collapse, as has occurred on so many occasions. A failure of a single shore will shift the center of gravity of the load and increase the load on the adjacent shores by as much as 50 percent (Figs. 2.133 and 2.134). In the case of a highly loaded formwork structure, such an individual failure may trigger the collapse mechanism.

Failure of a single shore may also cause uplift forces in adjacent shores. Where these are not rigidly connected to the forms, or otherwise laced to each other (Fig. 2.131) to form four-legged towers, they will fall down. This will initiate a progressive or domino-type effect, leading to a major collapse. In less dramatic cases, single shores have frequently fallen from high structures as a result of uplift, and caused serious injuries to workers or pedestrians. (See Chap. 1 for further discussion on progressive collapse.)

2.20.3 Mudsills

Shores and reshores supported on a lower completed floor may be assumed to have equal stable bearing. However forms for the lowest slab level are often supported on mudsills, which are not always bearing on solid ground. These continuous wood runners are frequently set on recently placed backfills with great probability of softening from flow of water. The source of this water may be from natural runoff, wash water from the forms, or from concrete truck mixers.

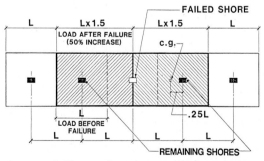

Figure 2.133 Formwork structures are very susceptible to progressive collapse. Failure of a single shore can increase the load on adjacent shores by as much as 50 percent. Shore load before failure: $W_\phi = \omega L$. Shore load after failure: $W_{-1} = 1.5 \omega L$.

FAILED SHORE

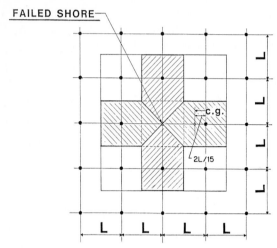

Figure 2.134 In flat slabs, failure of a single shore can increase load 25 percent on an adjacent shore. Shore load before failure: $W_0 = \omega L^2$. Shore load after failure: $W_{-1} = 1.25\ \omega L^2$.

Unequal settlement of the sills seriously affects the redistribution of loads among the various shores. Obviously, when the soil below some of the shores yields, the shores which do not settle receive more than their intended designed share of the load. Because of the low redundancy of formwork structures, this again triggers progressive collapse.

2.20.4 Lateral pressure of fresh concrete on forms

Hydrostatic concrete pressures on forms are important factors in many concrete formwork failures. While the pressures are dependent primarily on *the height of the fresh concrete column*, there are a number of other factors which heavily affect the amount of pressure generated in the forms. These are:

1. The rate of casting in feet per hour (meters per hour)
2. The temperature of the concrete
3. The level of vibration
4. Retarders or other admixtures used in the mix

Failures of form ties are also common occurrences, and are not even reported unless they are associated with injuries or deaths. When pressures are underestimated or where forms are not braced to resist these lateral pressures, the unfortunate consequence is failure of the formwork of one kind or another.

2.20.5 Formwork fires

There are too many formwork fires that start when temporary heat in enclosures causes buildup of high temperatures. With dried-out lumber, it is then a simple matter for a small gas-pipe to blow or a salamander to be overturned and ruin a construction project. It is no wonder that many articles have titles like "Forms are not for burning."

The fire damage to newly cast concrete can be devastating, with loss of strength of both concrete and reinforcement, not to mention the possible loss of life. Some of the precautions that should be taken are:

1. Avoid the use of heaters with open flames.

2. Do not refuel heaters while they are lighted.

3. Keep fuel tanks away from heaters.

4. Keep firefighting equipment readily available.

2.20.6 Shoring, reshoring, and backshoring

First we have to define these terms, since their functions within the formwork system vary. In the *Guide to Formwork for Concrete* of ACI Committee 347-78, we find these definitions:[11]

1. *Shores.* Vertical or inclined support members designed to carry the weight of the formwork, concrete, and construction loads above.

2. *Reshores.* Shores placed snugly under a stripped concrete slab or other structural member after the original forms and shores have been removed from a large area thus requiring the new slab or structural member to deflect and support its own weight and existing construction loads applied prior to the installation of reshores.

3. *Backshores.* Shores placed snugly under a stripped concrete slab or structural member after the original formwork and shores have been removed from a small area, without allowing the slab or member to deflect or support its own weight or existing construction loads from above.

The original shore must be installed "plumb" and "wedged securely" in order that each shore will carry its share of the load. It also must be "completely straight" and "undamaged." These exact requirements have not, unfortunately, been properly defined.

Any review of construction sites will confirm that the intent of these requirements is constantly violated. The American National Standards Institute (ANSI) Standard A10.9[12] did include an allowable construction tolerance of ⅛ in (3.2 mm) in 3 ft (0.92 m), which corresponds to ⅜ in

(9.5 mm) in 9 ft (2.74 m). (This requirement has since been omitted for reasons unknown to the writer.)

An interesting survey of construction sites in England in 1971 showed that 60 percent of the shores measured were found to be more than ⅖₆ in (7.9 mm) out of plumb in 3 ft (0.91 m). Over 10 percent were more than 1 in (25.4 mm) out of plumb in 3 ft (which amounts to 3 in in 9 ft; 76 mm in 2.74 m). There is no reason to believe that U.S. construction is any better.

The purpose of reshores is to distribute any further applied loads among several slabs that are stronger (older). Reshores should not be placed where they could cause tensile stresses where there is no reinforcement. Where reshores are not placed in the same location on each consecutive floor, they may cause serious cracking of the concrete.

There are two very important guideline requirements regarding the design of high-rise formwork shores which were addressed in the American Concrete Institute ACI-347 guide, and are often ignored by formwork designers and constructors:

1. Shores should be designed to carry the full weight of the concrete and formwork and construction loads of all floors above them *prior to the removal of the first story of shores supported by the ground* (emphasis is mine).

2. Shores in no case should be designed to carry less than one and one half times the weight of a given floor of concrete, formwork, and construction loads.

This last requirement may translate to a minimum shore load of *2½ to 3 times the weight of the slab.*

ANSI Standard A10.9 also contains valuable information regarding the requirements for design and construction of shores.

Some of the most common shoring failures are:

1. Height of screw jacks at the top of scaffolding legs is overextended (Fig. 2.135)

2. Eccentric loading on the top U bracket, causing bending of the hollow leg members

3. Use of bent or buckled bracing members — or even omitting them altogether

4. Failure to tie *high* towers together with diagonal bracing to resist horizontal loads

5. Use of double-tier shores (Fig. 2.136)

6. Failure to crossbrace individual shore legs to support vertical loads (Fig. 2.137)

7. Failure to scab *all four sides* at splices

8. Failure to use square ends with full contact at splices

Figure 2.135 Long shore extensions should be braced.

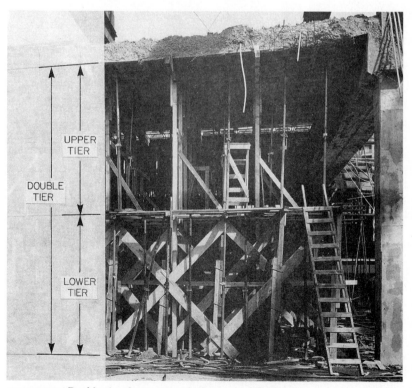

Figure 2.136 Double-tier shores are an invitation to collapse.

Figure 2.137 Single shores without crossbraces often loosen and topple. This not only results in loss of formwork support but also endangers workers.

9. Top of shores inadequately tied to stringers to resist uplift or other shifting loads

10. Use of initially bent and out-of-plumb shores (Fig. 2.138)

Understanding the complex interrelation between the shoring system and a recently completed concrete slab and column frame is essential to the proper design and construction of formwork — especially important in high-rise construction. The fact that the maturity of the concrete affects the loads in the shores must be recognized.

Major collapses related to premature removal of shores or other shoring failures are described in Sec. 2.6. These include such monumental failures as Bailey's Crossroads, Commonwealth Avenue, and Cocoa Beach.

2.20.7   Wall Form Failures

While major collapses get headlines, there are many minor failures which occur every day that go unreported. Many of these are simple wall form failures, which, contrary to the opinion of many, can result in serious accidents.

Figure 2.138 Formwork is often supported on out-of-plumb and bent shores that have been reused too many times.

Some of the causes for these failures are:

1. High rate of placing ("going too fast")
2. Improper or insufficient tying and/or the use of damaged ties
3. Missing or improperly connected hardware
4. Insufficient anchorage into previously cast concrete
5. Inadequate strongbacks, wales, or bracing (L/D greater than 50)

Formwork designers and constructors must remember that long 2 × 4s [longer than 6 ft (1.8 m)] are worthless as braces and accepting their effectiveness is wishful thinking. They at best serve as tension members, provided they are connected at their ends.

2.20.8   Newark, New Jersey—Garage collapse

In this collapse, about 120 yd³ (92 m³) of concrete in the roof of a three-story garage fell (Fig. 2.139) as the workers were screeding the completed surface of an 18-in (457-mm) slab designed to support 4 ft (1.22 m) of earth. The fall of the slab onto the second-floor slab, then 2 weeks old, also broke out the same area, 2000 ft² (186 m²), and then cracked the slab below. Steel pipe reshores punctured through the slab two tiers below the roof and triggered the collapse (Fig. 2.140).

Figure 2.139 Eighteen-inch roof slab for a three-story parking garage collapsed during screeding operation.

The concrete at that tier was supposed to have had type III (high-early-strength) cement in the mix, but, instead, normal cement (type I) had been used. The reliance of the contractor on the reported strength of labora-tory-cured cylinders was unwarranted, even though the laboratory was retained by the owner, because the *actual strength* of the concrete in this floor (placed in rather cold weather) was only *half the expected strength.*

The court found in this case that:[13]

> . . . a construction contractor who had agreed to build a structure in con-formity with plans and specifications is not relieved from doing so when he orders the right materials (i.e., those specified) but is shipped the wrong ones by his suppliers and unwittingly uses them.

Lessons

1. Stripping of forms must be done only after verification of the in-place concrete by *field-cured cylinders.*

2. Contractor shall always take *its own companion cylinders* and have them tested independently to verify actual in-place concrete strength.

2.20.9   Crystal Mall—Alexandria, Virginia

The collapse of the formwork of this low-rise concrete building started while the 8-in (203-mm) lightweight mezzanine floor slab was being cast

Figure 2.140 Steel pipe reshores, which buckled and punctured the slab two tiers below the roof, triggered collapse.

(Fig. 2.141). The support for the first-floor slab, which had been cast 6 days earlier, failed first. It failed in a funnel shape (Figs. 2.142 and 2.143), with its low point directly below a steel hanger which was designed to support the first floor from a second-floor girder to be posttensioned at a later time. Ordinarily, the lower decks support the freshly cast concrete above. Here, the shores between the basement and the first floor were expected to remain in place to support the total weight of the first, mezzanine, and second-floor slabs, not just until they achieved adequate strength, but until *after* the posttensioning of the second floor had been completed.

Apparently these shores were inadequate (they were quite high with insufficient diagonal bracing) and failed. Some of the shores on this project were jack-type shores which are assembled in two pieces lapped with bridle iron loops. This type of shore is incapable of resisting uplift forces which are often associated with collapses.

Figure 2.141   The collapse of the formwork for low-rise concrete building started when an 8-in mezzanine floor slab was being cast.

Figure 2.142   The first floor slab, cast 6 days earlier, failed first, in a funnel shape.

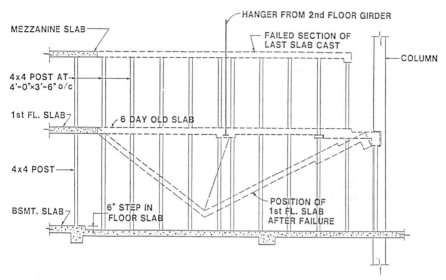

Figure 2.143 Diagram of formwork shows cause of collapse. Shores between basement and first floor were to remain in place until after posttensioning of second floor. Shores were inadequate and failed.

Lessons

1. Where an unusual sequence of construction is required, the contractor must be made aware of the exact procedure to follow, with a step-by-step written procedure reviewed by the engineer designing the formwork and shoring system.

2. Avoid the use of high unbraced shores. The shoring system *must be braced in both directions* for stability.

2.20.10   T. W. Love Dam—Cantilever form failure

In 1981 two workers were killed and 13 injured at the site of a dam construction project when the formwork supporting the mass concrete pour failed.

The formwork support system (cantilever forms) consisted of vertical and sloping strongback double channels reinforced by channel struts and an adjustable Patterson strut for alignment (Fig. 2.144).

The stability of the entire assembly depended on the "couple" reaction of a crimped anchor acting in tension and an angle standoff (kicker) acting as a compression leg. There was full agreement that the failed element was the anchor. The divergence of opinion centered on the question "Why?"

In 1983, a lawsuit was filed against the company which manufactured the she-bolts and crimped anchors and against the form company which allegedly designed the formwork for the general contractor for the project.

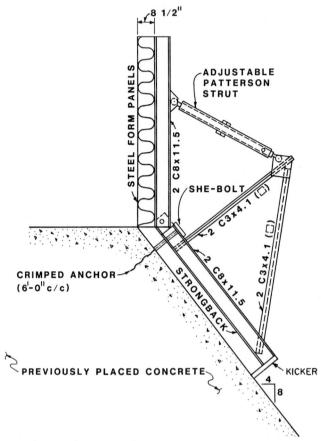

Figure 2.144 Formwork cantilever support system consisted of vertical and sloping strongback double channels reinforced by channel struts and an adjustable Patterson strut for alignment.

The findings of various investigations indicated that the formwork system was faulty in design as well as improperly constructed.

Subsequently, third-party claims were filed against the steel company which sold the steel to the anchor manufacturer, the company which machined the steel bars, and so on.

Some of the defendants alleged that the failure was caused by the contractor who overstressed the system during construction, that the concrete pour rate was excessive, and that one of the deceased workers was torquing the Patterson strut to replumb the wall (Fig. 2.145) when the system collapsed (Fig. 2.146).

The claims included the following:

1. Excessive rate of casting the concrete

Figure 2.145 Two workers were killed and 13 injured at dam construction project when formwork (shown here during installation) supporting the concrete pour failed.

2. Material failure of the anchor

3. Overtightening of the Patterson strut

4. No redundancy of the design

5. Excessive loads on the first anchor, which failed as a result of the end cantilevers

6. Failure of the bolts in bending

Figure 2.146 Cantilever formwork after collapse.

Nobody considered possibility 6 in spite of the fact that metallurgical tests referred to "bending fracture caused by unknown forces."

After considering all possible causes, we conducted a thorough investigation of the loads acting on this system. This investigation led us to conclude that, while every one of the above listed claims was a factor, they were only secondary ones. The actual mechanism of failure was bending of the anchor. Unfortunately, the designers of the formwork failed to consider that the loads acting on the slope tended to slide the entire system down the slope (force C in Fig. 2.147). The original design assumption, which disregarded the bending effect of force C, is shown in Fig. 2.148.

The dispute that arose after the collapse revolved around the fact that nobody claimed responsibility for the design of the forms. Because the design of formwork is often not regulated, and because it is regarded as part

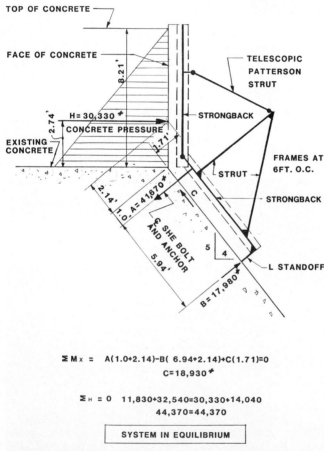

$$\Sigma M_x = A(1.0+2.14)-B(6.94+2.14)+C(1.71)=0$$
$$C=18,930\#$$

$$\Sigma_H = 0 \quad 11,830+32,540=30,330+14,040$$
$$44,370=44,370$$

SYSTEM IN EQUILIBRIUM

Figure 2.147 Designers of formwork failed to consider loads acting on the slope tending to slide the entire system down the slope.

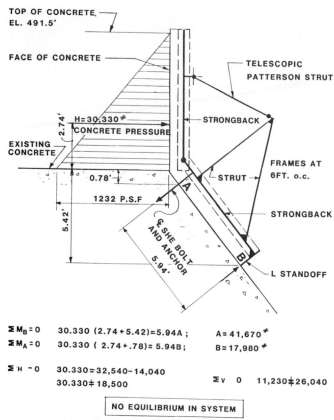

TOP OF CONCRETE, —
EL. 491.5'

FACE OF CONCRETE ——

H=30,330#
CONCRETE PRESSURE

2.74'

EXISTING
CONCRETE

0.78'

1232 P.S.F

5.42'

5.94'

TELESCOPIC
PATTERSON STRUT

STRONGBACK

FRAMES AT
6FT. o.c.

STRUT →

STRONGBACK

L STANDOFF

$\Sigma M_B = 0$ 30.330 (2.74 + 5.42)=5.94A ;   A= 41,670#

$\Sigma M_A = 0$ 30.330 (2.74 + .78)= 5.94B ;   B= 17,980#

$\Sigma H = 0$ 30.330=32,540–14,040
              30.330 ≠ 18,500 Σv 0   11,230 ≠ 26,040

NO EQUILIBRIUM IN SYSTEM

Figure 2.148 Original design assumption was in error because it disregarded the bending effect of the she-bolt anchor.

and parcel of the "methods and means" for which total responsibility is considered to lie with the contractor, formwork design is quite often left to untrained personnel lacking the required technical knowledge of the interactions among loads, stresses, structure, and stability.

In the case of the W. T. Love Dam, the design of the formwork and its supports was prepared by the formwork vendor, as a service to the contractor. After the collapse that same vendor referred to its so-called design as a "proposal," and refused to take responsibility for it.

Lessons

1. Avoid bending stresses on anchors.

2. Consider all the loads acting on a structure. Ascertain that the summations of all vertical and horizontal loads and bending moments are in equilibrium.

Figure 2.149 General view of the New York Coliseum collapse.

3. Provide an adequate factor of safety against failure of individual anchors, especially those acting in tension.

4. Provide redundancy in formwork design.

5. Control the rate of casting the concrete, so it stays within the design assumptions.

2.20.11 New York Coliseum formwork collapse

It is always difficult to find flaws in formwork construction after a collapse, since the rubble and debris are often totally disturbed during rescue operations.

In the New York Coliseum collapse (Fig. 2.149), however, the removal of the collapsed material uncovered drifted posts which were considered to be the most probably cause of the collapse. We discovered that two 4 × 4 in (102 × 102 m) posts "walked off" their 6 × 12 in (152 × 305 mm) beam supports spanning the escalator pit. The 4 × 4 (102 × 102 mm) wood posts fell approximately 3½ ft (1 m) and punched through the 4-in (102-mm) concrete pit floor (Fig. 2.150).

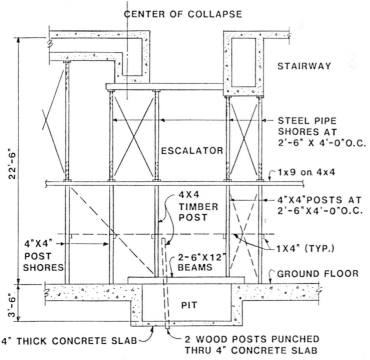

Figure 2.150 In New York Coliseum collapse, which killed one worker, two posts walked off their beam supports spanning the escalator pit. Double-tier shores were improperly used.

The structural design of the main exhibition floor consisted of 20-in-thick (508-mm) concrete waffle slabs. The immediate area of the failure had several deep girders framing out the main stairway and escalator opening.

The failure area covered about 10,000 ft² (929 m²) of formwork (Fig. 2.151), supported on two-tier shoring with a total height of 22 ft (6.7 m). The bottom tier consisted of wood posts spaced on a 2.5 × 4 ft (0.76 × 1.22 m) grid with a line of horizontal bracing in two directions at the top, with *some diagonal bracing*. The upper tier sat on top of a cap sill and consisted of adjustable pipe shores with flat cap plates for the support of the horizontal timber joists.

The same form design had been used on this project for some 60 previous castings of similar areas, some with the same 22-ft (6.7-m) height.

On the day of the collapse, May 9, 1955, some 700 yd³ (535 m³) of a scheduled 1000-yd³ (765-m³) pour had been placed by 2 p.m., when failure occurred without any warning. The workers were away from the deck at the time of the collapse. Most of the formworkers below the deck were also away at the time, except for one worker, who lost his life.

Placing of concrete was by power buggies traveling on timber duck-boards at reported top speeds of up to 12 mi/h (19.3 km/h), each carrying 12 ft³ (0.34 m³) of mix (Fig. 2.152).

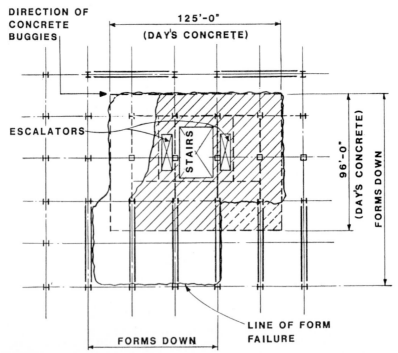

Figure 2.151 Layout plan showing the collapsed area. Note: 500 yd³ of concrete in place on May 9, 1955.

Figure 2.152 Concrete placing was by power buggies traveling on timber duckboards at reported top speeds of 12 mi/h, each carrying 12 ft^3 of concrete mix.

The exhaustive investigation determined that power buggies at high speeds, when acting in synchronism, will impose lateral thrust into the deck. The shoring system below must be designed to transfer these loads by means of diagonal bracing members. Proper bracing was lacking, causing the formwork to sway and fail. The horizontal forces transmitted to the deck by the fast-traveling buggies, stopping short on occasion, caused the posts to walk off the timbers (Fig. 2.150).

The job was completed using the buggies at slower speeds, adding diagonal bracings and nailing the wooden joists to the top caps.

Lessons

1. Shoring systems should be well-braced to resist lateral loads.

2. Special attention should be given to *high shoring systems.*

3. Avoid two-tier shoring where possible.

4. Power buggies should be operated at moderate speeds.

5. Where power buggies or motorized carts are employed, use an additional live load of 25 lb/ft^2 (1197 Pa) per ANSI A10.9.

2.20.12   Philadelphia LNG tank collapse

The magnitude and unpredictable effects of rainwater on temporary formwork construction became evident in Philadelphia on June 29, 1973, when 4.41 in (112 mm) of rain fell on the concrete berm which was being placed.

The concrete was to be the wall of one of two liquefied natural gas (LNG) storage tanks being built for the local gas company.

The tank's typical cross section is shown in Fig. 2.153. The diameter of the steel shell was 248 ft (76 m). The outer tank consisted of an 85-ft-high (26-m) prestressed wall with wire-wrapped reinforcing and pneumatically placed mortar surface over the wires. Supporting the fresh concrete 10 ft (3.1 m) away was a 5⁄16-in-thick (8-mm) carbon-steel welded liner. The 10-ft-wide (3.1-m) space was filled with 2000 psi (14 MPa) concrete. The casting of the concrete was being carried out in 2-ft (0.61-m) lifts in order to avoid high concrete pressures and resulting stresses in the steel liner. On the inside, the steel was braced against the inner prestressed concrete shell with a wooden shoring system consisting of double 2 × 8 in (51 × 203 mm) vertical shores with 4 × 4 in (102 × 102 mm) horizontal braces (Fig. 2.154).

As each concrete lift was placed, provision was made to keep hydrostatic pressure off the steel liner. Four 4-in-diameter (102-mm) galvanized pipe drains were being extended upward, and a cutting torch was used to make side outlets in these four pipes to drain away any rainfall between lifts. The flame-cut holes, which were about 3 × 4 in (76 × 102 mm), were then covered with a piece of light sheet metal to prevent concrete entry. In addition, every 10 ft (3.1 m) (five lifts), the joint between the concrete and the steel liner was caulked with mastic.

Just prior to the collapse, the rainwater penetrated the space between the steel liner and the hard concrete and developed full hydrostatic pressure. The amazing fact was that the actual width of the space (the *gap*, as it was referred to during the trial) was only 0.01 in (0.25 mm). But, as it turned out, that was all that was needed to fail the steel shell liner, with a loss of several million dollars. The separation between the concrete and the liner after the failure was more than 1 ft (0.31 m) (Fig. 2.155). The welded

Figure 2.153 Cross section of LNG tank. Design called for an interior high prestressed wall 8-in (203-mm) thick with wire-wrapped reinforcing and pneumatically placed concrete over the wires.

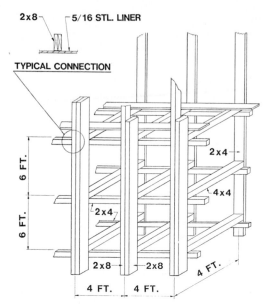

2x8   5/16 STL. LINER

TYPICAL CONNECTION

6 FT.

6 FT.

2x4

4x4

2x4

2x8   2x8   4 FT.

4 FT.   4 FT.

Figure 2.154 On inside of tank, steel was braced against inner concrete shell with a timber shoring system.

TORN STUDS

1-FT WIDE GAP
DEVELOPED AFTER COLLAPSE

Figure 2.155 Separation between the concrete and liner after failure was more than 1 ft (0.31 m). Welded steel studs used for bonding the liner and concrete actually tore from the liner under the high water pressures.

steel studs used for bonding the liner and the concrete actually tore from the liner under the high hydrostatic pressures.

In postfailure tests (Fig. 2.156), it was confirmed that even very narrow spaces are sufficient for developing a full hydrostatic head.

Our investigation determined that water accumulated in a particular area of the berm (referred to as *notch* in Fig. 2.153) which was approximately 2 ft (0.62 m) lower than the rest of the berm. There were claims that vertical pipes passing up through the concrete berm to serve as drains were inoperative during the morning of the failure because no holes had been cut in them at the level of the surface of the concrete (Fig. 2.157 shows such a cut in the vertical drain pipe).

The drainage system conceived (or "designed" as later alleged by the insurance company) for the construction of the tank totally failed. In addition to the question of holes not being cut, it was determined that the number and capacity of the pumps used to evacuate the water in the trenches were not adequate.

Because the insurance policy of the contractor had an exclusion for "design," the exact definition of this term became an issue in court.

Other than such dictionary definitions as:

Figure 2.156 Postfailure tests confirmed that spaces as narrow as 0.01 in (0.25 mm) are sufficient for developing full hydrostatic heads.

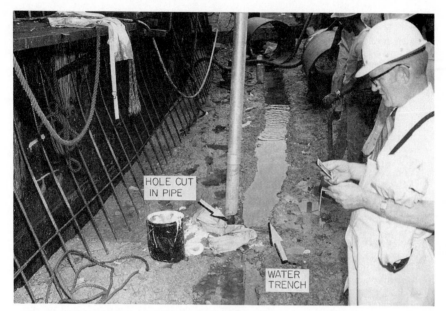

Figure 2.157 Vertical drain pipes were set in concrete. Water was to drain through trenches at top of berm into holes cut in the pipes.

"A mental project"

"A deliberate undercover project"

"A preliminary sketch or outline showing the main features of something to be executed"

"A decorative pattern"

codes and industry standards are surprisingly silent on this matter. To what extent the determination of the number of drainage pipes and the number of pumps required for temporary pumping is to be considered "design" was the issue.

Lessons

1. Temporary provisions for drainage of rainwater must be made by the contractor so as to avoid high water pressures on formwork and floods that may result in collapse.

2. Contractors should avoid the design of temporary structures such as formwork and shoring. This design should be left in the hands of qualified professionals, who are capable of obtaining the proper professional liability insurance coverage. These professionals should be retained by the contractors.

3. Even very narrow spaces are sufficient for the development of full hydrostatic heads and, therefore, should be tightly sealed.

References

1. Report of the Mayor's Investigating Commission, "The Building Collapse at 2000 Commonwealth Avenue," Boston, Mass., June 1971.
2. H. S. Lew, N. J. Carino, S. G. Fattal, and M. E. Batts, Investigation of Construction Failure of Harbour Cay Condominium in Cocoa Beach, Fla., NBSIR 81-2374, U.S. Department of Commerce/National Bureau of Standards Center for Building Technology, Washington, D.C., September 1981.
3. C. S. Culver, C. F. Scribner, R. D. Marshall, F. Y. Yokel, J. L. Gross, C. W. Yancey, and E. M. Hendrickson, Investigation of L'Ambiance Plaza Building Collapse in Bridgeport, Connecticut, NBSIR 87-3640, U. S. Department of Commerce/National Bureau of Standards Center for Building Technology, Gaithersburg, Md., September 1987.
4. J.-P. Levy, "The Pathology of Construction, the Causes of Cracking of Concrete," D. A. Sinclair, trans., Technical Translation TT-574, National Research Council of Canada, Ottawa, 1955.
5. D. Kaminetzky, "Rehabilitation and Renovation of Concrete Buildings," Proceedings, ASCE Workshop on Rehabilitation, Renovation and Reconstruction of Buildings, New York, N.Y., February 14–15, 1985, cosponsored by the National Science Foundation.
6. State of California Advisory Bulletin, Antioch High School, March 16, 1981.
7. Building Code Requirements for Reinforced Concrete (ACI-318-89), American Concrete Institute, Detroit, Mich., 1989.
8. H. S. Lew, S. G. Fattel, J. R. Shaver, T. A. Reinhold, and B. J. Hunt, Investigation of Construction Failure of Reinforced Concrete Cooling Tower at Willow Island, W.V., U. S. Department of Labor, OSHA/National Bureau of Standards Washington, D.C., November 1979.
9. Prestressed Concrete Institute, "PCI Design Handbook," 3d. ed., 1985, pp. 6–53. Tests conducted by Professor Fred Roll, in the Department of Civil Engineering, University of Pennsylvania, Philadelphia, Pa., 1971.
10. Guide to Formwork for Concrete, ACI 347R-88, American Concrete Institute, Detroit, Mich., 1988.
11. Ibid.
12. ANSI Standard A10.9-1983, "Standard for Construction and Demolition Operations Concrete and Masonry Work — Safety Requirements," American National Standards Institute, New York, N. Y. 1983.
13. Newark Garage Collapse, Court Decision, Superior Court of New Jersey, November 1944.

Chapter 3

Structural Steel

Structural steel is considered to be an extremely dependable construction material, the product of a closely controlled manufacturing operation. Steel with minimum specifications of 36,000 psi (248 MPa) [ASTM 36 ksi (248 MPa) grade] will test quite often at an average level of 39,000-psi (269-MPa) yield stress.

I frankly cannot recall a case where a failure has occurred because of some deficiency that could be traced to the strength of the steel itself. Even after spectacular collapses of dome roofs and failures of steel long-span and space structures subjected to extreme snow or water-ponding loads, in my experience not once was the strength of the base structural steel material itself found to be at fault.

Cases of brittle-fracture failures of structural steel have been recorded. These have been mostly in bridge structures which are exposed to low temperatures and impact loading. The most frequent such failures have been of welded thick plates using steel of acknowledged low toughness. The Wolftrap Center roof girder cracking (Sec. 3.5.1) is one such example.

We have investigated several cases of brittle fracture in buildings. These commonly have been traced to flaws but were triggered by unusual impacts and/or low temperatures.

Problems are frequent, however, with the connections of steel structures which will function only if designed and constructed properly. The most notable recent failure traceable to such a cause is the disastrous collapse of the Kansas City Hyatt Regency Hotel walkway (Sec. 3.3.1); all investigators agreed that the failure was caused by a defective connection.

Steel structures are coupled with two basic connection systems: high-strength bolts and welding. Rivets were extensively used in the past, but are no longer common.

Bolting is reliable and may be economically used for connections manufactured in the shop and erected in the field. Welding, on the other hand, is highly adaptable to field operations and is especially effective for rehabilitation and repair work. Its only drawback is the difficulty of achieving quality control in the field.

There have been very few recorded cases of failure in which high-strength bolts have been identified as the major cause. There have been, however, substantial numbers of failures where defective welding has been determined as the main cause of failure.

While there are many reasons why steel structures fail, these are the four most common and also the most likely to occur:

1. Insufficient temporary bracing during construction

2. Errors in design and/or construction

3. Deficient welding

4. Excessive flexibility and nonredundant design (Fig. 3.1)

Typical examples of insufficient temporary bracing during construction which resulted in total failure of the structure are covered in Sec. 3.1.

Errors in design and/or construction can lead to such spectacular structural failures as that of the Hartford Coliseum, Fig. 3.21.

Representative cases of deficient welding that led to major failures include those covered in Sec. 3.5.

There are many examples of failures resulting from designs that create structures that are too flexible or structures without redundancy. Typically, such structures will collapse under snow or water loads on their roofs. Two case histories (Secs. 3.7.1 and 3.7.2) reveal that open-web roof joist designs are often the culprit. Where the failure is initiated by buckling failure of a compression truss member or a welding failure because of the nonredundancy of the system, the load lacks an alternative path of transfer and the expected failure ensues. The code-mandated factors of safety are

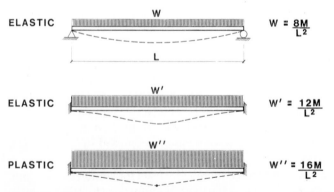

ELASTIC   W $W = \dfrac{8M}{L^2}$

L

ELASTIC   W' $W' = \dfrac{12M}{L^2}$

PLASTIC   W'' $W'' = \dfrac{16M}{L^2}$

Figure 3.1   Diagrams illustrate how redundancies increase the load-carrying capacity of steel beams.

only 1.5 for buckling and 1.65 for welding. Hence, a 50 percent overload or 40 percent reduction of weld strength will result in failure.

Examples of nonredundant designs that led to failure include those covered in Secs. 3.3.1 and 3.7.3. The introduction of redundancy in steel structures by incorporation of rigid connections will enhance their load-carrying capacity. Looking at Fig. 3.1, it is interesting to note that a single beam with hinged ends will support a load of only $8M/L^2$ (where M is the bending moment and L is the span), while a beam with fixed supports will carry loads of $12M/L^2$ up to $16M/L^2$, 1.5 to 2 times as great a load.

In addition, inadequate functioning of expansion joints can lead to major structural distress and failure. Unforeseen vibration modes resulting from wind forces can set up harmonics that can destroy a structure (Sec. 3.10.1).

Of course, many structural failures result from a combination of two or even more shortcomings.

Among important precautions the designer should take are these:

1. All loads used in the design should be included in the final design working drawings. It is not sufficient that general statements be made such as: "Live loads in accordance with local codes have been assumed in the design."

2. The live loads used should be tabulated in loading schedules which show all dead and live loads in the particular areas. These include wind loads and seismic loads (where applicable).

3. Particular attention must be paid to heavy load concentrations such as those from trucks, equipment, and machinery.

4. Because piping and machinery hanging from beams and truss chords are often critical and have resulted in buckling and yielding of structural elements (Sec. 3.9.2), these loads should be shown on the structural design drawings to the extent this information is available at the time of design. The design may be based on allowances such as "20 lb/ft^2 (958 Pa)" or "1000 lb (4448 N) concentrated load at bottom truss chord."

5. Do not forget to require that: "Mechanical shop drawings showing loads of machinery, equipment, and piping should be submitted to the structural engineer for evaluation before final approval of *structural shop drawings.*"

3.1 Failures Due to Insufficient Temporary Bracing during Construction

3.1.1 Supermarket warehouse — Plainfield, New Jersey

I visited this warehouse construction site 11 days after the building collapse, which occurred on December 26, 1969. I inspected the position of the various structural elements, looked for evidence of temporary cable guys and evidence of bolts within typical connections of beams to columns, and inspected the condition of column bases and anchorage to concrete work.

The structural steel erection had been in progress, with all major items in position but with no permanent connections. Some steel joists had been connected but most were in strapped bundles atop the steel girders when apparently a gusty wind caused a substantially complete fall of the entire steel assembly to the ground (Fig. 3.2).

The building was single-story with steel framed roof, 217 × 463 ft (66 × 141 m), with a two-story office area extension of 42 × 145 ft (13 × 44 m) at the south end of the rectangle. Columns, including the wall lines, were spaced at 36-ft (11-m) intervals in the short direction and 42-ft (13-m) intervals in the long direction. Columns were 8 WF sections with welded base plates, anchored into concrete piers or exterior-grade beams with two ¾-in-diameter (19-mm) bolts, 18 in (457 mm) long.

There was no indication of any foundation distortion or displacement or any damage to concrete except minor spalls where steel members fell and struck the concrete edges. A considerable number of bolts were pulled out or were found to have stripped threads. These anchor bolts required repair —extension by welding a new threaded length to them and, where they were missing, even inserting a replacement into a grouted drilled hole.

At the time of the collapse, the wind direction and speed at the site could only be approximately estimated. Winds in the morning had been very gusty from the northeast, with driving rain. Later in the day, the winds

Figure 3.2 Single-story steel frame warehouse collapsed during construction, with some steel joists connected but most still in strapped bundles atop girders.

swung around to the west as a cold front passed through. Such a change is always accompanied by strong wind gusts. Failure definitely came from an easterly wind which caused the center part of the long side to fall completely over and across the width of the building. This fall left one line of the columns standing in a sloping position (Fig. 3.3); the girders sat on top of the columns and could easily slide off the top cap where the bolt connection had not been made or had not been tightened. The center area then dragged the north and south parts inward with some drift eastward. The two east corner columns remained standing.

The girders were generally 21 WF. There was no rigid crossbracing or temporary guys across the width of the building to take the wind load directly into the ground. In fact, there was a series of cantilevers acting independently.

With the top of the steel girders at 25 ft (7.6 m) above the column base, the wind on the girders generated a lever arm of about 24 ft (7.3 m). The steel in the anchor bolts could be expected to fail in tension at 60,000 psi (414 MPa). Hence, a wind force of 4.5 lb/ft^2 (215 Pa) would be sufficient to cause bolt failure.

For a flat surface, wind pressure is $P = 0.003v^2$, where P is wind pressure in pounds per square foot and v is wind velocity in miles per hour.

Figure 3.3   Failure was the result of a gusty easterly wind which caused the center part of long side of building to fall over, leaving one line of columns standing but tilted over.

A pressure of 4.5 lb/ft² (215 Pa) therefore results from a wind speed of 39 mi/h (63 km/h). For I-beam sections, the pressure is higher than for flat surfaces and is close to $0.005v^2$; on an I beam, the 4.5 lb/ft² pressure would be caused by a wind of 30 mi/h (48 km/h). The overturning of the structure by a gusty wind could therefore be reasonably explained. No account was taken of the wind exposure of the joists stacked on the girders; when estimated, this exposure indicated that overturning could have occurred at even lesser wind speeds.

Had the steel been completely connected with crossbraced columns, portal-frame action would have substantially eliminated the moments transmitted to the column bases. The completed structure with a full roof deck and wall enclosures acts as a rigid box and wind loads are not transmitted into the columns.

Inspection of the steel indicated very good fabrication of the beams and columns.

Where steel joists 21 in (534 mm) deep were connected to the girders, failure was typically at the sole plate, which was either ripped off or torn at a bolt hole (Fig. 3.4). At these connections, where observable, only one bolt had been inserted (Fig. 3.5). Many of the joists were found partly disintegrated, the web rods having pulled away from the chords.

Figure 3.4 Where steel joists were connected to girders, failure was usually at the sole plate, which was either ripped off or torn at a bolt hole.

Figure 3.5 At the failed connections, only one bolt had been inserted.

Lessons

1. Brace the vertical columns of structural systems utilizing open-web steel joists with structural-steel rolled sections. This provides wind resistance simultaneously with the steel erection.

2. Avoid temporary "one-bolt" connections.

3.1.2 Arverne Nursing Home — Rockaway, New York

An eight-story hybrid structure was flattened during construction in February 1972 in a rainstorm and a relatively low intensity wind of about 15 mi/h (24 km/h). The entire structural-steel frame fell to the ground with several levels of precast-concrete hollow-core planks, which were totally demolished (Figs. 3.6, 3.7, and 3.8).

The design of the structure called for a structural-steel frame supporting precast slabs (Fig. 3.9). It became clearly evident at the beginning of my investigation that the designers expected the lateral resistance of this structure to wind forces to be provided by longitudinal masonry walls and rigidity to be provided by reinforcing bars within concrete topping layers on the planks.

Figure 3.6   This eight-story steel and precast-concrete building collapsed during construction in a low-intensity wind.

Figure 3.7   This is what the structure looked like prior to collapse.

Figure 3.8   The entire structural-steel frame fell to the ground with several levels of precast-concrete planks which were totally demolished.

Since both of the provisions were not yet present in the structure at the time of collapse, there was nothing to resist lateral wind forces and, hence, nothing to prevent the ensuing collapse. The minimal cable crossbracings were completely inadequate. Welds to the bottom plates in the precast planks failed (Fig. 3.10) and anchor bolts pulled out of the foundations.

A question that arises in such cases — and an obvious dispute — is "Who is responsible for temporary bracing when wind resistance is to be provided *later* by other trades?" Because temporary bracing can be a major cost item, there is also a question of whose contract it is in.

Lessons
1. During construction of standard steel frames, the wind resistance provided by moment connections, shear trusses, and bracing should be installed floor-by-floor as steel construction progresses upward.

2. For hybrid construction, the design engineer should include a typical temporary bracing system in the drawings as a part of the contract documents. At a minimum, the design engineer·should require the contractor to engage a licensed professional engineer* to design the bracing

---

* An individual licensed to practice engineering in the state where the construction takes place.

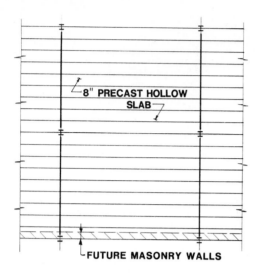

PLAN

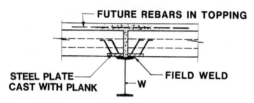

FUTURE REBARS IN TOPPING

STEEL PLATE
CAST WITH PLANK

W

FIELD WELD

TYPICAL SECTION

Figure 3.9 Lateral resistance of the structure to wind forces was to be provided by longitudinal masonry walls, and rigidity by rebars inside concrete topping layers on the planks.

system and submit it for review. This option, while relieving the engineer of record* from liability for temporary construction, always entails claims for extra costs by the contractor because the extent and type of the bracing system is not clearly shown to the contractor prior to bidding.

3. Give consideration to the inclusion in the design of a permanent steel spandrel bracing system which will eliminate a temporary system.

3.1.3   Staten Island Racquet Club—Steel frame

This failure occurred during construction. The building consisted of prefabricated steel rigid frames supporting lightweight purlins. Our examina-

---

* The licensed professional engineer who files the construction plans with the local building department.

Figure 3.10 Welds to bottom plates embedded in the precast planks failed.

tion of the site (Fig. 3.11) 3 days after the collapse showed that four frames of the multibay structure had toppled in the wind and collapsed (Fig. 3.12). The 8-in (203-mm) easterly block wall, which was approximately 17 ft (5.2 m) high, fell over as a single mass. The anchor bolts connecting the frame column legs to the concrete piers (two bolts per column) were pulled out. There was no distress in the frames and the splice connections.

Our analysis of the design of the frames concluded that both the frame design and fabrication were proper. The toppled block wall rotated about its grade beam base under the wind load because the temporary top re-

Figure 3.11 General view of racquet club steel structure which failed during construction, after wind loads.

Figure 3.12 Welded rigid steel frames fell flat on the ground after the failure.

straint was not adequate. The long slender purlins were not capable of acting as compression members to restrain the frames.

Lessons

1. Steel structures must be temporarily braced during construction to resist wind loads, especially when connected to *high walls.*

2. A few anchor bolts are not enough to provide lateral wind resistance for tall unbraced structures.

3.2 Other Failures during Construction

3.2.1 Industrial park warehouse — Spring Valley, New York

In July 1970, a two-story building under construction suddenly collapsed without warning. The structure was approximately 132×80 ft $(40 \times 24$ m), with six bays of 22 ft (6.7 m) in the long direction and four bays of 20 ft (6.1 m) in the short direction.

The building was designed as a structural-steel frame with precast plank [8-in (203-mm) Flexicore] floors and a 2-in (51-mm) cast-in-place topping. On the east, an earth retaining wall 20 ft (6.1 m) high was founded on a narrow 3-ft-wide (0.91-m) continuous footing (Fig. 3.13).

I found that local newspaper accounts were grossly inaccurate and misleading, a common occurrence after building collapses. For example, the local newspaper had this to say:[1]

AFTER WORKERS LEAVE, BUILDING COLLAPSES

After construction workers doffed their hard hats and left the site of a 45,000 square foot [4180 m²] industrial building at the Industrial Park on Monday, all was quiet.

Instead of the rumble of cement mixers and clash of steelwork going up, the

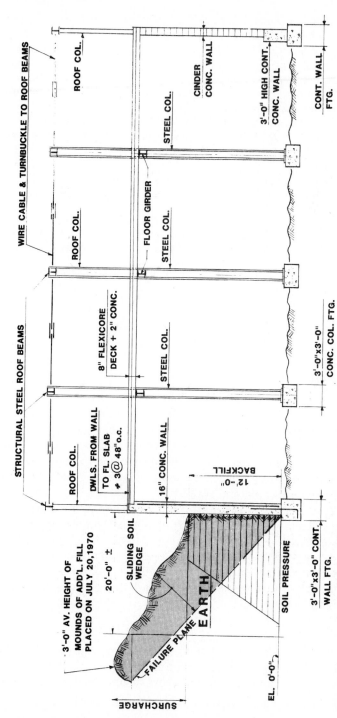

Figure 3.13 Cross section, before collapse, of structural-steel frame warehouse designed with precast plank floors and a 2-in cast-in-place topping.

215

sounds of children at play in the Village apartments, 100 yards [91 m] away from the construction project, could be heard.

But at 5 p.m., scarcely one half hour after the workers left the job, other sounds were heard — the crack of concrete breaking up, the groaning of steel under great stress, then a roar as the two-story building collapsed.

Soon after, the Village Administrator viewed the wreckage. On Tuesday morning, the entire inspection staff of the village was on the scene. He ordered a 24-hour watch on the building in case a 50-foot [15.2 m] long concrete retaining wall, cracked by the collapsing steelwork, should show signs of caving in.

Neither the Administrator nor any of his staff would speculate on the cause of the building collapse. Experienced construction workers told the *News,* however, that the weight of newly poured concrete floors on the second story probably triggered the collapse.

Our investigators, who were at the site several days after the collapse, found the facts about the cause quite different. They reported as follows:

Owner of the facility requested us, by phone, to make inspection of the collapsed building. Visited site, met with Owner, inspected collapse with him from 8:30 a.m. to 12:00 p.m. Returned 1 p.m. to 4:30 p.m. to finish sketches and notes.

At the time of pouring of 2-inch (51 mm) concrete topping for deck, the general earth fill against wall was 11 feet (3.4 m) +/−. Earth fill slid against wall to bring fill up to another foot (0.31 m). At 4:30 p.m., July 20, Owner stopped fill. At 5:30 p.m. collapse started.

While it may be clear that a 20-ft-high (6.1-m) earth retaining wall based on a 3-ft-wide (0.91-m) footing is incapable of supporting full-height earth lateral loads by itself, it should also be obvious that the structural-steel frame is not sufficiently rigid to retain this wall. Number 3 dowels spaced 48 in (1.22 m) on center could not possibly provide the necessary joint rigidity to resist the rotational movements. Figure 3.14 shows cross sections with schematics of lateral stability.

Figure 3.15 shows the structure after collapse. Note the total destruction of the cinder-block wall on the west. The photograph clearly shows the lateral motion which occurred.

I have investigated several similar cases of imbalance. Fortunately, these other cases did not end up in total collapse, but rather as excessive joint deformations and movement of the tops of the columns by as much as 2 to 3 in (51 to 76 mm).

Lessons

1. Where only one side of a structure is retaining earth, provide stability either by a properly designed retaining wall or by stiffness (moment connections) designed into the structure to support the wall laterally.

2. Be aware that soil surcharge may be critical in developing additional

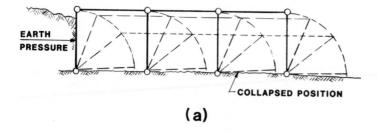

(a)

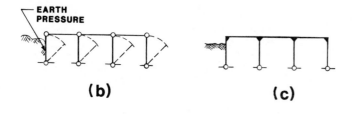

(b) **(c)**

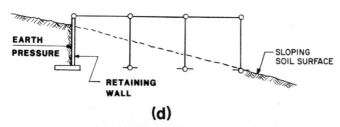

(d)

Figure 3.14 Simple schematic drawings show (a) collapse mechanism and (b) unstable/hinged, (c) stable/rigid, and (d) stable/retaining wall configurations.

lateral loads, even if the soil is not directly against the wall, but within the soil failure zone (Fig. 3.13).

3.3 Failures Due to Errors in Design and/or Construction

3.3.1 Hyatt Regency Hotel—Kansas City, Missouri

This shocking and fatal failure occurred on July 17, 1981 (Fig. 3.16). The walkway, a relatively simple hanging structure (Fig. 3.17) incorporating steel rod hangers and steel box beams, failed and fell to the first floor lobby level, taking the lives of 113 people and seriously injuring many others.

The high loss of life caused by the failure of a simple detail gives credence to the philosophy of relating factors of safety of structures to the possible

Figure 3.15   Warehouse after collapse and total destruction of cinder-block wall. The lateral motion which occurred is clearly shown.

Figure 3.16   A hanging walkway of simple design failed, taking the lives of 113 people and injuring many others. The shocking death toll caused by the failure of a small detail gives credence to the philosophy of relating factors of safety of structures to the possible consequences of failure.

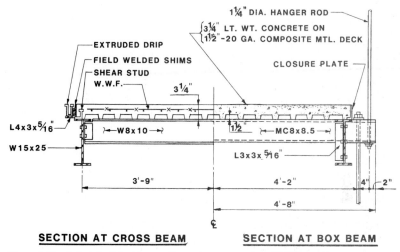

SECTION AT CROSS BEAM   SECTION AT BOX BEAM

Figure 3.17   A sketch of the hung walkway in cross section (NBS report).

consequences of failure. Hence, high-density occupancy structures should be designed using higher safety factors than low-density areas.

This failure also brought to light the question of assignment of responsibility for various phases of the building process. In particular, the question of responsibility for shop drawings and detail development was raised. Who is responsible for design and detail of connections of steel structures? The problem becomes even more acute when design details are changed in the final structure as was the case at the Hyatt Regency.

The fact is that neither the original design detail nor the as-built detail (Fig. 3.18) could safely support the required dead and live loads. To the trained eye, both details are obviously inadequate and violate fundamental rules of good engineering practice. It is not permissible to transfer loads to an edge of a flange or a channel section without any reinforcing plates or stiffeners.

A clear view of the failed assembly can be seen in Fig. 3.19. The 1¼-in-diameter (32-mm) hanger rod which supported the second-floor walkway is shown through the totally bent flanges of the channel box beam. The upper hanger rod, which carried twice as much load, had already failed and pulled through upwards. The remaining bent flanges of the channel are clear evidence of this passage.

The National Bureau of Standards simulated the failure by building a full-scale mock-up and testing it. The conclusion of the test was that dynamic effects were insignificant compared to the static loads.

The factor of safety for the original design was found to be equal to 60 percent of that required by code.

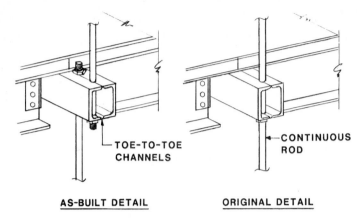

AS-BUILT DETAIL   ORIGINAL DETAIL

COMPARISON OF INTERRUPTED AND CONTINUOUS HANGER ROD DETAILS

Figure 3.18 Neither the original design detail nor the as-built detail could safely support the required dead and live loads. Both details are inadequate to the trained eye and violate fundamentals of engineering practice.

The as-built system (with split hangers) had an ultimate capacity slightly greater than the dead load acting at the time of the collapse.

Lessons

1. Typical details of critical connections which are not AISC standard connections should be designed and detailed by the engineer of record and shown in the contract documents.

2. Thoroughly check new designs which depart from accepted standard.

3. Modifications by contractors of design details must have prior written approval by the engineer of record.

4. Provide redundancy when possible to avoid progressive collapse.

5. For hanging structures prefer the use of back-to-back channel detail, with and without web stiffeners (Fig. 3.20a). Where it is necessary to go to a box-beam design (i.e., to transfer torsion), use crossplates (Fig. 3.20b).

3.3.2 Hartford Coliseum—Hartford, Connecticut

The view of the remarkable collapse of the space roof of this stadium (Fig. 3.21) will not be easily forgotten. This monumental structure, supported on four columns (Fig. 3.22), suddenly collapsed (in January 1978) under a snow and ice load, luckily without a single fatality. There had been nearly 5000 spectators in the arena for an evening college basketball game just 5

Figure 3.19 The fourth-floor walkway hanger-rod and channel/box-beam assembly was identified as the most probable cause of failure initiation. Part of the hanger rod which supported the lower walkway remained in place, but the upper rod on which the walkway was suspended pulled upward through and between the toes of the channels.

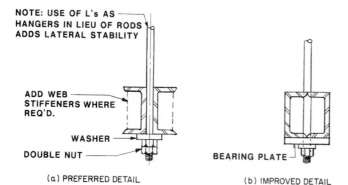

(a) PREFERRED DETAIL   (b) IMPROVED DETAIL

Figure 3.20 (a) Preferred detail showing back-to-back channels, with and without stiffeners. (b) Improved detail showing toe-to-toe with bearing crossplate.

(a)

(b)

Figure 3.21 Early-morning total collapse of Hartford Coliseum space roof structure following a heavy winter storm luckily took no lives. Five hours earlier, nearly 5000 spectators watched a college basketball game. (*a*) General view; (*b*) failure at end of Coliseum.

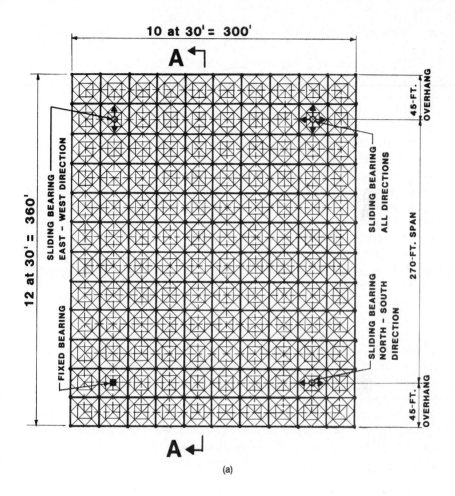

10 at 30' = 300'

12 at 30' = 360'

A

A

SLIDING BEARING EAST – WEST DIRECTION

FIXED BEARING

SLIDING BEARING ALL DIRECTIONS

SLIDING BEARING NORTH – SOUTH DIRECTION

270-FT. SPAN

45-FT. OVERHANG

45-FT. OVERHANG

(a)

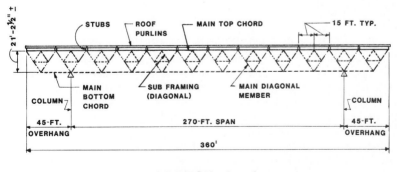

21'-2½" ±

STUBS   ROOF PURLINS   MAIN TOP CHORD   15 FT. TYP.

COLUMN

MAIN BOTTOM CHORD

SUB FRAMING (DIAGONAL)

MAIN DIAGONAL MEMBER

COLUMN

45-FT. OVERHANG   270-FT. SPAN   45-FT. OVERHANG

360'

SECTION A - A

(b)

Figure 3.22 General roof plan (*a*) and typical main truss of huge space structure (*b*).

hours before the early Sunday morning collapse. When I visited the site several days later, buckled mounds of steel were all over the area.

The enormous steel space structure, loaded within the limits of the local code, collapsed after large deformations were noticed yet not heeded. It was an example of structural instability of compression members which were not sufficiently braced. Only 20 lb/ft^2 (958 Pa) of snow was estimated to be on the roof at the time of the collapse.

Normally, the trusses in roof structures are braced against lateral buckling by the roof structure itself, roof skin or membrane, subdividing beams, secondary beams, and the like. In this case, the roof skin and secondary structures were remote from the space structure itself. These elements did not sit directly on top, in full contact, but sat instead on stub columns. This design eliminated the bracing elements which are commonly present. This fact weakened the structure enormously.

In addition, as often happens in cases such as this, there were very complicated connections, with members facing in many directions and with high overstresses due to eccentricities.

During the investigation, it was also determined that the actual dead load of the roof structure was about 15 lb/ft^2 (718 Pa) heavier than that assumed by the design engineer.

There were also signals which went unnoticed or unheeded. Very large deformations [up to 8½ ins; (216 mm)], were ignored and were considered "normal" for such large structures. These were "cries for help" from the structure itself!

For example, the contractor could not fit the windows below the girders. This should have been an alarm signal to check deformations and to determine if they were of the order of magnitude expected by the designer. The design error could have been noted, identified, and corrected at that time, at relatively low cost.

The following newspaper story vividly describes the emotions associated with this collapse:[2]

WHY THE ROOF CAME TUMBLING DOWN

HARTFORD, CONN. — The ironworkers knew. . . . In the downtown shot-and-a-beer bars, they would gather after work that summer of 1972 and tell each other they would never set foot inside the Hartford Civic Center Coliseum when it was completed, that the web of steel was a death trap. . . . They did as directed, thrusting the 1,400-ton [12.5-MN] space frame into place 85 feet [26 m] above the ground.

They looked along its twisting, bowing, sagging metal, and they knew . . . One of the ironworkers . . . remembers, as others did, having brought the twisted metal to the attention of superiors, and remembered nothing having been done.

Everybody knew it, he said.

Everybody couldn't be wrong. You work on a lot of buildings and you say, "Jeez, I hope the thing stays up until we come down." But this was no joke. He was dead serious. So was . . . another of the ironworkers. He told his children: When the Civic Center is finished, I don't want you ever to go in there when there's snow on the roof.

The ironworkers knew.

Yet the space frame was considered a marvel of architectural engineering. It was "computer designed." Who were laborers, blue-collar workers, to doubt a computer? So the steel spider web rose, majestically, on the skyline of Hartford. And, at 4:19 on a Wednesday morning, last January 18, there was a low rumble that grew until it shook the earth around it—the death scream of agonized metal.

The roof came crashing down.

The ironworkers knew.

Why did it take until the roof collapsed—three years after the coliseum was opened and six hours after a crowd of 4,800 had filed out of the building from a college basketball game—for everyone but the ironworkers to know?"

Lessons

1. Check to verify the existence of roof skin or other members that will laterally brace roof compression chords.

2. Large deformations are usually signs of possible distress. Do not ignore them! Investigate immediately and take proper precautions, such as evacuation and/or shoring of the structure.

3. Avoid complicated details which are usually difficult to execute properly and are invariably very costly.

4. Increase the factor of safety when designing structures such as sports complexes where the consequences of failure may be catastrophic.

3.4   Failures Due to Incorrect Fabrication and Detailing

3.4.1   Midwestern university basketball arena

The collapse of the roof of an arena in February 1969 was a monumental failure of a long-span roof. The roof collapsed during construction (Fig. 3.23), fortunately with no fatalities.

The framing of this structure consisted of roof joists spaced 8 ft (2.4 m) on center, with an overall span of 180 ft (55 m). The joists were supported at each end on open-web structural-steel main trusses, which, in turn, were connected to steel columns. Figure 3.24 shows the arena before the collapse.

There were several differing opinions on the cause of the failure, including lateral buckling, impact from erection equipment, and lack of tempo-

Figure 3.23 University basketball arena, which collapsed under construction, is an example of a monumental failure of a long-span roof.

rary bracing. However, the remarkable phenomenon here is that all parties involved—owners, general contractors, structural-steel fabricators, and erectors—jointly agreed to rebuild the entire structure so it could be readied in time for the upcoming basketball season. The determination of liability was postponed until after reconstruction, at which time the costs for the repair and rebuilding were assigned to the various parties in accordance with a mutually agreed-on formula.

Figure 3.24 The arena before the collapse.

Generally, the basic structural design was proved to have been adequate and required no major changes. Some structural refinements were agreed on and incorporated into the reconstruction. We recommended a revision to the erection procedure for the long-span joists. Subassemblies consisting of two roof joists and their connecting bracing members were fabricated on the ground and lifted into position as a single unit (Fig. 3.25). In this way, sufficient lateral and torsional stability was provided and undesired deformations were controlled as the members were put in place.

As is frequently the case, our investigation revealed many deficiencies, including:

1. There had not been good compliance with the shop drawings during fabrication. This was concluded from actual measurements of lengths of joists in comparison with joists which had not been erected.

2. There were other variations from the original design. For example, the original design called for standard-type end supports; namely, the joists were to be supported *on top of* the carrying trusses. However, what were actually supplied were *eccentric-face* connections for the full depth of the joists to the trusses. (The top bearing connection is not sensitive to fabrication inaccuracies, while the full-depth connection is.)

Figure 3.25 Revised erection procedure for the long-span joists in the rebuilt structure consisted of subassemblies and bracing members fabricated on the ground and lifted into position as a single unit. Pairs of joists were fabricated into box beams.

The major cause for the failure was established as the incorrect fabrication and detailing of the joists. The development of the bearing detail at the point of intersection between the joists and the supporting trusses resulted in 10- to 12-in (254- to 305-mm) eccentricities, since the top chord of the joist was 4 to 5 in (102 to 127 mm) off the end of the joist. This resulted in an overstress of the top chord of the joists and the end panel which was above the allowable stress level. Thus, the stress of the top chord of the truss was *at or above the yield stress* of the material. There was also lack of torsional stiffness of the jack trusses (supporting trusses), which were unable to accommodate the rotation of the long-span joists at the supports. This connection was sufficiently rigid to transfer rotational stresses from the joists to the jack trusses. While the eccentricity from the working point of the diagonal to the face of the joist was approximately 4 in (102 mm) (Fig. 3.26), this eccentricity created a torque with an effective eccentricity of about 10 in (254 mm) to the real support line (centerline of the supporting truss).

As part of the reconstruction, flexibility was incorporated into the four main columns at the corners to permit roof expansion and contraction without overstress of the structural-steel columns.

Lessons

1. Careful end-bearing details of long-span trusses are required to avoid large eccentricities.

2. Where eccentricities are unavoidable, design the supporting jack trusses to accommodate the rotations without overstress and possible failure.

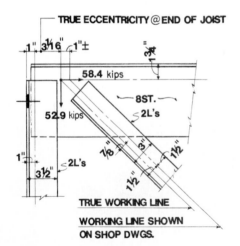

Figure 3.26 Detail at end of joists (top chord) shows eccentricity of critical connection.

3.5   Failures Due to Deficient Welding

3.5.1   Wolftrap Center — Fairfax, Virginia

A recent failure as a result of brittle fracture occurred at the famous Wolftrap Center. One of the main roof girders constructed of structural steel actually split vertically, with a gap as wide as 1½ in (38 mm) (Figs. 3.27 and 3.28). From the scaffolding, I could see that the gap was at its maximum width on the bottom of the girder and extended for most of its depth.

The roof structure remained suspended and did not fall. There was sufficient redundancy in the structure, and alternate paths were available for load transfer. Therefore, total collapse was averted.

Figure 3.27   A main roof girder of welded steel plates failed as a result of brittle fracture, actually splitting vertically (see arrow).

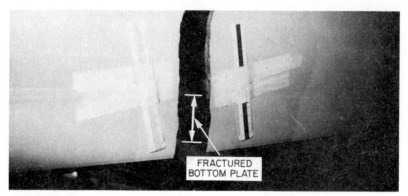

FRACTURED
BOTTOM PLATE

Figure 3.28 Blow-up of 1½-in (38-mm)-wide crack with attached scales shows fractured bottom plate.

Bridge engineers have been confronted for many years with brittle fracture of welded girders consisting of heavy plates. They have been taught to realize that structural steel has a property referred to as *fracture toughness,* in addition to yield strength and ultimate strength. For proper perform-ance of bridge structures, especially those in cold climates, the steel must have adequate fracture toughness. This requirement is critical for welded bridge girders. Such members are also subject to vibration and shock and, therefore, are excellent candidates for brittle fracture.

However, the fact that was highlighted by the Wolftrap failure is that enclosed building structures, too, may be the victims of brittle fracture, even without the effects of vibration and shock. It is important to re-member that during construction, many steel structures are exposed to the weather (and often, low temperatures) prior to enclosure and heating.

The Wolftrap roof structure was analyzed and found to have stress levels below the allowable for the required dead and live loads. Because brittle failures often emanate from minor flaws within the metal or the welds, flaws which are not visible to the naked eye, they were not detected during fabrication or erection. Close metallurgical examination after the failure detected that the backup steel plate used for the full-penetration weld was not itself welded continuously. This created a flaw large enough to generate the fracture.

Although other criteria are involved in the selection of a steel for a given welded structure, resistance to low-energy brittle fracture is particularly important because many fabrication and service failures have resulted from inadequate attention to this property.

The steel industry has not brought this fact to the attention of both engineers and contractors, who, as a matter of common practice, do not specify minimum toughness requirements for structural steel used for buildings.

In 1948, the American Bureau of Shipping pioneered the practice of controlling the notch properties of tonnage structural steels by requiring the specification of composition (particularly maximum carbon), grain size, and heat treatment. Up to ½ in (13 mm) thickness, only a tensile test was required, but as thickness increased, tight composition, fine-grain practice, and a normalized heat treatment were added.

Corrective actions. At the Wolftrap girder the portions of the cracked plates were removed and new plates were welded. As an extra precaution the entire girder was posttensioned with cables after the new plates had been welded.

Lessons
1. Specify minimum fracture toughness for structural construction of welded thick steel plates. The tougher the steel, the more resistant it is to brittle fracture.
2. Minimize the risk of brittle failure by:
   a. Avoiding introducing notches, either from design or from weld or base metal imperfections
   b. Maintaining welding temperatures below the ductile/brittle transition for the particular parts of the structure
   c. Avoiding excessive restraint by heavy-section thickness or design of connections
   d. Avoiding high rates of loading (impact)
3. Provide redundancy in the structure so that in the event one element fails, the load is transferred to other members, thus avoiding total collapse.
4. Avoid unnecessary use of high-strength steel, since high-strength steels, which have lower fracture toughness, are susceptible to brittle fracture.
5. Improve quality control (inspection) to detect material and welding flaws, especially at critical details.

3.5.2 New York broadcasting headquarters

On the afternoon of June 1, 1983, the entire west facade of the new wing of the CBS broadcasting headquarters building, located on the west side of Manhattan, collapsed during construction (Fig. 3.29). The wall fell suddenly and the debris covered the roof setback at the sixth floor, causing damage to the sixth floor's framing.

I arrived at the site the next morning and examined the collapsed area. A careful review indicated that the precast spandrel beams hung from the structural-steel overhangs (Figs. 3.30 and 3.31) had separated at the steel hangers. The weakest point was established at the connection of the support plates at the bottom of the steel overhanging beams (Figs. 3.31 and 3.32).

Figure 3.29   A newly built facade of a large Manhattan office building suddenly collapsed during construction. Luckily, no pedestrians were injured as the debris fell on the roof setback and caused damage to framing of the floor below.

The welds connecting the plate to the beam had failed and fractured. The broken weld surfaces were immediately coated with Krylon to prevent rusting.

An important observation was the fact that, where these plates were continuously welded across the flanges, they did not fail. Since the plates were supported from and welded to the bottom flanges of the beams, the welds were subjected to a combination of shear and tensile stresses.

It was established by analytical computations and by full-scale laboratory testing that the welds as constructed (without returns) were not adequate to safely support the required load on the plates.

The question arose regarding responsibility for detailing the connection and especially for the required weld return. Was it the designer's responsibility? Or was it, in fact, the structural-steel detailer's, who is an agent of the steel contractor?

Figure 3.30 A side view of the failed support system shows the precast beam and the remaining hanger rod at welded plates which did *not* fail.

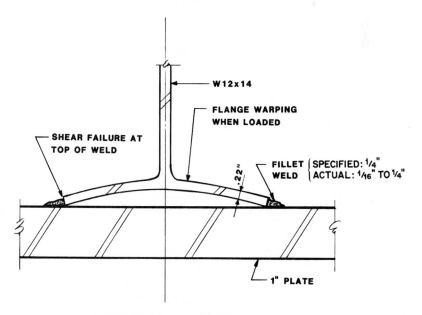

W12x14

FLANGE WARPING
WHEN LOADED

SHEAR FAILURE AT
TOP OF WELD

FILLET | SPECIFIED: 1/4"
WELD | ACTUAL: 1/16" TO 1/4"

.22"

1" PLATE

HANGER PL. ASSEMBLY
(UNDER LOAD)

Figure 3.31 The weakest point was established at the connection of the support plates at the bottom of the steel overhanging beams.

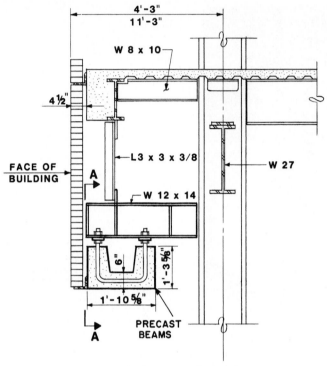

TYPICAL MASONRY WALL

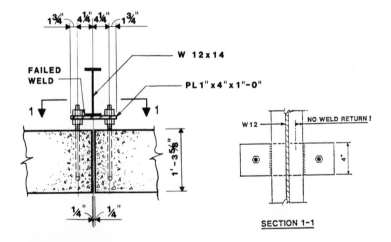

ELEVATION A - A

Figure 3.32   The precast spandrel beams hung from the structural-steel overhangs separated at the steel hangers.

As in many weld failures, it was found here that several welds were undersized, reducing the factor of safety even further. The existence of undersized welds pointed to another deficiency, namely, inadequate controlled inspection,* which did not detect the low-strength welds.

Lessons

1. Avoid, where possible, welded underflange plates subjected to pure shear combined with bending tension.

2. Provide weld returns to prevent fracture at the end of the welds.

3. Provide close-range, thorough weld inspection — both in the laboratory and at the site.

3.5.3 General manufacturing plant — Long Island, New York†

As a result of manufacturing errors in welding of the cold-formed flanges of steel open-web roof joists for a manufacturing plant in Long Island, New York, several joists cracked and fractured (Fig. 3.33) and later collapsed.

There was no way by which the possible defects in other joists still supporting the roof could be detected, identified, and repaired. The simplest solution would have been to replace the entire roof. However, because it was an operating plant, enormous costs would have been incurred to stop production. Taking into account that the roof structure had performed properly over many years with only isolated joist cracking, we determined that the best method of repair would be:

1. Reinforcing the joist bottom chords by means of added steel plates bolted to the existing chords.

2. Adding heavy steel diagonals to provide the necessary redundancy.

This method was adopted (Fig. 3.34). The design was such that, in the event of a failure of one joist, the adjacent joist would pick up its share of the increased load. This way, the domino effect, or progressive collapse, would be avoided. There could only be a localized failure which could be repaired easily.

The computer models we examined confirmed the possible future behavior of the structure after repair, assuming failure of a single joist (Fig. 3.35). Full-scale, in-place load tests (Fig. 3.36), including deformation measurements, were conducted by our engineers and reconfirmed the analytical studies.

---

* Inspection required by the New York City building code for various construction activities, including welding.

† Dov Kaminetzky, "Are Our Codes Adequate for Snow Loads?" American Society of Civil Engineers Convention, New York, N.Y., May 11–15, 1981, Reprint No. 81–166.

(a)

CRACK

(b)

Figure 3.33 (a) Welding flaws in cold-formed flanges of steel open-web joists cracked and fractured. Several joists later collapsed. (b) Bottom view of bottom chord.

Lessons.   Minimize stress concentration in cold-formed steel by avoiding:

1. Abrupt changes in section in highly stressed areas
2. Concentrations of numerous heavy welds in a small area with no provision for relief of shrinking forces by peening or stress relief
3. Unnecessarily large welds

Unless these three conditions are followed, the integrity of the structure may be determined, not by design stress, but by the notch resistance of the

Figure 3.34 Heavy steel diagonals were added to provide necessary redundancy for the repair. Bottom chord plates (arrows) were also added. The new design ensured that, in the event of failure of one joist, the adjacent joist would pick up its share of the increased load, thus avoiding the possibility of a progressive collapse.

BRIDGING TYPE

NO.	LOAD TRANSFER %	CROSS−BRIDGING ARRANGEMENT	CROSS−BRIDGING SIZE	REMARKS
I	0	NONE	NONE	MEMBER 1027 OMITTED
II	39.0 lb/ft²	⋉⋊	1½ X 1½ X ⅛	MEMBER 1027 OMITTED
III	66.0 lb/ft²	⋉⋊	3 X 3 X ½	MEMBER 1027 OMITTED
IV	71.0 lb/ft²	⊠ ⊠	2 X 2 X ¼	MEMBER 1027 OMITTED
V	76.0 lb/ft²	⊠ ⊠	3 X 3 X ¼	MEMBER 1027 OMITTED

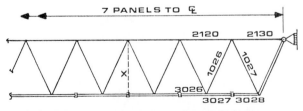

Figure 3.35 Computer models confirmed the likely future behavior of the structure after repair, assuming failure of a single roof joist. In the diagram, crossbracing is added for redundancy (X = location of crossbracing).

Figure 3.36 A full-scale, in-place load test using water pool was performed after the repair and confirmed the analytical studies.

joints. Problems such as those at the manufacturing plant can be minimized if the designer consults with fabricating shop personnel to take advantage of their welding experience.

3.5.4   College science building—Brooklyn, New York

On February 28, 1971, at about 12 noon, several steel girders failed under load while supporting the upper five floors and roof during construction of an addition to a college science building (Fig. 3.37). Severe damage to the project resulted.

The girders were fabricated by welding of heavy A441 steel plates. Large openings in the web sections were included to accommodate the passage of plumbing and mechanical lines.

According to the design engineers who inspected the project after the failure, many welds connecting the diagonal web members to the flanges were undersized. Ultrasonic inspection indicated several joints to be defective. Joints tested by radiography included structural discontinuities exceeding the permissible limits of the American Welding Society (AWS) specifications. The design engineers concluded that the major factors causing the failures were weld-fusion defects and toe or underbed cracking and the low fracture toughness of the high-strength A441 steel.

Some investigators concluded that the major reason for the failure was weld-shrinkage cracking from the large transverse weldments.

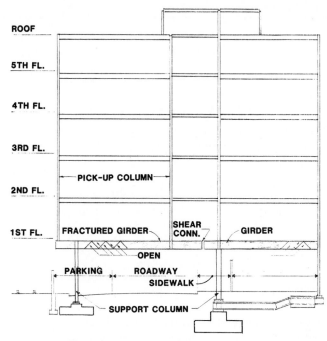

Figure 3.37 Typical section through building showing location of fractured plate girder.

Later research revealed that this failure was, in fact, caused by *lamellar tearing* (not widely known by that term at the time of the failure).

The following is a vivid description of this failure by Dr. Robert D. Stout, a metallurgical investigator:

> The building in Brooklyn was put up with a very peculiar design—those design engineers, you know—where they had two diagonals meeting a horizontal tension flange and they filled up that whole area with weld. *It was just "welded to death" there and so when they put in the final passes it pulled apart.* [Emphasis mine.] One of the diagonals pulled out and there is some evidence of lamellar tearing on the one side but all the rest of the fracture was brittle. [Figure 3.38.] What happened was that they were doing something with concrete up above, adding load, and what that did was propagate this lamellar tearing . . . a complete failure of that member by brittle fracture because it was cold. It was in December and the building was still not closed in; it was unheated and so on. They got a brittle failure that propagated but you could see that a lot of it had torn by lamellar tearing first and then it propagated.

The design was indeed unusual, consisting of a combination open-web truss and a plate girder. In effect, the plate girders were pierced to provide for penetration by heating and ventilating ducts, thereby resulting in a composite-type plate-girder–structural-truss combination (Fig. 3.39).

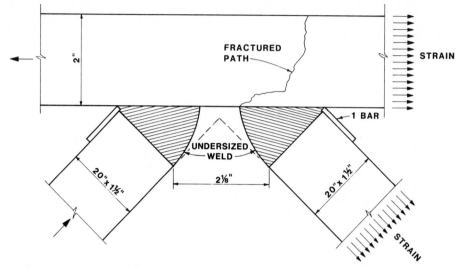

Figure 3.38 Joint detail shows fracture that later research revealed was from lamellar tearing. (See Sect. 3.6 for additional information.)

The girders which cracked in several locations (Figs. 3.40 and 3.41) were temporarily shored (Fig. 3.42) and had to be entirely removed and replaced by other girders.

Remedial work involved the construction of a new girder below the existing cracked girder which would carry the entire load. (*Note:* the design change reduced the clear headroom below the new girder.) At some of the unfailed girders, new gusset plates were added to transfer the loads from the diagonals to the flanges without having to rely on the welds — bypassing the welds (Fig. 3.43).

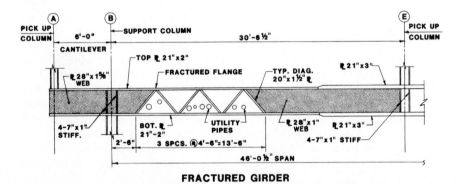

FRACTURED GIRDER

Figure 3.39 Design was unusual, consisting of a combination open-web truss and a plate girder. The plate girders were pierced to accommodate heating and ventilating ducts, resulting in a composite-type plate-girder–structural-truss combination.

Figure 3.40 Close-up photograph shows cracked plates.

Figure 3.41 Close-up photograph shows heavy welding at cracks.

Figure 3.42   The cracked girders were shored until replaced.

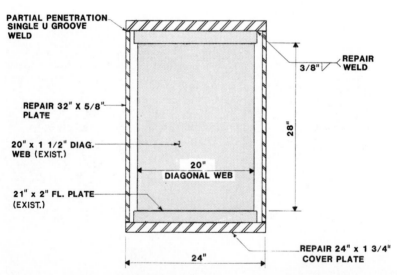

PARTIAL PENETRATION
SINGLE U GROOVE
WELD

REPAIR
WELD
3/8"

REPAIR 32" X 5/8"
PLATE

28"

20" x 1 1/2" DIAG.
WEB (EXIST.)

20"
DIAGONAL WEB

21" x 2" FL. PLATE
(EXIST.)

24"

REPAIR 24" x 1 3/4"
COVER PLATE

Figure 3.43   Section detail of a proposed repair to structure in Fig. 3.39. At some of the new girders, gusset plates were added to transfer loads from the diagonals to the flanges without having to rely on the welds. (Existing plates are shown shaded. New plates are cross-hatched.)

Lessons

1. Where possible, avoid complicated details involving heavy welding of thick plates.

2. Avoid large multipass welds and/or specify low-yield-strength welding electrodes. Welds should be balanced around the midthickness of the section and peened (except first and surface passes).

3. Specify steel with minimum notch toughness for welded *thick* steel plates. These steels should especially be used for structures subjected to high restraint and low service or construction ambient temperatures.

3.6 Lamellar Tearing

One of the problems that has baffled structural and civil engineers over the last 30 years is brittle fracture in structural steel. In particular, *lamellar tearing* has come to the forefront and forced intensive research to try to solve and eliminate this problem.

This phenomenon of lamellar tearing is actually an old one. It seems that occurrences have become more frequent lately because of the increased use of welding in building structures, as well as the growing size of the weldments themselves.

Three of the most well-known cases of lamellar tearing occurred in the steel compression ring of the Performing Arts Center in El Paso, Texas; the 52-story Atlantic Richfield twin towers in Los Angeles; and the 20 electric power transmission towers at the U.S. Bureau of Reclamation's third powerhouse at Grand Coulee Dam, Colorado.

While the general problem of fracture and fatigue in structural-steel members, bridges in particular, is well-known, the problem of lamellar tearing has been intensified and brought to light mainly because of the occurrences in the early 1970s which caused heavy expenses and delays to the completion of the structures involved.

In general, lamellar tearing can be described as a failure of steel in the dimension *perpendicular to the plane of rolling*. It most commonly occurs in steel at heavy, restrained, welded joints [usually with plates thicker than 1½ in (38 mm), although some instances of lamellar tearing have been recorded with thinner plates]. The phenomenon manifests itself in the form of microscopic cracks. These are most often internal and may be detected by ultrasonic inspection. In some instances, these cracks may also be seen at the surface. Tearing then develops soon after welding.

The general explanation for lamellar tearing is that critical expansion from the heat of welding occurs in the weld. When the weldment at the restrained joint cools, the resulting contraction can tear the steel members.

What engineers have come to understand now is that structural steel is not as homogeneous a material as they were originally led to believe — and that there are characteristic differences in the *strength* of steel in the $x, y,$ and z directions. In fact, structural steel is weakest in the direction perpendicular to rolling.

There is now a strong consensus among engineers that structural-steel codes should be modified to recognize this reality and that the codes should include qualitative restrictions, e.g., size of weld, proportion of length of weld to base metal, and interrelations between direction of rolling and direction of stress action. Such code modifications will help educate the general engineering public to the importance of better connection details which, while being more costly, may contribute significantly to a reduction in lamellar tearing failures (Fig. 3.44).

At the El Paso Arts Center, which was designed as two intersecting crescents, the compression ring consisted of a box girder 4.6 × 2 ft (1.4 × 0.61 m) in cross section. Its 25 sections were joined together by massive, full-penetration welds which connected 2½- to 3-in-thick (63- to 76-mm) flange plates to 1-in-thick (25-mm) diaphragm plates. The design of this compression ring had been changed so it utilized fillet welds across the top and bottom of the girder, thus eliminating full-penetration welds in the girder side plates in their *through-thickness dimension.*

At the twin towers of the Atlantic Richfield Plaza in Los Angeles, delayed cracks appeared at the column spandrel connections. The columns and spandrels at these elevations were fabricated of very thick members [1½ in (38 mm) and heavier]. The cracks appeared within the base metal and within the welds.

Lamellar tearing was only one type of a number of cracking failures at these locations. In spite of the fact that the steel used was ASTM* A-36, it did not behave in a ductile manner, which is extraordinary for this mild steel. Approximately 10 percent of the spandrel/beam flange connections were affected.[3]

The *Engineering News Record* reported:[4] "The problem in the Grand Coulee Towers is surprising because the plates are not thick, ranging from ⅜″ to 1⅟₁₆″ (10 mm to 27 mm), however, the tower joints are restrained." In this case, initially the problem was considered to be minor, but later magnetic-particle testing revealed additional cracks. This forced a complete redesign of all the towers, including costly dismantling of those already erected.

---

* American Society for Testing and Materials.

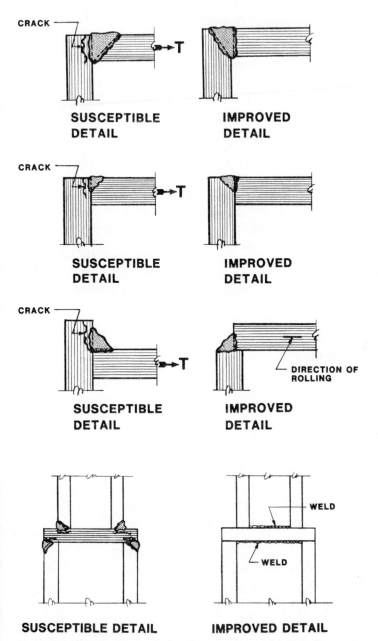

CRACK

SUSCEPTIBLE
DETAIL

IMPROVED
DETAIL

CRACK

SUSCEPTIBLE
DETAIL

IMPROVED
DETAIL

CRACK

SUSCEPTIBLE
DETAIL

IMPROVED
DETAIL

DIRECTION OF
ROLLING

WELD

WELD

SUSCEPTIBLE DETAIL   IMPROVED DETAIL

Figure 3.44 Details showing in four instences how risk of lamellar tearing may be minimized by careful detailing. (Per AISC) (T = tensile force caused by weld contraction).

3.7   Failures Due to Excessive Flexibility and Nonredundant Design

3.7.1   IRS warehouse—Philadelphia

So-called *dead-level* roofs are getting lighter and spans longer. As a result, ponding—the retention of water due solely to the deflection of flat roof framing—is increasingly becoming a problem. Drains and leaders are normally located at the columns, which renders them inefficient for drainage as they become high points of the roof.

A 12,000 ft² (1115 m²) section of this 425 × 500 ft (129 × 152 m) structure collapsed (Fig. 3.45) during a violent rainstorm, pulling down the expansion joint ends of the beams. It was a lightweight structure, weighing 15 lb/ft² (718 Pa) and consisted of open-web steel joists supported on castellated steel girders (Fig. 3.46). We found that the structure was extremely flexible and did not comply with AISC ponding requirements (see lessons). In addition, the castellated girders did not have their webs stiffened, leaving them susceptible to web rippling and buckling (Fig. 3.47).

The sensitivity of light steel structures to failures can be demonstrated as follows:

Average steel roof weighing		15 lb/ft²	(718 Pa)
plus a live load of		30 lb/ft²	(1436 Pa)
Total		45 lb/ft²	(2154 Pa)
Allowable stress	=	24 ksi	(165 MPa)
Yield point	=	36 ksi	(248 MPa)
Factor of safety 36/24	=	1.5	
Without additional redundancies, ultimate capacity	=		
45 × 1.5	=	67.5 lb/ft²	(3232 Pa)
Existing loads	=	15 lb/ft²	(718 Pa)
Collapse load	=	52.5 lb/ft²	(2514 Pa)
or 10 in (254 mm) of water			

Thus, 10 in (254 mm) of water will be sufficient to collapse an average structural-steel roof building without taking into account ponding effects. When these latter effects are considered, the depth of water required to collapse the roof can be less than 10 in.

Lessons

1. Check roof structure for ponding criteria. Use the AISC specification for roof stability: The roof system shall be considered stable and no further investigation will be needed if $C_p + 0.9C_s \leq 0.25$ and $I_d \geq 25S^4/10^6$, where

Figure 3.45 Internal Revenue Service warehouse section that collapsed was adjacent to an expansion joint. A 12,000-ft^2 section of the dead-level lightweight roof collapsed during a severe rainstorm, a result of ponding.

Figure 3.46 The roof was a lightweight structure, weighing 15 lb/ft^2 and consisting of open-web steel joists supported on castellated steel girders. Shown here is the roof frame under construction.

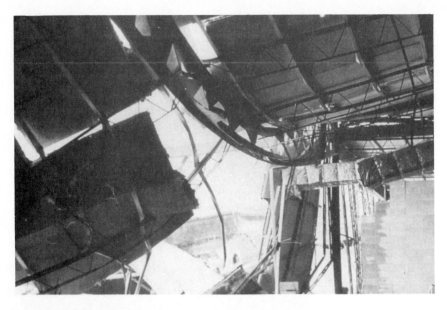

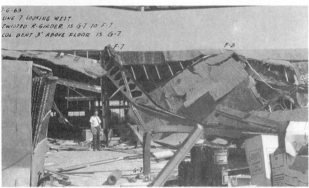

Figure 3.47 Castellated girders did not have webs stiffened, leaving them susceptible to web rippling and buckling.

$$C_p = \frac{32L_sL_p^4}{10^7I_p} \qquad C_s = \frac{32SL_s^4}{10^7I_s}$$

and L_p = column spacing in direction of girder, ft (length of primary members)

L_s = column spacing perpendicular to direction of girder, ft (length of secondary members)

S = spacing of secondary members, ft

I_p = moment of inertia of primary members, in⁴
I_s = moment of inertia of secondary members, in⁴
I_d = moment of inertia of the steel deck supported on secondary members, in⁴/ft

2. Provide sufficient slope toward the point of free drainage or adequate drains to prevent accumulation of water. If possible, locate drains at midspan.

3.7.2 Warehouse—Jamaica, Queens, New York

A similar, more recent roof collapse as a result of ponding of rainwater occurred on July 3, 1984. This flat, level roof was of light-gauge steel beam construction, far too flexible to comply with the AISC code ponding requirements. Also, there was insufficient drainage provided within the roof.

The section of the roof which collapsed was adjacent to an expansion joint, which was the weakest link because the roof beams deflected severely under accumulation of rainwater and turned into catenaries. The lateral thrust could not be resisted at the expansion joint, causing the edge beam to buckle laterally away from the joint (Figs. 3.48 and 3.49). This was my

Figure 3.48 This is another flat-roof failure that can be attributed to its flexible construction and inadequate drainage provisions. The section of the roof that collapsed was next to an expansion joint, the weakest link.

Figure 3.49 As the roof beams deflected severely under accumulation of rainwater, they became catenaries. The lateral thrust could not be resisted at the expansion joint, causing the edge beam to buckle laterally away from the expansion joint.

first site observation, which was subsequently confirmed by structural analysis.

Lessons

1. Always check to ascertain that roof beams and girders comply with AISC ponding requirements for stiffness.

2. Give special attention to beams adjacent and perpendicular to expansion joints.

3. Provide adequate number and size of drains and locate them at low points of the roof *away from columns*.

3.7.3 Lightweight canopy — Poughkeepsie, New York

A lightweight canopy which served as a marquee for a theater was framed in steel and hung by two steel rods (Fig. 3.50). The 1-year-old structure collapsed on a cold and windy day in January 1963, at which time it was covered by about 1 ft (0.31 m) of snow. One person was injured in the accident.

About a week before the collapse, the canopy had been raised on its outer end by adjusting the turnbuckles which were provided for the purpose. This had been done because the canopy was not draining properly.

Our investigation showed that failures occurred on all four points of attachment of the canopy to the building, at both the third-floor and the canopy-level connections. These were determined to be not tensile failures, but rather fractures due to flexure.

At our request, a metallurgical examination of several of the parts re-

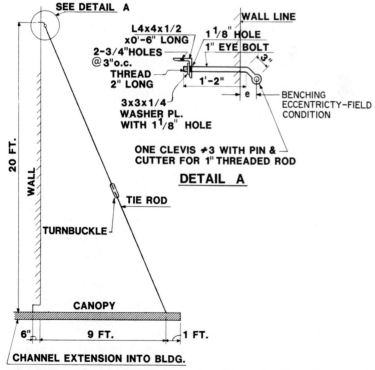

Figure 3.50 Detail of lightweight canopy, framed in steel and hung by two steel rods, which collapsed on a cold and windy day after a heavy snowfall. Eyeblolt fractured.

moved from the canopy was conducted at the Polytechnic Institute of Brooklyn (now Polytechnic University). The examined parts consisted of the two joints between the anchor bar and the eye bar extending to the canopy. The anchor bar at each joint terminated at a transverse fracture. The eye bolts terminated at a flame cut.

The joints consisted of an eye bar and an anchor bar connected by a bolt through the respective eyes with 3 × 4 in (76 × 102 mm) plates placed on the outside of the eyes. The anchor bar was 1 in (25 mm) in diameter.

The fracture surfaces generally showed a grainy condition when observed under a binocular microscope.

Office studies and review of the photographs indicated that the construction did not follow the architect's plans and details, which were identical to those found in the engineer's design notes. This fact was confirmed by the fractured specimens themselves.

The engineer's design, based on his computations, was for a total weight of 25 lb/ft² (1197 Pa), with an assumed snow load of 40 lb/ft² (1915 Pa), for a total of 65 lb/ft² (3112 Pa). The outer support of the marquee consisted of two diagonal hanger rods, 1 in (25 mm) round or ¾ in (19 mm) round upset to 1 in (25 mm) at ends which were threaded to receive an adjusting turnbuckle. The upper end was connected by a clevis, a steel forging with a socketed thread on one end and a circular hole in each of two prongs at the other end. The intent was to thread the hanger into the socket and bolt the prongs into a bent eye bolt embedded in the wall. In turn, the eye bolt would project through the wall and connect to a 4 × 4 × ½ in (102 × 102 × 13 mm) angle welded to a 6 × ⅛ in (152 × 3.2 mm) steel strap 4 ft 6 in (1.4 m) long, which was to be bolted to four timber floor beams.

The actual loading of the marquee was approximately 35 lb/ft² (1675 Pa). Since failure occurred with an additional load of only about 1 ft (0.31 m) of snow, the maximum *total actual load* was not over 50 lb/ft² (2394 Pa), or *77 percent* of the total design load. We determined that overloading, therefore, was not a factor in the collapse.

The actual loads produced combined tensile unit stresses of 8080 psi (55 MPa). The ASTM A36-62T steel used provided for a minimum yield of 36,000 psi (248 MPa), with an allowable stress of 24,000 psi (165 MPa). This would provide a factor of safety of 4.5.

There was, therefore, no deficiency in the original design. Rather, the important fact was that the construction differed from the intended design in a way which increased the stresses in the structure. A straight rod with a loop rod connector was used in lieu of an eye bolt. The ensuing eccentric load resulted in bending which theoretically could have been as high as 190,000 psi (1310 MPa) initially, but which readjusted and reduced such stresses by actual bending deformation, still leaving an eccentricity and an unacceptable connection.

Lessons

1. Avoid determinate structures without built-in redundancy, the safety of which depends on a single eye bolt or a single tie rod — or else design with very high safety factors.

2. There is no substitute for a thorough on-the-job inspection during and after completion of construction. Such information should reveal unauthorized substitutions.

3. Avoid bending in tension members such as rods or bolts. The stresses generated by bending are very high compared to the intended direct tensile stress and can result in fracture at relatively low tensile loads.

3.7.4 Library—Plainfield, New Jersey

This case history (Fig. 3.51) is included primarily to demonstrate that large deflections in a structure are *not always* signs of distress, but may occur in an adequately designed steel structure.

The Plainfield library is a hybrid structure consisting of a concrete waffle-slab roof supported on steel girders with very long spans (58 ft;

Figure 3.51 Library in Plainfield, New Jersey, is a hybrid structure consisting of a concrete waffle-slab roof supported on long-span steel girders. While the building itself was not overstressed, its doors, windows, and other architectural members were distressed.

17.7 m). For architectural reasons, the depth of construction was chosen to be very shallow for the long spans involved. The framing plan (Fig. 3.52) showed shallow, heavy 14-in (356-mm) deep WF members rolled from high-strength steel. This translated into a ratio of span to depth (L/D) of 50, which considerably exceeds the maximum L/D ratio of 30 recommended for roofs. (L is the total clear span and D is the depth of the member.) The deflection was so great (1½ in; 38.1 mm) that doors were locked in and would not open and glass windows were cracking.

What is of particular interest here is that we concluded that the structure itself was not overstressed—but, rather, the architectural members *were, indeed*, distressed.

Lessons

1. Avoid ratios of span to depth exceeding 30 for roof construction.

2. Where shallow construction is dictated, provide sufficient camber for dead loads and provide generously for proper allowance for live-load deflections by introducing ample clearances at tops of walls, doors, windows, and other architectural finishes. Fig. 3.52 shows where adequate clearance should be provided.

3.8   Failures Due to Roof Overloads

3.8.1   Stepped-roof structure—Elwood, Long Island, New York

A common case of roof overload at steps in the roof elevation—and subsequent collapse—occurred after a heavy snowstorm on Long Island. In fact, there were several snowfalls during the same week and subfreezing temperatures. At the line of change in roof elevation, there was approximately 5 ft (1.52 m) of differential level between the low roof and the high roof (Fig. 3.53).

There was also drifting snow as a result of winds. On arrival at the collapse site, I removed a section of snow 12 × 12 in (305 × 305 mm) for weight measurement. The total weight of the snow on the low roof was determined to be 57 lb/ft² (2730 Pa).

The ANSI code and local codes in snow zones call for low roofs at stepped areas to be designed for heavier snow loads (Fig. 3.54). This fact, unfortunately, is not always known by designers. This inevitably results in understrength roofs and eventual collapse after heavy snowfalls.

The design live loads for roofs in the area was 20 lb/ft² (958 Pa) at the time of construction. The 57-lb/ft² (2730-Pa) snow load that accumulated was nearly 3 times the original design level. The actual result of the snow load was the buckling of the first steel-pipe compression diagonal (Fig. 3.55), which is an indication of overstress as a result of overload. This pipe

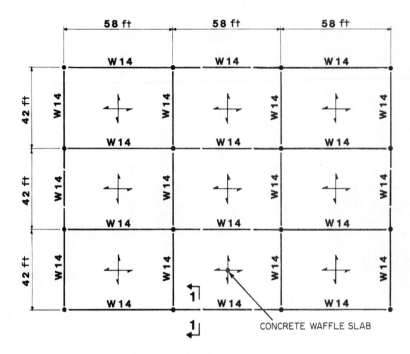

ROOF FRAMING PLAN

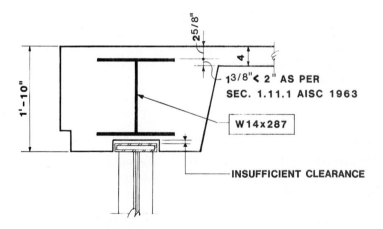

SECTION 1-1

Figure 3.52 Roof framing plan shows shallow and long, heavy, 14-in (356-mm)-deep WF members, which translates into a ratio of span to depth of 50, nearly double the recommendation L/D ratio of 30. Section detail shows insufficient clearance, resulting in inoperable doors and cracked window panes.

Figure 3.53 There was about a 5-ft (1.52 m) differential level between the low roof and the high roof, allowing heavy buildup of drifting snow.

buckled into an S shape (Fig. 3.56). The undeformed steel open-web joist can be seen in Fig. 3.57. The computed stresses are indicated in Fig. 3.55.

It is common for determinate structures such as simply supported steel joists to fail in a catenary shape, which results in the lateral movement of the supports (Fig. 3.58).

Lessons

1. For stepped roofs, design the lower roof elevation for increased loads.

2. It is recommended that roofs at party lines (common property lines) should be treated similarly, since future taller construction on the adjacent property will result in snow drifting and a buildup of load.

3.8.2 Parking deck — Nyack, New York

In early November 1967, a 40 × 40 ft (12.2 × 12.2 m) section of an earth-supporting roof slab over an underground parking garage collapsed into the lower area (Fig. 3.59). The collapse was accompanied by a tremendous noise which was mistaken for an explosion by those nearby. A broken gas pipe (Fig. 3.60) which protruded above the debris was initially considered as having played an important part in this "explosion-triggered" failure.

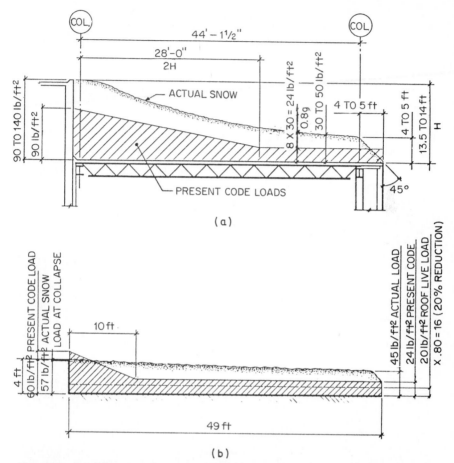

Figure 3.54 Load diagrams showing high snow loads at roof drops. (*a*) Approximate snow depths and load intensity at time of collapse. Failure was from imposed load exceeding the capacity of the joists; there were no errors in design or construction, and stress at steel chord was at yield point. (*g* = ground snow load, psi. *G* = 3*g* or 15*H*, whichever is smaller. (*b*) Snow depths at time of collapse.

The collapsed shape of some columns which were found to be bent like pretzels (Fig. 3.61) also added credibility to the theory of explosion.

Several weeks later, while we were still investigating the collapse, another section of the structure collapsed, this one 40×60 ft (12.2×18.3 m). This time there was absolutely no indication that a gas explosion was in any way involved.

The original design was for a load of 410 lb/ft² (19.6 kPa) which consisted of 3½ ft (1.1 m) of soil weighing 350 lb/ft² (16.7 kPa) and a superimposed live load of 50 lb/ft² (2.4 kPa). The load which was actually in place at the time of the collapse consisted of soil *varying in depth from 10 to 12 ft (3.1 to 3.7 m)* (Fig. 3.62), for a superimposed load of 1100 lb/ft² (52.7 kPa).

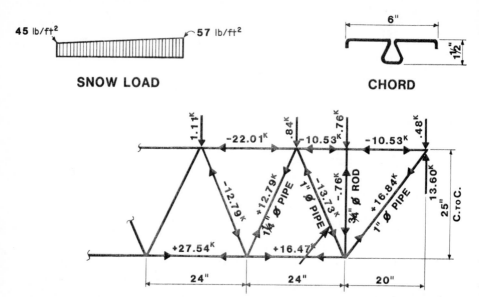

45 lb/ft²   **57 lb/ft²**

SNOW LOAD

6"

1½"

CHORD

Figure 3.55   Loads in joist truss members are shown. The snow accumulation resulted in a load 3 times the design load, causing buckling of the first compression diagonal (arrow).

WELDING DID
NOT FAIL

Figure 3.56   The welding of the first compression diagonal did not fail. The steel-pipe diagonal buckled into an S shape.

Figure 3.57   Undeformed steel open-web joist.

Figure 3.58   Determinate structures such as simply supported steel joists typically fail into a catenary shape, which results in lateral movement of supports (arrow).

Figure 3.59 The collapse of a large section of earth-supporting roof slab over an underground parking garage was mistakenly thought at first to have been caused by a gas explosion.

Figure 3.60 This broken pipe misled investigators to conclude that explosion was the cause of the collapse.

Figure 3.61 Pretzel-shaped column also threw investigators off the right trail.

This translated into an overload factor of approximately 2.6. With a factor of safety of 1.5 for structural-steel designs, there was no mystery why this structure came down. Some columns actually buckled under the load (Fig. 3.63). Furthermore, our investigation found no trace of explosion residues. We therefore attributed the sound of an "explosion" to the noise resulting from air pressurizing in the enclosed space below the deck as it collapsed.

An additional interesting aspect of this failure was the "stepping" of the structural frame of the deck to accommodate the final sloping grade of the soil fill. This was conceived by the structural engineer as a method by

Figure 3.62 Roof load in place at time of collapse consisted of soil at a depth of up to 12 ft.

Figure 3.63 Buckled columns should have alerted everybody to the serious overload condition.

which skewed connections would be avoided, thus minimizing fabrication and erection costs. The stepping resulted in soil depths far in excess of 3½ ft (1.1 m) which, when combined with overfilling by the contractor, resulted in the overload (Fig. 3.64).

The collapsed sections of the structure were reconstructed by raising the level of the deck framing, reinforcing deficient beams, and adding diagonal bracings (Figs. 3.65 and 3.66).

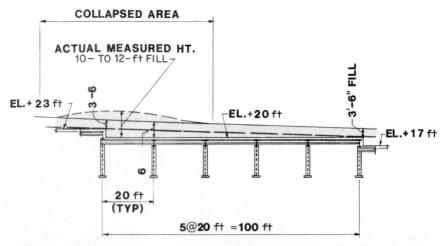

Figure 3.64 Stepping of structural-steel frame to accommodate sloping grade resulted in soil depths far above design. When contractor also overfilled, collapse was certain.

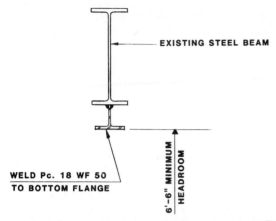

EXISTING STEEL BEAM

WELD Pc. 18 WF 50
TO BOTTOM FLANGE

6'-6" MINIMUM
HEADROOM

Figure 3.65 Typical detail for strengthening overstressed beams. Beams are jacked prior to welding. End connections of overstressed beams are checked and reinforced, if needed.

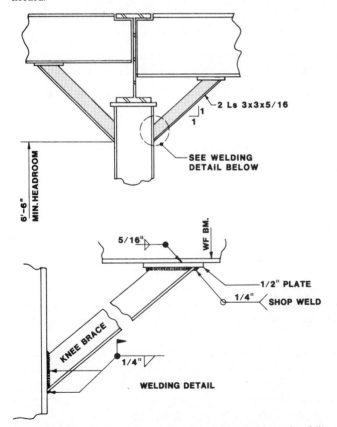

2 Ls 3x3x5/16

SEE WELDING
DETAIL BELOW

6'-6"
MIN. HEADROOM

5/16"

WF BM.

1/2" PLATE

1/4" SHOP WELD

KNEE BRACE

1/4"

WELDING DETAIL

Figure 3.66 Detail of knee braces added for additional lateral stability.

The sensitivity to failure of a soil-covered steel structure can be demonstrated as follows:

Garage roof (self-weight)	=	60 psi	(2.9 kPa)
3.5 ft (1.00 m) of soil	=	350 psi	(16.7 kPa)
Total dead load	=	410 psi	(19.6 kPa)
Live load	=	50 psi	(2.4 kPa)
Total dead and live loads	=	460 psi	(22.0 kPa)
Ultimate capacity = 460 lb/ft^2 × 1.5	=	690 psi	(33.0 kPa)
minus existing dead load	=	410 psi	(19.6 kPa)
		280 psi	(13.4 kPa)

Conclusion. An additional 2.8 ft (0.85 m) of dry soil weighing 100 lb/ft^3 or 2.3 ft (0.70 m) of wet soil weighing 123 lb/ft^3 are sufficient to collapse the roof [allowable stress = 24 ksi (165 MPa), yield stress = 36 ksi (248 MPa), 36/24 = 1.5 factor of safety].

Lessons

1. Pay special attention to avoid soil overloading during construction when earth-supporting structures are covered with mounds of soil.

2. Use *actual* soil weight, not theoretical weight, when stepped construction is used for sloping soil fills.

3. Immediately cut off all electrical power (Fig. 3.67) and other utilities after a collapse to avoid injuries and possible fatalities resulting from electrocution.

3.9   Failures Due to Improperly Designed Expansion Joints

3.9.1   New York City stadium

As part of the complete rehabilitation of a 50-year-old New York sports stadium in 1972, the two main expansion joints of the stadium were upgraded and rebuilt. The stadium is horseshoe-shaped and has sloped, stepped seating. These joints utilize slotted holes to allow for movement (Fig. 3.68).

Late in 1981, when cracks were noticed in the concrete slabs adjacent to the joints (Figs. 3.69 and 3.70), we suspected that these expansion joints were not functioning as planned. The alarm was finally sounded after a large piece of concrete fell, penetrating a hung ceiling, and crashing to the spectator seating below. Fortunately, no one was injured.

Figure 3.67 Close-up of failed connection. Typical torn electrical line clearly indicates the importance of immediate cutoff of power following structural collapse.

Temporary scaffolding and protective boards were immediately installed in accordance with our directions and we launched a thorough investigation. This investigation involved analytical computer models, monitoring of an instrumented truss (using electrical strain gauges, Fig. 3.71), movement gauges (Fig. 3.72), and material strength tests. The instrumentation program continued during both winter and summer periods so that a complete temperature cycle could be studied.

The accumulated data established clearly that the expansion joints were actually locked and were not operating at all. The cracks which had developed adjacent to the joints (Fig. 3.70) were actually acting as de facto joints and were moving with temperature changes. We recommended that the joints be released and reconstructed properly using slide plates with totally free movement to accommodate temperature changes, thus avoiding stresses and resulting cracking and damage. The repair method is described and illustrated in Fig. 3.73.

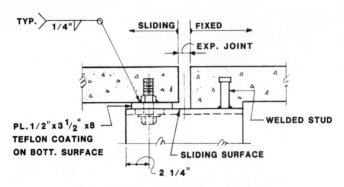

ESSENTIALS OF THE DESIGN DETAIL

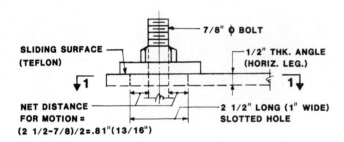

SLOTTED HOLE DETAIL

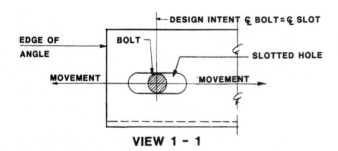

VIEW 1 - 1

Figure 3.68   Details of original main expansion joints of New York stadium. Note how movement of joint is restricted to 13⁄16 in (20.6 mm) by small slotted hole.

Figure 3.69 Cracks in the concrete slabs adjacent to the expansion joints indicated that the joints were locked and not functioning as planned. Wood pieces were used for temporary support.

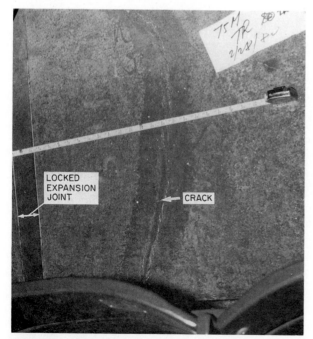

Figure 3.70 Locked expansion joint is seen next to the crack in the concrete deck.

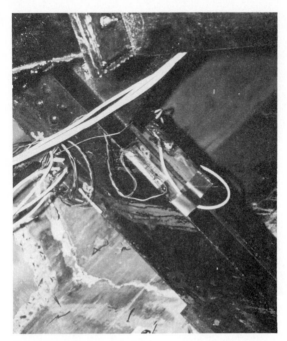

Figure 3.71 Electrical strain gauges were used in the stadium investigation.

Figure 3.72 Movement gauges were employed in stadium to confirm locked joint.

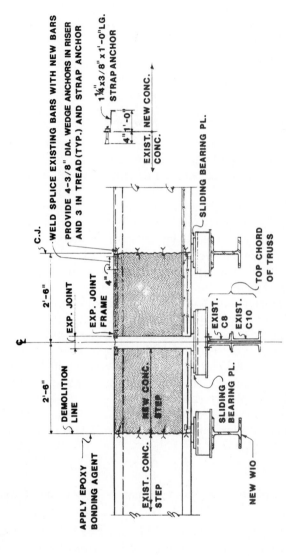

Figure 3.73 Reconstruction of expansion joints used sliding plates with totally free movement to accommodate temperature changes, thus avoiding stresses and resulting cracking and damage.

Lessons

1. Expansion joints using slotted holes to accommodate movements are likely candidates for failure. The reason is that it is impractical to properly install the bolts within the slotted holes so as to avoid lock-in conditions. Bolts may lock when installed too close to the edge of the slotted holes (Fig. 3.74) or when the slot is skewed in relation to the direction of movement (Fig. 3.75). Surprisingly, sometimes the bolts are even welded for "extra strength."

2. Design expansion joints so that in outdoor structures exposed to ambient temperatures, such as sports stadiums, each section has a maximum length not exceeding 100 ft (30.48 m).

3.9.2 Nassau Coliseum — Long Island, New York

This long-span structure (Fig. 3.76) had two basic problems which were evident even before the structure was completed.

The first problem involved thermal expansion which was blocked because of immobilized slotted holes in the single expansion joint provided. Coupled with this serious deficiency was another: some of the steel roof joists were subjected to heavy mechanical loads for which they were not designed. These loads were not shown on the structural drawings — so they were overlooked. Since the steel joists were not designed, detailed, or fabricated to support these loads, they failed. The buckled compression members are shown in Fig. 3.77.

One single expansion joint had been designed and constructed using the slotted-hole solution (which rarely works properly, since the bolts are placed on opposite sides of the slots, thereby locking them). We found that the 2½- to 3-in-long (63.5- to 76.2-mm) slotted holes even had friction bolts installed in them, so that when fully tightened, the joint did not work at all.

Because the Coliseum was scheduled to open shortly, we were forced to propose a reasonable solution that could be speedily implemented. Therefore, the friction bolts were removed and replaced with new bolts with sufficient clearance to allow for movement on each side of the slot and the buckled members replaced.

Lessons

1. Avoid use of slotted holes to accommodate movement in expansion joints.

2. Very large structures (usually over 200 ft long; 61 m) require expansion joints in sufficient number and adequate width to allow for the required movement as well as for the proper performance of sealants.

3. *Mechanical* loads must be shown on the *structural drawings* so that they will be accounted for and provided for in the final structure and its connections.

Figure 3.74 Photograph clearly shows locked joint with bolts at extreme ends of slotted hole.

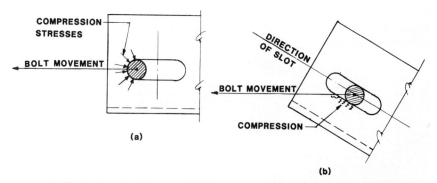

Figure 3.75 Expansion joints using slotted holes, like original joints in New York stadium, are likely candidates for failure because of their tendency to lock. (*a*) Joint is locked because of incorrect bolt placement. (*b*) Joint is locked because of incorrect slot installation.

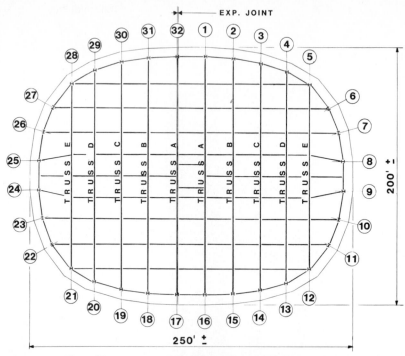

EXP. JOINT

⟨30⟩ ⟨31⟩ ⟨32⟩ ⟨1⟩ ⟨2⟩
⟨29⟩ ⟨3⟩
⟨28⟩ ⟨4⟩
⟨5⟩
⟨27⟩ ⟨6⟩
⟨26⟩ ⟨7⟩

TRUSS E TRUSS D TRUSS C TRUSS B TRUSS A TRUSS A TRUSS B TRUSS C TRUSS D TRUSS E

⟨25⟩ ⟨8⟩
⟨24⟩ ⟨9⟩   200' ±
⟨23⟩ ⟨10⟩
⟨22⟩ ⟨11⟩
⟨21⟩ ⟨12⟩
⟨20⟩ ⟨13⟩
⟨19⟩ ⟨18⟩ ⟨17⟩ ⟨16⟩ ⟨15⟩ ⟨14⟩

250' ±

Figure 3.76 Plan of long-span roof for Nassau Coliseum, a structure with two basic problems that were evident even before completion.

BUCKLED MEMBER

Figure 3.77 Buckled steel truss members at Nassau Coliseum.

272

Figure 3.78 During construction of Shea Stadium, structural-steel girts fractured and snapped at end connections during high winds. Photograph shows two of the failed girts on the right.

3.10   Failures Due to Vibration

3.10.1   Shea Stadium — Queens, New York

During construction of this stadium, structural-steel girts (beams) fractured and snapped (Fig. 3.78) during windy conditions. The unusual behavior of these members was even more puzzling because the beams were not supporting any load (other than their own weight) at the time of failure.

The two controlling features of these girts were:

1. The girts were erected with their flanges in the vertical direction and the web in the horizontal plane (Fig. 3.79).

2. There were deep cuts in the top flange and the webs at the end of the girts (Fig. 3.80).

Our analysis disclosed that the girts vibrated at a low frequency as a result of the wind. The natural frequency of the girts coincidentally was identical to the wind-gust frequency — resulting in resonance. As a result, the girts literally vibrated like violin strings and snapped at their ends (Fig. 3.78), leaving the connection clips attached to the columns.

Figure 3.79   Girts were erected around the stadium with flanges in vertical direction and web in horizontal plane.

Figure 3.80   Ends of girts and unreinforced deep cuts in top flange and web are shown. Peeling paint was indicative of incipient failure (arrow).

Lessons

1. Reinforce deep cuts at ends of steel beams by welding plates, angles, and the like.

2. Avoid blocking both flanges, especially where the web is horizontal. When blocking is a must, reinforce the web with welded plates.

References

1. *Rockland Leader,* July 22, 1970.
2. B. Keidan, "Why the Roof Came Tumbling Down," *The Philadelphia Inquirer,* May 28, 1978.
3. Reader's comment, *Engineering News Record,* July 19, 1973.
4. *Engineering News Record,* July 26, 1973.

4

Masonry

Masonry construction is one of the oldest methods of building. While masonry structures such as the pyramids in Egypt have survived thousands of years, many newer masonry structures fail before reaching their first birthday.

Masonry can be defined as "an assembly of stones and mortar." Theoretically, concrete structures are considered to be masonry structures; however, in this book, concrete construction is treated in a separate chapter (Chap. 2). This chapter covers the following topics:

1. Brick
2. Block
3. Cavity walls
4. Natural stones
5. Curtain wall
6. Precast concrete
7. Water penetration

4.1 Brick Failures

Brick masonry was used in ancient Egypt and Rome. The bible refers to bricks and mortar in *Genesis,* Chap. 11, Verse 3: "And they said to one another: 'Come, let us make brick, and burn them thoroughly' . . . and they had brick for stone, and slime had they for mortar."

Two components combine to form brick masonry: brick units and mortar. The failure of brickwork results from defects in the brick units themselves, the mortar between the units, and/or the bond in the interface between the two. The failure of the bond between the mortar and the brick units is often the cause of structural failure of the joints and the resulting water penetration through the joints. There are numerous cases of non-performance of brick masonry where both the mortar and the brick units were of acceptable quality, yet the bond between the two failed.

4.1.1 Brick units

Clay as it exists in nature is the basic material used in the production of brick units or, as they are commonly called, *bricks*. This material varies considerably in its chemical and physical properties. During the manufacture of bricks, clays from different sources are mixed and thus the brick produced is not always uniform.

Clay appears in three principal forms with similar chemical ingredients, but different physical properties:

1. Surface clays

2. Shales (clays that have been subjected to high pressure)

3. Fire clays (clays that are mined at greater depths and have more uniform properties and low oxide content)

Bricks are manufactured by the burning (baking) of clay or shale which has been molded in special forms. The units are baked at a high temperature (most building bricks are burned to the incipient stage at a temperature of 1500 to 2100°F (815 to 1149°C). The period of time during which the bricks are in the kiln — referred to as the baking time — determines the final properties of the brick. These properties include the color, absorption, tensile and compressive strengths, coefficient of saturation (COS), and the initial rate of saturation (IRS).

While compliance with the requirements of strength is important, the most critical property of brick which affects its performance and the rate of its deterioration is the COS.

ASTM specifications outline in detail the specific requirements for brick. However, mere compliance with these minimum requirements does not in all cases ensure quality brick.

4.1.2 Mortar

The major components of mortar are cement, sand, lime, and water. The proportions of these materials will determine the properties of the hardened mortar and its performance. The various types of mortars are identified with a letter such as M, N, K, S, and O. These mortars are described in detail in ASTM C 270.[1]

Type M, for instance, has a minimum compressive strength at 28 days of

2500 psi (17.24 MPa), while type N has a strength of only 750 psi (5.17 MPa). Portland cement–lime mortars of these types have the sand/cement ratios of 3:1 and 6:1, respectively. Masonry cements are proprietary mixtures with great variations in composition, and, therefore, their performance cannot be accurately predicted.

Because of the many ingredients involved in the manufacture of mortar and the fact that most mortars are mixed in the field, there are many reasons for mortar failure. For example, the sand component itself may vary by its gradation, fineness, void content, cleanliness, and color. Many cases have been recorded where beach sand containing salt had been used in mortar, resulting in corrosion of metal embedments. The destructive effects of chlorides on metal embedments such as anchors and reinforcing have been well-recorded (refer to Chap. 6).

The effect of the many admixtures which are used in masonry mortars has to be carefully studied prior to their use. The effect of air-entraining agents and other additives on the performance of mortars is not always clearly known, and such additives should be treated cautiously, or not used at all when in doubt. There is at least one instance of a catastrophic failure attributable to mortar which contained a special "high-bond" additive used to increase its bond strength.

4.1.3 Bond

One of the most important factors in the performance of brick masonry is the bond strength between the brick units and the adjacent mortar. This bond manifests itself by the ability of the mortar to adhere to the face of the brick. The tightness of this contact area is the property which will determine the ability of the finished wall to flex under load without separation at the contact lines. These separations, which are actually cracks through the contact surfaces, are the avenues through which water penetrates the brick wall. Once this penetration occurs, the deterioration process starts. The freeze-thaw cycle then takes over, cracks widen, and the failure process is underway.

The bond strength depends on the following factors:

1. Mortar type and its cement content
2. Brick material and its initial rate-of-saturation property
3. Workmanship

Several bond tests that we have conducted proved that bond strength is not necessarily directly proportional to the compressive strength of the mortar. The compatibility of the two (mortar and brick), is the determining factor. In certain instances, mortar of 750 psi (5.17 MPa) compressive strength yielded a higher bond strength than a mortar with a 2500 psi (17.24 MPa) compressive strength.

As a result of many investigations and field and laboratory tests, we recommend that preconstruction tests be performed in accordance with ASTM E-514.[2] At the conclusion of these tests (performed with various mixes using the actual brick and mortar to be used in construction), the suitable brick and compatible mortar type will be selected.

Coefficient of saturation.   *It has been our experience that brick which has a COS of more than 0.75 may cause deterioration problems.* It is, therefore, important to establish the value of this property prior to a firm commitment to purchase the particular brick:

$$\text{Coefficient of saturation} = \frac{\text{easily filled pores}}{\text{total fillable pores}}$$

$$= \frac{W_{24\text{h/cold}} - D}{W_{5\text{h/boil}} - D}$$

where W = weight in water and D = weight dry. Typical durability requirements per ASTM C 62[3] for grade SW (severe weathering) are as follows: minimum compression strength, 3000 psi (20.69 MPa); maximum water absorption, 17 percent; maximum saturation, 0.78.

Suction (initial rate of saturation).   *It is also important to check the suction characteristics of the brick to verify the need of prewetting the brick before laying.*

4.1.4   Workmanship

The so-called tightness of the joints depends to a great extent on the quality of the workmanship that went into building them. Proper tooling of joints will ensure high-quality bond and reduce the avenues through which water may enter the brick wall. The pressure applied by tooling forces the wet mortar into the brick on both sides of the joint, resulting in tighter bond. Proper tapping also will improve the resulting bond.

Full bedding of the mortar joints not only will increase bond strength but also will make it impossible for water to penetrate the joints.

4.1.5   Metal embedments

Metals, mostly steel, are embedded in brick masonry to enhance its strength properties. There are several forms in which metals are used in brick construction: horizontal reinforcing, vertical reinforcing, and miscellaneous steel embedments.

Horizontal reinforcing. Horizontal reinforcing with prewelded wire mesh embedded in the mortar joint is used primarily to increase the overall tensile strength of the joints.

Vertical reinforcing. Vertical rebars are sometimes installed in the mortar joints to serve as column reinforcing. Occasionally, the vertical rebars are embedded within mortar cast in the hollow cores created within the bricks during manufacture. Where corrosion occurs in rebars within these cores, the volumetric expansion ultimately causes distress in the form of cracks in the brick units.

Miscellaneous steel embedments. Steel embedments such as clip angles, shelf angles, channels, ties, bolts, and rods are used to support, anchor, and attach brick walls to backups and structural frames. Where brick panels are prefabricated in the shop and then erected and lifted into place, clip angles are used for attachment. When corrosion of the steel is triggered by one factor or another, distress ensues because of the volumetric change resulting from iron oxide expansion.

4.1.6 Structural factors

In order to perform properly, a masonry wall must be adequately stable and sufficiently strong to resist all the loads it is expected to support. Bearing walls, for example, support heavy loads of floors, roofs, trusses, and beams.

While masonry walls are often erroneously considered to be merely envelopes or enclosures, they are always acted on by forces such as compression, shear, and lateral loads (Fig. 4.1).

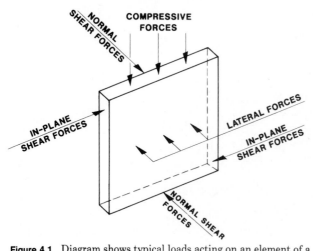

Figure 4.1 Diagram shows typical loads acting on an element of a wall.

Sometimes bearing masonry walls are mistakenly considered to be merely partitions dividing occupied spaces. In such cases, the removal of these "partitions" results in a catastrophe similar to the one shown in Fig. 4.2, which occurred in downtown Manhattan.

Stresses and damage mechanisms in masonry are as follows:

1. Wind movements
2. Foundation movements
3. Shortening of the frame
   a. Elastic shortening
   b. Creep
   c. Shrinkage

Wind movements. Wind loads on structures result in movements, lateral deflections, and stresses. Consequently, stresses are transferred from the frame to the brick walls.

This frame/facade interaction can be relieved by the introduction of what are commonly referred to as *soft joints.* These joints are, in fact, separations between brick units which allow for movement. They are also referred to as *expansion joints* or *movement joints.* These joints are filled with compressible fillers such as neoprene, premolded fillers, or other foam or elastic materials.

Figure 4.2 This parking deck supported on brick bearing walls collapsed when a bearing wall, thought to be merely a partition, was demolished.

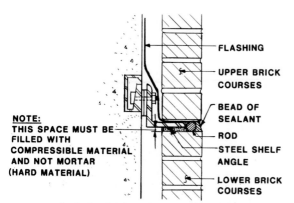

NOTE:
THIS SPACE MUST BE FILLED WITH COMPRESSIBLE MATERIAL AND NOT MORTAR (HARD MATERIAL)

FLASHING

UPPER BRICK COURSES

BEAD OF SEALANT

ROD

STEEL SHELF ANGLE

LOWER BRICK COURSES

Figure 4.3   Locked soft joint full of mortar in brick facade.

Even when soft joints (Fig. 4.47) are required by the design, they are often improperly constructed. Contractors sometimes assume that a joint full of mortar with a bead of sealant is, in fact, a satisfactory soft joint (Figs. 4.3 and 4.4). Where relief in the form of soft joints exists, the stresses will dissipate without distress.

Without such soft expansion joints, high stresses may result in cracks and spalls within the brick masonry.

Figure 4.4   This joint was designed as a soft joint. Unfortunately, behind the sealant the joint was full of mortar. Our inspector classified it as "no soft joint."

Figure 4.5 Diagonal cracks in corner brick wall caused by foundation settlement.

The highest stresses resulting from wind loads will occur at the base of the structure, with additional cracks forming in diagonal patterns.

Foundation movements.* As a result of settlement of the foundation of the structure, brick masonry supported on slabs, beams, or other structural elements will exhibit high stresses and cracks. The latter are similar to those developed as a result of wind movements; however, the distribution of the cracks is different. Cracks developed as a result of foundation movements telegraph upward throughout the height of the structure. Cracks caused by wind movement increase in magnitude from the top of the building downward.

As with other distress caused by structural causes, when foundations settle, the masonry walls affected act as deep beams (stiff diaphragms) in trying to resist the ensuing stresses. Unfortunately, because these walls are often not designed to do so, they will crack — most commonly in diagonal patterns (Figs. 4.5 and 4.6).

Shortening of the frame. The structural frame may shorten as a result of several causes. Frames shorten simply as a result of *elastic deformation* caused by compression forces generated in the frame. These deformations occur directly after the application of the loads.

Creep, on the other hand, is a phenomenon which is time-dependent. Creep deformations increase in time without the addition of any loads. Creep may continue for many years and is associated with concrete structures only.

Shrinkage is similar to creep; however, it is prevalent only during the first 4 to 5 years of the life of a concrete structure, after which it ends.

The shortening of the frame coupled with no shortening of the skin causes a shift of loads from the frame to the skin, thereby causing stresses

---

* Dov Kaminetzky, "Preventing Cracks in Masonry Walls," *Architectural Record*, November 1964, pp. 210–214.

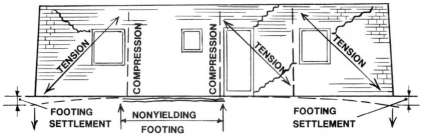

Figure 4.6 Diagonal tension cracks due to differential settlement. The center of the wall is based on nonyielding footing on rock.

and sometimes undesirable distress in the brick masonry. This condition, which is exacerbated in those instances where the skin actually expands, may be relieved by incorporation of soft joints.

The movement of the frame in relation to the brick face and the resulting pressures on the shelf angles will inevitably cause rotation of the angles with potential chipping and spalling (Fig. 4.7).

4.1.7 Brick expansion

It is commonly known that brick naturally expands when it absorbs water and when it is subjected to increased temperatures. It is *not* generally known that brick also expands when it has been improperly fired during manufacture. When its expansion is not constrained, no stress is generated. On the other hand, when expansion is restricted, the deformation is accompanied by stresses. When these stresses exceed the limit of strength, the brick fails by cracking or spalling.

Water absorption. As clay brick units cool after firing in the kiln, they absorb moisture from the air and expand. Where brick is exposed to environmental effects such as rain and temperature variations, the expansion will continue for several years. Expansion rates of 0.02 percent/year are not uncommon. Since this effect is often cumulative, the expansion may add up to 0.06 to 0.07 percent in a few years.

The practice of using softer mortars has often been suggested as a way to accommodate moisture expansion in brickwork. Unfortunately, tests have indicated only slightly less expansion with softer mortars within the first few days.

Temperature. Brick expands with rising temperatures and contracts with dropping ones, like most construction materials. The average coefficient of

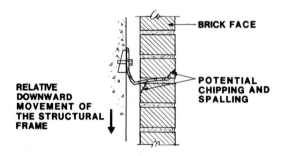

RESULT: SPALLING

Figure 4.7 The potential for chipping and spalling of brick at the shelf angles, resulting from movement of the frame, is clearly shown here.

thermal expansion for brick is 3×10^{-6} per °F (5.4×10^{-6} per °C), which is about half the expansion for steel or concrete.

A fact which is not widely known about brick is that residual expansion often occurs, so that actually the brick "grows" in time.

Actual surface temperatures of brick exposed to the sun are considerably higher than the ambient temperature. While the ambient temperature may be 90°F (32°C), the brick itself may be at 140°F (60°C). Brick sills, for example, are particularly subjected to high temperature, which often has serious effects, especially when the sills are restrained at their ends.

Insufficient baking. When brick units are underbaked in the kiln, they later expand excessively while in service. These improperly baked brick units are often recognizable by their lighter color.

Cases where such brick units were laid and incorporated into building walls have been recorded. The walls in these instances expanded excessively, even cracking the restraining concrete columns.

With the expansion of the brick, we always witness the concurrent shortening of the structural frame as a result of elastic shortening, creep, shrinkage, or wind forces (see Sec. 4.1.6). This combination further increases the distress and damage.

4.1.8 Construction defects

Nonconformance by the masonry contractor may result in serious consequences. When key anchorage elements such as metal ties or metal anchor slots have been misplaced or even totally left out, walls quite often will shift. Where steel shelf angles have not been properly attached or are otherwise loose, they will rotate or deflect, causing the brick to crack, spall, bow, or buckle.

Failure *may and does* occur in any element or component of brick walls.

Material failure may take place as a result of deterioration of the brick units or the mortar bonding them together; corrosion of the metal anchorage system; tearing or disintegration of the flashing membrane; or deterioration of coping stones, window sills, and all other wall components.

Missing ties. Where dovetail slots have not been embedded in concrete columns or beams, the mason sometimes will take the easy way out and omit the ties entirely. The proper action to correct for missing slots is to install brick ties into the concrete by attaching metal rails sometimes called *Gripstays*, or ties expansion-bolted into the concrete.

We have investigated cases where certain portions of the walls were heavily anchored with ties while other sections had none — no ties at all! With further research we found that the masons on these projects apparently did not want to carry the ties in their aprons all day so they typically used them up in the early hours of the workday.

Missing dovetail slots.   These slots are usually specified to be installed in the concrete by the concrete subcontractor. They are nailed into the side of the formwork, and the metal slot form filler is removed after the concrete has hardened. These slots are installed when brick or block walls are anchored to or are backed by concrete columns or beams. They are erroneously or carelessly omitted when poor coordination exists among the various trades on a project.

Missing or loose shelf angles.   Shelf angles are often either short or totally missing at the corners of the building (Fig. 4.8) and at the returns of brick piers. Where shelf angles are missing at the corners, the direct result is

Figure 4.8 Missing shelf angle at the corner of this high-rise building caused brick to crack. Cavity filled with mortar did not help matters.

buildup of compression forces, leading to corner cracking. When adjustable inserts have not been embedded in the spandrel beams and are missing, the miscellaneous-iron erector often assumes that they are not needed and the angle is left out. The error is then perpetuated by the brick mason who "does not ask questions" and builds two- or three-story heights of brick on one inadequate shelf. Good coordination among the trades, supplemented by alert field inspection, will eliminate these errors.

Missing bolts.   Bolts are often missing when there is no alignment of the holes in the steel shelf angles with the adjustable inserts embedded in the concrete. This problem may be corrected by either providing new holes in the shelf or drilling an expansion bolt into the concrete. Unfortunately, the bolts are sometimes left out completely and may escape the inspector's eye.

Missing or torn flashing.   Where water penetrates through exterior walls, we frequently find the culprit to be missing, damaged, or otherwise inadequate flashing. Flashing is a barrier to water flow. When flashing is punctured by careless construction, an open path for water inflow is created. Punctures, tears, open splices, and loose flashing are all construction defects that can be easily avoided by good workmanship and alert inspection procedures (Fig. 4.9).

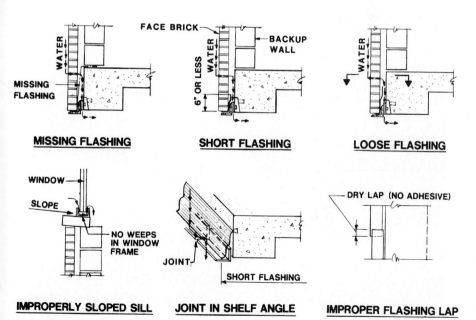

Figure 4.9   Sketches show the many avenues of water penetration through brick walls, resulting from a variety of deficiencies.

Improperly connected shelf angles. Where the faces of beams are placed too far away from their intended design location, the brick load cannot be adequately transferred to the spandrel beams by means of shims. Steel bars may be used for this purpose; however, bars must *not* be used in the flat position shown in Fig. 4.10, which will bend the flat bar, but in the correct vertical position as shown in Fig. 4.11.

4.1.9 Glazed brick

This special brick, manufactured by application of a glazing finish on the brick prior to baking, has had more than its share of problems. During the 1950s and 1960s, glazed brick was very popular and was used extensively for exterior facing. Unfortunately, its performance in cold environments was not always successful. As a result of freeze-thaw cycles, many of these walls exhibited massive spalling and cracking. The present theory, which is still not completely proved, is that the glazing serves as an unintended barrier, preventing trapped water in the brick from escaping to the exterior face from behind the glazing.

Cases of enormous distress are recorded, especially of walls which are not heated during the winter. Typical cases are brick garden walls, which are exposed on both faces. Unheated enclosed spaces such as stairways and storage rooms are also often victims of this phenomenon.

Figure 4.10 Condition of steel bars connecting shelf angles and spandrel plates, as found during investigation. Bars were improperly welded in flat position.

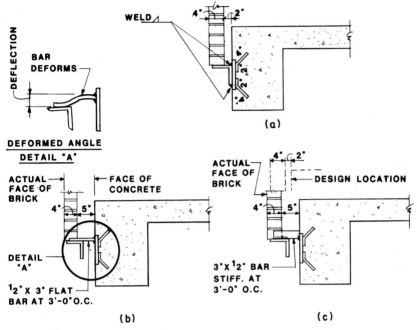

Figure 4.11 Sketches depicting (*a*) the design intent, (*b*) improper field solution, and (*c*) correct alternative for welding shelf angles to spandrel beams.

4.1.10 Parapet walls

Parapets in exterior walls are perpetual problems. The parapet is exposed to the environment on both its faces and at its top (coping). It is exposed to rain and wind and to high and low temperatures without any help and protection from air conditioning and heating. The parapet is thus subjected to the greatest number of freeze-thaw cycles and therefore is typically the first element to show deterioration.

The one basic tool to combat temperature effects in these cases is the introduction of expansion joints along the length of the parapets and at the corners. Customarily these joints are placed at 15- to 20-ft (4.6- to 6.1-m) spacings. These joints reduce the effective length of the walls, thereby minimizing pressures and distress. It is common practice that specifications require that the joint must be totally soft and not blocked with mortar.

When such joints are not provided or when they become locked, the parapets will bow, buckle, and cause cracking in various modes:

1. The corners are where the expansion of parapets is manifested most severely. The expansion of the parapet sometimes will carry portions of the lower wall with it (Fig. 4.12).

Figure 4.12 Because of lack of vertical expansion joints in this long parapet wall, the corners pushed out and cracked, with horizontal separation lines and 45° cracking at the exterior corner.

2. Cracking and bowing of the parapet wall at its top (Fig. 4.13) is a frequent occurrence.

3. Sometimes the parapet will bow upward when dowels connecting the roof parapet have been cut to avoid the cap flashing and the expansion joints are blocked with mortar (Figs. 4.14 and 4.15).

Figure 4.13 This parapet wall bowed and cracked in place as a result of temperature expansion without relief joints.

(a)

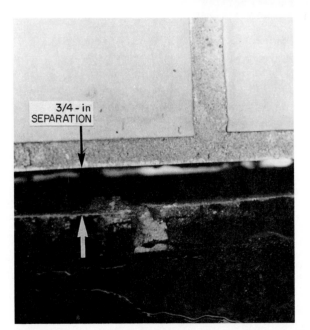

(b)

Figure 4.14 This parapet wall buckled upward (a) [see separation in enlargement]. (b) The investigation revealed that the vertical dowels were improperly cut at the cap flashing.

Figure 4.15 Probes of the expansion joints of the wall in Fig. 4.14 showed that, behind the exterior sealant, the joints were locked and full of mortar.

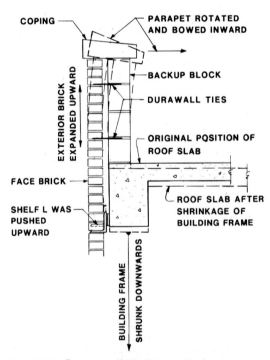

Figure 4.16 Parapet wall tilted inward as a result of the differential movement of expanding brick and shrinking of the structural frame. A horizontal soft joint would have prevented this failure.

4. Where the building frame shrinks and the exterior brick expands excessively, the parapet will bow inwardly as the block backup bearing on the frame goes down while the brick expands upward (Fig. 4.16).

4.1.11 Other design problems

The problems of omission of soft joints in design and of improperly designed masonry supports were discussed earlier. Other design deficiencies which unfortunately recur frequently are:

Lip brick design.   In order to conceal the shelf angle for aesthetic purposes and reduce the thickness of the soft joints, architects sometimes resort to the use of "lip brick" (Fig. 4.17). This specially manufactured brick is hard to produce and its success is always highly questionable. During my long experience I have not seen even one successful use of lip brick. Because of construction and manufacturing tolerances, the lip ends up with undesired contact at the toe of the steel angle or harmful pressure against the brick unit directly below (Fig. 4.18).

Low shelf angles.   Because of dimensional difficulties or certain window details, the designer may place the shelf angle at a low elevation in relation to the soffit of the spandrel beams (Fig. 4.19). In such cases, where the

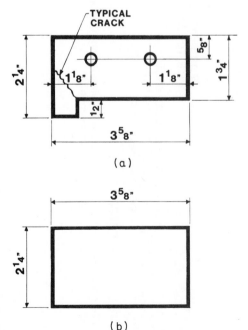

Figure 4.17   Side view of the geometry of lip brick (*a*) compared to standard brick (*b*).

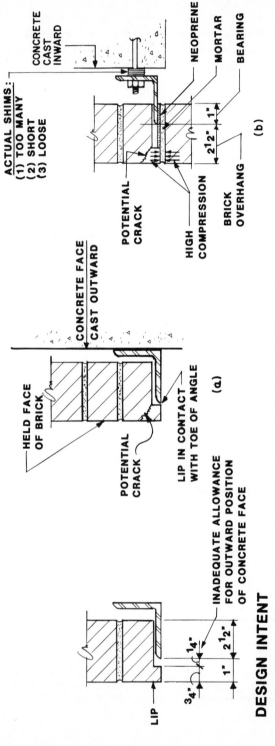

DESIGN INTENT

Figure 4.18 The use of lip brick often results in failure because the lip breaks after (*a*) contact with the shelf angle or (*b*) pressure on the lower brick.

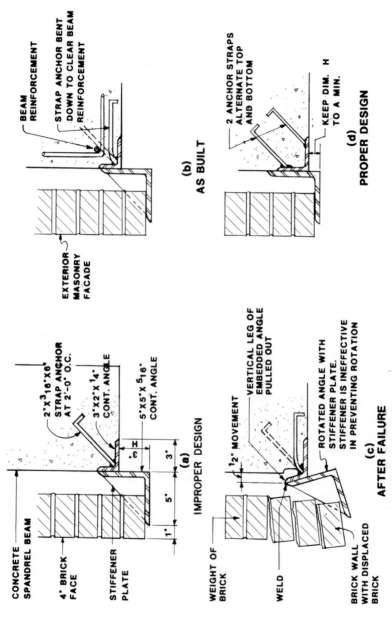

CONCRETE SPANDREL BEAM

4" BRICK FACE

STIFFENER PLATE

2"x3⁄16"x6" STRAP ANCHOR AT 2'-0" O.C.

3"x2"x1⁄4" CONT. ANGLE

5"x5"x5⁄16" CONT. ANGLE

H

3"

3"

5"

1"

(a)
IMPROPER DESIGN

BEAM REINFORCEMENT

STRAP ANCHOR BENT DOWN TO CLEAR BEAM REINFORCEMENT

EXTERIOR MASONRY FACADE

(b)
AS BUILT

WEIGHT OF BRICK

1⁄2" MOVEMENT

VERTICAL LEG OF EMBEDDED ANGLE PULLED OUT

WELD

ROTATED ANGLE WITH STIFFENER PLATE. STIFFENER IS INEFFECTIVE IN PREVENTING ROTATION

BRICK WALL WITH DISPLACED BRICK

(c)
AFTER FAILURE

2 ANCHOR STRAPS ALTERNATE TOP AND BOTTOM

KEEP DIM. H TO A MIN.

(d)
PROPER DESIGN

Figure 4.19 Where the shelf angle is designed too low in relation to the spandrel beam, (*a*), the shelf tends to rotate and even fail (*c*), especially where metal strap anchors are improperly (*a*) designed and (*b*) built. (*d*) Proper design.

297

dimension H is greater than 1 in (25 mm), the shelf will rotate, causing distress in the brick facing [3 in (76 mm) was used for H in the example of Fig. 4.19]. The steel stiffeners provided by the designer in this case were ineffective and only wishful thinking, since they could not prevent rotation of the entire angle. Proper anchorage into the concrete must also be provided as illustrated in Fig. 4.19*d*. It is clear that even if the contractor had followed Fig. 4.19*a* perfectly, and did not bend the strap anchors, the shelf would certainly have rotated and failed.

4.2   Concrete-Block Failures

Concrete block is a general term which applies to small precast masonry units which are assembled together and bonded by means of a cementitious material (often carelessly called *mortar*) to form a wall unit.

Block walls are used as bearing walls, enclosure walls, and backup walls. In the last capacity, they are built behind the brick face wall to increase its flexural capacity and to serve as an additional water barrier.

The material cast in metal forms can be standard stone concrete, lightweight concrete, cinder concrete, or other varieties of lightweight concrete. Block units are manufactured as solid (minimum 75 percent solid and maximum 25 percent hollow) or hollow units (less than 75 percent solid).

The bond between block units and the mortar within the joints is as critical as in brick walls. Where block walls serve as the main structural element, the bond at the interface between the block and the mortar will determine the performance of this wall.

Even where block walls serve as backups in cavity walls or other composite walls, tight mortar joints at the contact surfaces between the block and the mortar will prevent air movement through the wall, thereby cutting the flow of water through the block wall and into the occupied space. *Water cannot flow through a wall where airflow is cut off.* Of the four boundary lines of a wall panel (top, bottom, and two sides), the top line (horizontal) is the weakest, while the bottom line (horizontal) offers the greatest resistance to water penetration because it is compressed.

It is extremely critical that these four boundary lines be tight against air and consequently against water flow. It is equally important that they be secured by mechanical means. This may be achieved by the installation of metal dovetail slots in the concrete columns at both sides and also at the soffit (bottom) of the slab above. Dovetail metal anchors are then embedded in the mortar joint during the laying of the block walls to provide the necessary mechanical bond.

We cannot underestimate the importance of workmanship in building a block wall. *Proper tooling of the joints* will increase its strength and its resistance to water flow and will greatly extend the life of the wall.

4.3   Cavity Walls

The idea behind the cavity wall is to construct a structurally strong wall, yet create a space (a cavity) which will serve as a drainage ditch to first collect whatever water succeeds in penetrating the brick face wall (first line of defense) and then lead this water back outside through specially installed ports (weep holes).

Unfortunately, this idea, while perfectly logical, does not often work as planned. As with any other complicated system, the possibilities of failure multiply with the number of components. This could be compared to a chain constructed of multitudes of links of different sizes and metals, constructed by different workers at different times.

The various components of the cavity wall are:

1. Brick face (brick and mortar)

2. Backup (block and mortar)

3. Cavity

4. Metal anchors (between brick and block walls)

5. Metal anchors (between block wall and concrete surfaces)

6. Dovetail metal anchor slots

7. Metal reinforcing (horizontal)

8. Weep holes

9. Membrane flashing

10. Sealant (at toe of shelf angle)

11. Backup rod for sealant

12. Metal shelf angle

13. Bolts (connecting the shelf angle to the structure)

14. Adjustable metal anchor inserts

15. Metal shims (for adjustment of shelf angles)

These 15 components, together with their subcomponents, yield at least 20 variables. When metal studs are used as backup, the number of variables increases further.

4.4   Natural-stone failures

Ancient civilizations such as the Egyptian, Greek, Roman, and Israelite built structures and heavy fortifications from quarry-cut natural stone. Some of these earlier structures even used cut-stone units for heavily loaded bearing walls.

More recently, human-made materials such as reinforced and pre-stressed concrete panels and structural steel are used for support of heavy loads because of their ease of construction and the advantages of high allowable stresses both in compression and tension.

Cut stones, therefore, are used increasingly as facing materials only, because of their aesthetic features. Stone units, after being cut to a predetermined size, are incorporated into building facades in one of the following ways:

1. As facing units with backup structures — usually a concrete-block wall, a cast-concrete wall, or a metal-stud backup wall.

2. As facing units which are cast into the backup concrete wall to form a monolithic final element.

3. As facing stone laminates which are bonded together with a precast-concrete element or a steel frame to form a composite section, which is then erected as a facing unit with or without a backup wall.

While stone facades serve as well as any other enclosure to protect building interiors from environmental effects such as temperature, rainwater, wind, dust, and noise, they are not immune to some of the typical construction problems and failures which frequently haunt and harass other facade systems. These include:

1. Failure of the stone by cracking, spalling, and bowing. These failures may be induced by an improper vertical support system (shelf angles or other cast brackets), defects in the stone material itself, or the selection of a type of stone *which cannot* be used in an alien environment.

2. Failure of the anchorage system as a result of corroded or otherwise defective connectors. These failures may result in the collapse of an entire stone unit.

3. Water penetration through the stone itself or its joints.

4. Bowing or bulging of units as a result of temperature exposures or differential temperature between the outside and inside faces of the stone units.

5. Bowing of stone units as a result of expansion caused by absorption of water at the exterior surface.

6. Bowing of stone units as a result of stresses imposed on the stone facing by the structural frame. These stresses can be induced by effects such as elastic or plastic shortening of the building frame (Fig. 4.20) or stresses induced by lateral loads such as wind. Where reinforced concrete is used as the main structural frame, the effect of shrinkage and creep of the

Figure 4.20   Buckling of stone units resulted from shortening of the frame.

columns may be a dominant factor, especially if properly designed soft relief joints are not incorporated into the structure.

7. Sealant failures by tearing of the joint in tension (Fig. 4.21) or squeeze-out of sealant material in compression (Fig. 4.22).

Several case histories in which unexpected failure has occurred will be described later in this chapter. The causes underlying the failures will be identified, and how these situations could have been avoided in the first place and how they were later rectified by corrective measures will be described. Sometimes, the total removal of the stone facade and the subsequent substitution of a metal facing or a more suitable stone facing system was the answer (Fig. 4.23).

The use of rather thin units of stone bearing on narrow shelf angles results quite often in insufficient bearing, even when tolerances are met. The tolerances involved are those of the building frame, steel supports, the stone units themselves, and the vertical lines of erection of the stone.

Auxiliary metal support systems (Figs. 4.24, 4.25, and 4.26) will improve bearing and load transfer. However, all minute details of slotted adjustment holes, both horizontal and vertical, must be compatible with the ability of the metal accessory to properly transfer the loads of the stone to

Figure 4.21 Sealant tore in this vertical expansion joint because it could not accommodate horizontal movement.

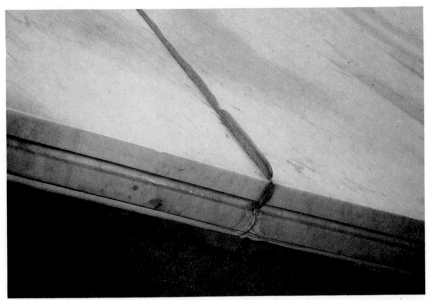

Figure 4.22 Squeezed and failed joints resulting from horizontal expansion of marble units.

Figure 4.23   The failed marble has been removed from the entire face of the building. Aluminum face panels have already been installed on left.

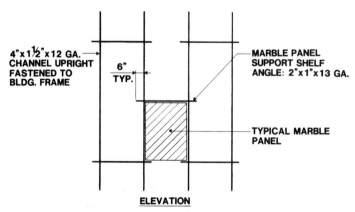

4"x1½"x12 GA.
CHANNEL UPRIGHT
FASTENED TO
BLDG. FRAME

6"
TYP.

MARBLE PANEL
SUPPORT SHELF
ANGLE: 2"x1"x13 GA.

TYPICAL MARBLE
PANEL

ELEVATION

Figure 4.24   Independent auxiliary metal support system will minimize undesired stresses in stone facing panels.

the structural support. Excessive use of horizontal or vertical shims must be avoided (Figs. 4.27 and 4.28), since this practice inevitably results in undesirable movements and rotations. The capacities of bolts will then be exceeded and they consequently will bend, with resulting rotation of the shelf angle stone supports.

Figure 4.25   Vertical metal channels are attached to frame and connected with adjustable clips to shelf angles.

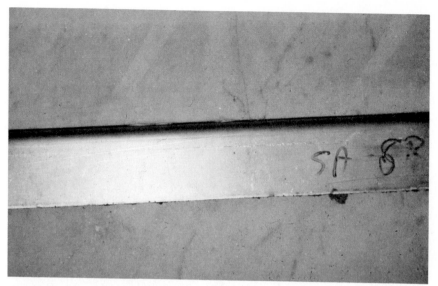

Figure 4.26 Back of the shelf angle shows soft-joint provision to allow for future movement.

Figure 4.27 Excessive horizontal shims (eight could be counted) used for adjustment. These shims could not transfer lateral loads.

Figure 4.28 Excessive vertical shims viewed from bottom of shelf angle. This caused rotation of the shelf and dangerous movements of stone panels.

Marble, as a structural material, also goes back to the ancient Greek and Roman periods. Many marble works of art, such as Venus de Milo, Michelangelo's David, and structures like the Taj Mahal, survive to this day because marble is a natural material with great durability when kept dry (its mechanical properties vary, however, with different varieties).

Each design should be prepared in accordance with the specific properties of the material actually used. Table 4.1 includes typical properties of marble, limestone, travertine, and granite, which are the stone materials most commonly used for stone facing.

Thermal expansion of natural stone is an important factor when a temperature gradient exists across the panel thickness (Fig. 4.29). This effect may be magnified in composite sections of dissimilar materials, sometimes stone laminated on precast-concrete backup panels. The designers should assure themselves that the stone selected has a coefficient of thermal expansion and modulus of elasticity compatible with that of the precast backup panels.

Some stone materials are not as durable as their suppliers claim. Tests of samples exposed to the environment and unexposed samples have demonstrated considerable differences in compression, tensile, and flexural strengths; temperature coefficients; and elastic modulus values. It is, therefore, extremely critical that tests of this nature be conducted *prior to* the incorporation of a particular stone material into the structure. Laboratory tests for marble, for instance, have demonstrated that after many

TABLE 4.1 Properties of Stone

Property	Marble	Limestone	Travertine	Granite
Compressive strength				
ksi	6–17	8–16	5–9	17–20
MPa	41–117	55–110	34–62	117–138
Flexural strength or modulus of rupture				
ksi	1.1–2.8	0.8–1	0.7–1.8	1.0–2.0
MPa	7.6–19.3	5.5–6.9	4.8–12.4	6.9–13.8
Modulus of elasticity E				
ksi	2×10^3–15×10^3	4×10^3–8×10^3	5×10^3–8×10^3	5×10^3–7×10^3
MPa	14×10^3–103×10^3	28×10^3–55×10^3	34×10^3–55×10^3	34×10^3–48×10^{-3}
Coefficient of thermal expansion				
per °F	2×10^{-6}–12.3×10^{-6}	2.2×10^{-6}–3.4×10^{-6}	7×10^{-6}–10×10^{-6}	2×10^{-6}–5×10^{-6}
per °C	3.6×10^{-6}–22.1×10^{-6}	4×10^{-6}–6.1×10^{-6}	12.6×10^{-6}–18×10^{-6}	3.6×10^{-6}–9×10^{-6}

thermal cycles, a residual thermal expansion of up to 20 percent of the original expansion may be realized. This permanent growth of the stone units must be accounted for when soft and other types of expansion joints are designed. Results of tests performed on marble in the course of one investigation are shown in Table 4.2.

The table shows that both flexural strength and modulus of elasticity in bending decreased after exposure. The standard deviation and coefficient

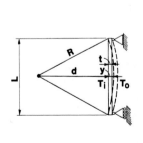

$\triangle T$	y, in
10° F	.025
20° F	.049
30° F	.074
40° F	.096
50° F	.121
60° F	.144
70° F	.168
80° F	.192

Figure 4.29 When interior and exterior surfaces are at different temperatures, stone panels will bow. $\Delta T = T_0 - T_i$, where $\Delta T =$ thermal gradient across panel thickness, $T_0 =$ outside temperature, and $T_i =$ inside temperature. $y = R - d$, where $y =$ deflection, $R = t/\Delta T \alpha$, $d = \frac{1}{2} \sqrt{4R^2 - L^2}$, thickness $t = 1$ in, and $\alpha = 7.4 \times 10^{-6}$.

TABLE 4.2   Results of Tests on Marble

Material	Property	Average Value
Original marble not exposed to weather	Flexural strength	1400 psi (9.7 MPa)
	Modulus of elasticity in bending	3.5×10^3 ksi (24×10^3 MPa)
Exposed marble (5 years), taken from wall	Flexural strength	200 psi (1.4 MPa)
	Modulus of elasticity in bending	1.7×10^3 ksi (12×10^3 MPa)

of variation of each property also increased after exposure. These data indicated that a change in the marble had occurred and that the material has become less homogeneous after exposure to environmental effects.

The factor of safety most commonly used for stone materials is 10, with a factor of 6 used for units subjected to wind stresses.

Relieving angles must be provided as often as practicable. The detail of stacked units (Fig. 4.30) has its shortcomings, since with this system any minor misalignment may cause movement of units, especially when:

1. Compression buildup is permitted to occur.

2. Units are too thin. While *thick* stone units can be stacked successfully, applying this system to thin panels, 2 in (51 mm) or less, should be avoided.

3. Insufficient anchors are provided to tie the stone back to the backup or structural frame.

The complete assembly of an independent metal support system such as shown in Figs. 4.24, 4.25, and 4.31 will avoid compression buildup when installed properly. However, even here bowing and bulging of units may occur when insufficient width of soft joints is provided in actual construction (Fig. 4.32). These assemblies must be fully secured in place prior to stacking of subsequent upper units; failure may ensue if this precaution is not carefully adhered to.

The interaction of the frame and the more rigid (stiffer) masonry enclosure is critical to the behavior of the entire structure. The relationship between the two also brings new problems. Any change in relative position of the two systems imposes shears in the facade (the stiffer system), which is quite often also the more brittle component, as compared with the more flexible and yielding frame.

Elastic shortening of the steel columns from loadings applied after the wall enclosure is complete can seriously affect stone facing when no relief is provided in the horizontal joints. The common old detail of lead pads at

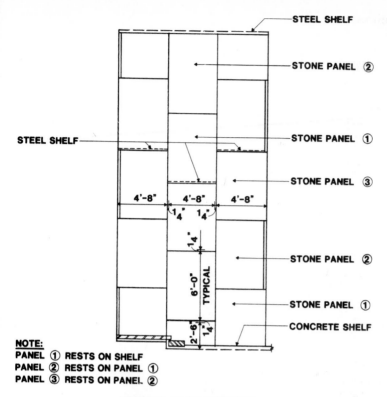

STEEL SHELF

STONE PANEL ②

STEEL SHELF

STONE PANEL ①

STONE PANEL ③

4'-8"   4'-8"   4'-8"

1¼"   1¼"

¼"

6'-0" TYPICAL

STONE PANEL ②

STONE PANEL ①

CONCRETE SHELF

2'-6" ¼"

NOTE:
PANEL ① RESTS ON SHELF
PANEL ② RESTS ON PANEL ①
PANEL ③ RESTS ON PANEL ②

TYPICAL FACE SUPPORT STACKED ARRANGEMENT ELEVATION

Figure 4.30 When stone panels are stacked one on top of the other, three or four in a row, a *highly unstable* system results. Such a system should not be utilized for stone panels, especially thin panels.

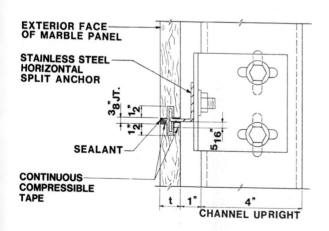

EXTERIOR FACE OF MARBLE PANEL

STAINLESS STEEL HORIZONTAL SPLIT ANCHOR

⅜" JT.

1½"

1½"

5/16"

SEALANT

CONTINUOUS COMPRESSIBLE TAPE

t   1"   4"

CHANNEL UPRIGHT

SPLIT-EDGE DETAIL - AS DESIGNED

Figure 4.31 Design intent was to leave the joint ⅜ in (9.5 mm) wide.

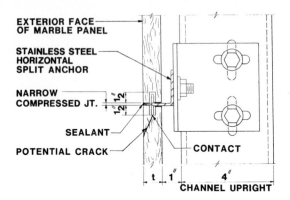

EXTERIOR FACE
OF MARBLE PANEL

STAINLESS STEEL
HORIZONTAL
SPLIT ANCHOR

NARROW
COMPRESSED JT.

SEALANT

POTENTIAL CRACK

CONTACT

t $1''$ $4''$

CHANNEL UPRIGHT

SPLIT-EDGE DETAIL - AS BUILT

Figure 4.32   Actual joint width provided was insufficient, thus negating the soft joint.

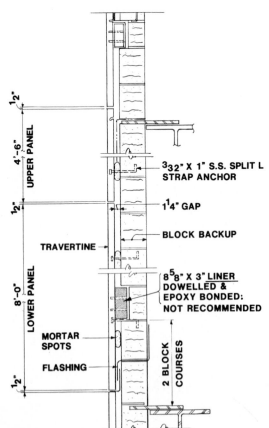

$\frac{1}{2}''$

$4'-6''$

UPPER PANEL

$\frac{1}{2}''$

$\frac{3}{32}''$ X 1" S.S. SPLIT L STRAP ANCHOR

$1\frac{1}{4}''$ GAP

BLOCK BACKUP

TRAVERTINE

$8\frac{5}{8}''$ X 3" LINER DOWELLED & EPOXY BONDED: NOT RECOMMENDED

$8'-0''$

LOWER PANEL

MORTAR SPOTS

FLASHING

2 BLOCK COURSES

$\frac{1}{2}''$

Figure 4.33   Epoxy-bonded laminated liners attached with short dowels to stone units serving as the main support are unreliable for this purpose and often fail.

every second story for stone covering supported by steel framing seems to have been overlooked in the rush of construction starting in the 1930s.

When very high facades are set in thin, rigid, high-strength mortar joints, compression resulting from thermal movements may cause warping of the building face and constitute a dangerous condition.

There are several additional factors here in action. The delayed shrinkage and time-dependent creep of the loaded columns and the elastic shortening from loads added after the walls have been built, especially where lower floors are started before the framework has been topped out, impose compressive forces into the sheet of facing material.

Some stone materials also exhibit loss of flexural strength from exposure to weather and freeze-thaw cycles. This loss may amount to 40 percent or more of the material's original strength.

Details which incorporate liner plates "glued" by epoxy and short dowels to the face stone and then, in turn, supported on the block backup wall (Fig. 4.33) lack positive means of support from the facing to the backup. Also, the so-called glues (some of the epoxies and especially polyesters) are

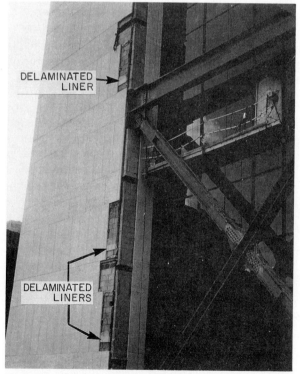

Figure 4.34 A half-ton stone unit fell to the sidewalk during construction of this high-rise building. The stone liner attached with adhesive separated and failed.

questionable as a reliable method for providing the sole main structural support for sizable stone units. The stone facade shown in Fig. 4.34 illustrates a case where a large travertine unit supported by a glued liner fell to the ground after installation into the backup wall. Stone panels were subsequently through-bolted for proper anchorage as a corrective measure.

4.4.1 Guidelines for stonework

The following guidelines are suggested for avoiding distress and other problems in stone masonry facing:

1. *Prevent physical contact between the back face of stone walls and the structural frame (columns, floor slabs, and end beams);* expansion may push the stone out and cause damage to the stone.

2. Specify mortar joints to be tooled for proper moisture resistance. The forced close contact between the mortar and the stone will cut off the path of water penetration (Fig. 4.35).

3. It is difficult to maintain fully intact a long vertical mortar joint. Long vertical joints should be sealed with elastic sealants.

4. Use the preferred detail at corners as depicted in Fig. 4.36. This detail is also more economical.

5. Pierced and torn flashing should be resealed.

6. Steel lintels (simple span or continuous) carrying stonework should have ample stiffness or support the superimposed load with a deflection of less than $L/360$, where L is the clear span of the steel lintel. This require-

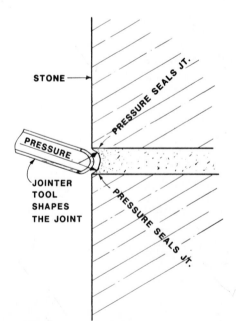

STONE

PRESSURE SEALS JT.

PRESSURE

JOINTER
TOOL
SHAPES
THE JOINT

PRESSURE SEALS JT.

Figure 4.35 Tooling of mortar is critical for the success of nonmovement joints.

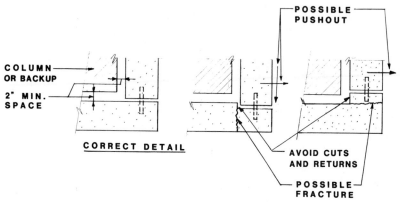

Figure 4.36 Proper corner detailing avoids cracks.

ment is especially critical where the stone is separated into several units (Fig. 4.37).

7. Dowel pins or disks should not be embedded solidly when installed through a relieving angle at an expansion joint or horizontal anchors. *Compressible* material should be installed at *both ends of the dowel pin holes* to allow for thermal expansion of the stone wall. Otherwise, mortar will restrict any motion and lock and negate the action of the soft joints (Fig. 4.38).

8. *Cavities in self-draining walls* should be kept *clear of mortar droppings* during construction. *Make sure the cavity is not poured full of grout.*

9. The use of *calcium chloride* or other salts which act as accelerators or retarders is not permitted in *setting stone.*

10. Make certain an adequate number of setting pads are placed in the horizontal joint under heavy stones to sustain the weight until the mortar has set (Fig. 4.39). Setting pads should *never be placed near the edges* of the stone units and should not be left there. In one such case, where plastic

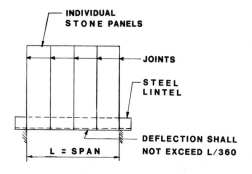

Figure 4.37 Steel lintels supporting individual stone units must be sufficiently rigid to avoid deflection. They should also be fireproofed when spanning more than 4 ft.

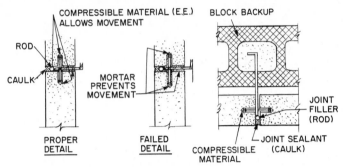

Figure 4.38 Soft joints should be designed and constructed to allow free movement. *Note:* Compressible material should be installed at both ends of dowel pins or discs to allow for thermal expansion and other movements of the stone wall.

shims were set at the corners of limestone blocks, the corners chipped as a result of cumulative compression stresses (Fig. 4.40).

11. Large panels may be set on soaked soft wood wedges so as to support the stone in proper alignment until the mortar has set. Wedges should be removed when dry — without breaking them off.

12. During construction, make sure walls are adequately protected by temporary covers at night and during rains. This will help prevent staining and efflorescence. Failure to do this will result in trouble, dissatisfaction, and expense (Fig. 4.41).

13. Design for wind-loading effect on full-size floor-to-floor panels (Fig. 4.42). The maximum allowable tensile bending stress f customarily used for stone units is 125 psi (0.86 MPa). This represents an approximate safety factor, S.F., of 8 to 10. Often modulus of rupture values of 1000 to 1250 psi (6.9 to 8.6 MPa) are used. Actual properties of the stone should be checked; values may vary from material to material (see Table 4.1). Use the following formula:

$$t = h \sqrt{0.006w}$$

which is based on

$$M = \frac{wh^2}{8}$$

$$S = \frac{t^2}{6}$$

Figure 4.39 Setting pads are used to temporarily support stone panels until the soft mortar is set and hardens. *Warning:* Setting pads must not be placed near the edges of the stone and must not be left there.

Figure 4.40 Corner spalling and cracks in limestone panel. Setting shims were placed at corners of panels and left there after grouting.

and

$$f = \frac{M}{S} = \frac{1250}{10} \ 125 \text{ psi}$$

with S.F. = 10 and modulus of rupture = 1250 psi (8.7 MPa), where M = bending moment and S = section modulus.

14. Full stone thickness should be provided to resist wind loads (Fig. 4.43a). Thickness of stone may be reduced with use of backup. The wall should be checked as a composite with backup and facing acting in unison where two or three units are used per story height (Fig. 4.43b).

METAL ANCHOR ——————————— WEIGHTS OR
POSITIVE CONNECTION

KERF IN STONE ——————— PROTECTIVE MEMBRANE
COVERING

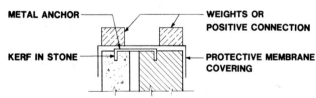

Figure 4.41 When work on walls is temporarily halted, stone should be tied to backup for stability and properly braced and covered to prevent damage by wind and rain.

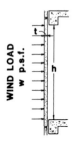

WIND LOAD *w* **p.s.f.**

Figure 4.42 Example of typical wind design for stone facing of full story height (h = clear span, ft, and t = panel thickness, in).

15. Provide expansion joints in parapet walls. Parapet walls are subjected to extremes of moisture infiltration and thermal changes which cause movements differing from those in the wall below, which is exposed to the outside environment on one side only.

16. Mild steel or even galvanized steel and other potentially corrodible materials should not be used for stone anchorage and support and should not be installed within the stone. Thus, dowels, straps, ties, rods, dovetail anchors, and similar connectors which are to be mortared or otherwise engaged in holes or slots in stone should be of *stainless steel, brass, bronze, or other noncorrodible material.* A large variety of accessories is available for tying the stone to its backup. Bimetallic corrosion should be prevented by using metals which are far apart in the electrogalvanic series. The term *anchor* refers generally to the straps, rods, dovetails, and other connections between stone and structure. Anchors as noted above are intended to maintain stones in their positions rather than support their weight. Certain anchors such as bolts (including their nuts and washers) and brackets and shelf angles used for support of panels should also be made of stainless steel or other noncorrodible material.

17. Use properly designed and anchored angle and plate supports (Fig.

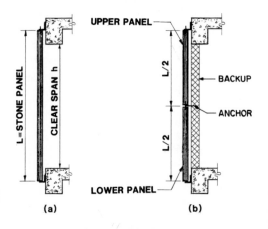

UPPER PANEL

L = STONE PANEL

CLEAR SPAN h

L/2

L/2

BACKUP

ANCHOR

LOWER PANEL

(a)   **(b)**

Figure 4.43 (*a*) Stone should be thick enough to resist wind loads with no backup. (*b*) Where several stone panels are stacked within a story height (not recommended), the backup wall or the auxiliary metal system should be designed to resist the wind load acting on the stone facing.

FLOOR TO FLOOR

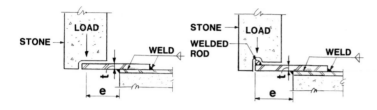

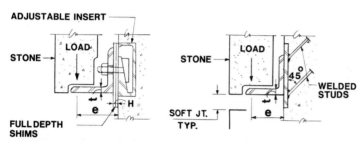

Figure 4.44 Properly designed shelf supports should be capable of transferring the stone weight to the structure. Thickness of plates and angles should be designed with full consideration of the eccentricity of the loads. Stainless steel is recommended for shelfs, anchors, plates, bolts, nuts, and washers. When H exceeds ⅜ in (9.5 mm), use welded shims.

4.44). In support systems for stone utilizing steel angles attached to the building's frame, sufficient strength of support members is of paramount importance.

4.5 Curtain Walls

A curtain-wall system is an enclosure using metal and glass elements intended to form a single watertight and airtight facing. Unfortunately, this system quite often is neither watertight nor airtight. When curtain walls are constructed in such a manner as to yield a flexible structure overall, deformation will be excessive. These deformations and movements break the seals between metal and glass elements and between the metal sections themselves. The integrity of the sealing material and its ability to adhere to both glass and metal is also a frequent source of problems. The interaction of the structural frame and the curtain wall, as with other masonry enclosures, also imposes undesirable stresses on the curtain-wall components. The problem areas are, therefore, as follows:

1. Flexibility of curtain wall
2. Defective sealant system
3. Frame and facade interaction
4. Corrosion of metal elements
5. Connections of the curtain wall to the structure

Preconstruction mock-up tests in wind tunnels are a reliable and cost-efficient method of confirming the strength of the panel and its ability to resist the required wind loads. These tests should be coupled with water-penetration and air-infiltration tests which are performed with specially fabricated boxes. Wind-driven rain at 62.5 mi/h (100 km/h), for example, is simulated with compressors attached to the test boxes. Each suspect avenue of infiltration and penetration is then sealed and opened sequentially to locate leaks.

4.6   Precast Concrete

Precast-concrete panels are sometimes referred to as "human-made" cast-stone units. These units have many advantages over natural stone, since they may be fabricated in large panels and in a variety of shapes, designs, colors, and surface finishes. With the advent of reusable elastomeric form liners, specially textured architectural concrete surfaces can be created. The appearance of the textured concrete surfaces may be enhanced with admixtures.

However, all the problems associated with concrete production in general and with precast elements in particular are also present in the design, manufacturing, and erection of precast-concrete wall units.

Especially critical are the *connections* used to secure the precast units to the structure. These connections must be flexible enough to allow for movements and rotations so as to avoid stresses and resulting distress and failure. The use of neoprene washers and bearing plates or other flexible pads is recommended to achieve the desired freedom of movement.

The use of welding for attaching precast units to the supporting structure and for connecting units one to another is absolutely forbidden, since fully welded connections will not permit movements. Where precast-concrete panels are installed in such a fixed fashion, increase in temperature, movements of the structure itself, shrinkage of the concrete of the panels, wind loads, vibrations, and similar effects often will result in distress in the panels.

Quality control in the production of the units is essential to the *performance and appearance of precast panels.*

Tolerances must be adhered to carefully since small variations in locations of reinforcing mesh can have serious consequences. Because of the relatively small thickness of precast panels, these variations can greatly decrease the effective depth of the panel and its ability to resist bending stresses. Placing reinforcing bars or mesh too close to the surface of the panel can be detrimental, too, since corrosion of the steel embedment within the panel depends greatly on the amount of concrete cover on the rebars. The protection of the steel within the units should also be considered. At a minimum, rebars should be galvanized. In severe environments, or in especially thin concrete panels, using stainless steel mesh or coating the rebars with epoxy should be considered. Asymmetrical placement of rebars in the panels can also be damaging, because temperature differentials and/or shrinkage of the concrete will result in warpage.

When precast wall panels are used also as bearing walls rather than merely as facing units, their strength properties are of great importance.

4.7   Water Penetration through Walls

One of the most misunderstood mechanisms is the entrance of water into occupied spaces through the boundaries of buildings. Insurance claims arising from water penetration through exterior walls amount to billions of dollars per year.

Water is one of the most damaging agents of destruction. It corrodes metals, decays wood, and disintegrates concrete. Water penetrating walls will cause damage not only to the walls themselves and the structural elements behind them, but also will cause cracking, spalling, mortar deterioration, and rust staining. Heavy damage is often also caused to interior finishes, floors and ceilings, and to the contents of the space itself.

There are five major areas where water can and does infiltrate the building envelope:

1. Roof

2. Parapets

3. Exterior walls (including windows, doors, and other openings)

4. Foundation walls

5. Ground or basement floor

Figure 4.9 illustrates some of these paths.

One of the simplest problems to address is that of the material itself: the material either is not inherently water-resistant or is defective. These are the easiest cases to first identify and then solve and rectify.

The material may be porous, torn, punctured, or cracked. It may be porous and permeable as a result of a manufacturing defect or be torn and punctured because of an accident. It may be also cracked as a result of overstresses. In such instances, the water will penetrate through the material itself or through its defects. We have tested and documented the fact that separations and cracks as narrow as 0.01 in (0.25 mm) are sufficient to permit the free flow of water. Therefore, fine cracks or separations which are hardly visible to the naked eye may be responsible for serious leaks.

Much more difficult are those instances where the water penetrates through the boundaries and contacts between different materials. The

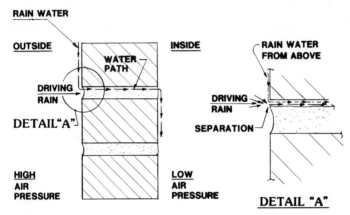

Figure 4.45 Water penetration through contact surface between mortar joint and brick. Occurs especially in driving rain.

contact surface between masonry units (brick, block, or stone) and the mortar connecting them is the most common avenue for water inflow (Fig. 4.45). Since the bond between mortar and masonry depends on the tightness of the joint, the workmanship factor is very important.

Often the water will penetrate at the top of masonry openings such as windows (Fig. 4.46) or doors, where the flashing is not continuous, or through joints in the shelf angles.

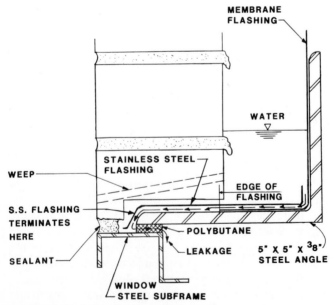

Figure 4.46 Typical leakage over window frame.

The most common masonry wall construction is the cavity-wall type, and the primary reason for its popularity is that it is expected to be ideal for preventing the entry of water into buildings.

Essentially the cavity wall (see Sec. 4.3) consists of two walls (wythes) separated by a gap of 2 to 3 in (51 to 76 mm). The exterior wall, most commonly made of brick, is tied to the interior wall (the backup wall, usually constructed of concrete block, cinder block, or metal-stud wall) with rust-resistant metal ties. At floor levels, a membrane flashing is installed to cut off the water flow (Fig. 4.9), with weep holes inserted to drain the water (Fig. 4.47).

4.7.1 Flashing

Flashings are dams or barriers constructed within walls or at junctures between walls and on roofs and floors, in order to prevent water entry. These dams are sometimes constructed of metal, such as stainless steel, copper, or aluminum, or fabric membranes. Flashing also must be provided over windows, below window sills, below coping stones, and at junctures between penthouse walls and the main roof (Fig. 4.48).

Since it is impossible to exclude *all* water infiltration, a properly designed and constructed flashing system will serve as a second line of defense against the penetration of water. Whatever water does infiltrate will be contained with the help of the flashing and then evacuated through weep holes.

In order to be effective, flashing must be:

1. Fully continuous, without tears or holes.

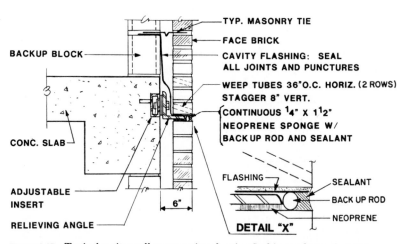

Figure 4.47   Typical cavity-wall cross section showing flashing and weep-hole arrangement with horizontal soft joint.

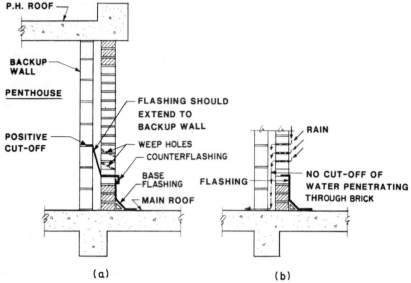

Figure 4.48 Flashing at juncture between penthouse walls and roof is a frequent cause for many leaks. (*a*) Proper flashing. (*b*) Improper flashing.

2. Lapped a minimum of 6 in (152.4 mm) at all splices.

3. Sealed at all splices with adhesive.

4. So designed as to create a "tub" not less than 12 in (305 mm) deep.

5. Set in mastic and fully adhered to the spandrel beams. Details of unsatisfactory flashing installations are shown in Fig. 4.9.

As early as 1936, the NBS conducted tests on various wall panels. Water was applied first by spray at the top of the wall and penetrated by forces of gravity and capillarity only. Later water was applied under 10 lb/ft² (478.8 Pa) air pressure to stimulate wind-driven rain. These tests clearly demonstrated that *all* walls of class B workmanship with unfilled interior joints and flush-cut face joints *leaked,* whereas most walls of class A construction having solidly filled interior joints and tooled face joints showed little or no leakage.

In commenting on these tests, it is interesting to quote from the *Brick Engineer's Digest*:[4]

> There are a few, if any, construction problems that have been more erroneously treated than that of moisture penetration through walls. The sales literature, and the method of attacking the problem, of the vendors of the many panaceas for leaking walls remind one of the patent medicine quack who promises to cure everything from corns to appendicitis with a bottle of

colored water. No amount of theory or irrelevant demonstrations will eliminate the obvious necessity of flashing and caulking in a masonry building.

It is evident that the importance of flashing and caulking were well-
recognized more than 50 years ago. "And there is nothing new under the
sun."(*Ecclesiastes,* Chap. 1).

4.8   Case Histories

4.8.1   Imperial House, Virginia

On my first visit to this relatively new project located in Richmond, Virginia, cracking and spalling of the brick facing was evident throughout the
building facade. Spalling was consistently at shelf angle locations (above
and below the steel angle) and at the corners of the building (Figs. 4.49 and
4.50).

An intensive investigation was launched, and exploratory probes were
opened at the corners (Fig. 4.51). These probes revealed that the steel shelf
angles did not extend fully around the corners, and thus full support for the
brick was not provided (Fig. 4.52). In addition, the first support bolt was
installed approximately 24 in (610 mm) from the corner, resulting in a
long and weak cantilever section of the shelf angle (Fig. 4.53).

Another interesting fact that surfaced during the investigation was that
most of the damage was confined to the more brittle, light-colored brick;
the dark-colored brick was generally unaffected (Fig. 4.54).

Figure 4.49   Spalling of the face brick above and below the relieving angle.

Figure 4.50 Cracking of the face brick at corner of building.

Figure 4.51 Close-up of probe at corner. Flashing covers relieving angle.

Figure 4.52 Probe shows condition at cracked corner. Note shelf angles are *not* continuous around corner, and mortar droppings in cavity. Flashing is *not* lapped at corner shelf, and left leg of angle is tilted with crack in brick at toe of angle.

Lessons

1. Shelf angles must be continuous around building corners.

2. The first bolt supporting shelf angles should be placed not farther than 9 in (228 mm) from the end of the angle.

4.8.2 Federal Plaza office building

Shortly after completion of this prestigious New York City building in 1976 (Fig. 4.55), many of the corners of the stone piers began chipping and spalling.

Probes through several of the piers revealed that the plastic shims used for leveling the stone units had been left in place and not removed after

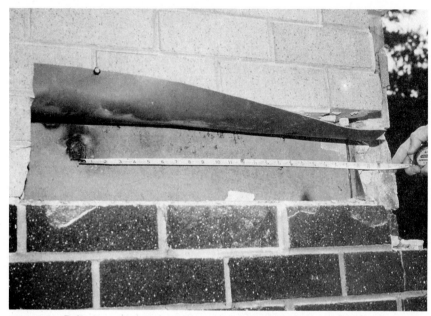

Figure 4.53   Relieving angle was exposed after lifting of metal flashing. First bolt supporting the shelf is located 24 in (610 mm) from corner, which is excessive; 9-in (228-mm) distance is standard. Excessive length caused shelf angle to cantilever, deflect, and rotate.

Figure 4.54   Spalled tan brick with dark brick intact.

Figure 4.55   General view of Federal Plaza building.

setting. Where these shims had been placed very close to the edge of the stone panel, the stone units cracked and spalled.

Our investigation consisted of both an elastic stress analysis and laboratory testing. The analysis indicated relatively high compressive stresses (vertical). These stresses in combination with the existence of the plastic shims close to the edges contributed to the distress in the stone piers.

The laboratory study consisted of a test of a stone pier section removed from the building. This stone was placed in a compression machine with the plastic shims inserted at the edges. Upon loading of the stone, it cracked (Figs. 4.40 and 4.56) in exactly the same failure mode as the piers at the building itself. This result confirmed the conclusion arrived at earlier as to the cause of the distress.

Lessons

1. To avoid spalling, shims used for setting stone units should never be placed near the edge of the stone.

2. Where possible, shims should be removed after grouting.

4.8.3 Coney Island nursing home

The striking fact about this three-story brick building was that on first glance one could see the building "crying," i.e., shedding many small flakes of red brick chips. These chips were on the ground all around the structure

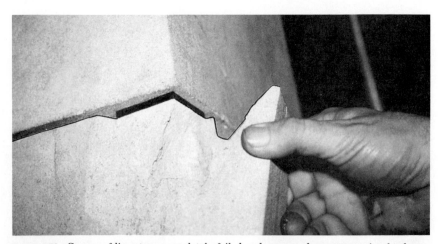

Figure 4.56 Corner of limestone completely failed and separated as compression load was increased. Separation was along 20° angle. Removal of spalled pieces reveals plastic shim located too close to stone edge. Shims should have been placed away from corner or at least removed after grouting of joints.

Figure 4.57 Brick chips spalled from brick face and fell to the ground along exterior building walls. Brick units were defective and could not withstand the freeze-thaw cycles.

(Fig. 4.57), indicating that the problem was throughout the entire periphery and not limited to a particular area.

Large areas of brick had their exterior faces completely missing or delaminated and about to fall (Fig. 4.58). A light pull of the hand could peel off the brick exterior surface (Fig. 4.59).

The first order of business was to remove typical brick units for laboratory testing. The most important indicator in such cases is the coefficient of saturation, earlier discussed in Sec. 4.1.3. As expected, the laboratory results confirmed that the COS was considerably higher than 0.75.

Another interesting fact that helped to confirm our conclusion was that two types of brick were used in the construction of the facade of this building. Around the windows, the architectural treatment called for a panel of brick of light color. This brick, which was of completely different manufacture, with a low coefficient of saturation, performed perfectly without any distress (Fig. 4.60).

Lesson. Insist on certification by the brick manufacturer attesting to the fact that the coefficient of saturation of the actual brick to be used (not old tests of earlier brick) is lower than 0.75.

Figure 4.58   While brick units have spalled, the mortar is intact (see head and bed joints). Full-depth weep joints have been provided for positive drainage of the cavity behind.

Figure 4.59   Brick face in imminent spalling condition is being removed just by the touch of the hand.

Figure 4.60 Light tan brick was not damaged at all, proving the manufacturing defect of the red brick.

4.8.4 The Bicentennial Building, Washington, D.C.

After an entire section of the facade of this brick building bowed extensively to the danger point, structural-steel strongbacks were installed to temporarily secure the brick walls against total collapse (Fig. 4.61).

The 10-story concrete framed building was large, about 230 × 160 ft (70 × 50 m). It was designed and constructed with solid brick walls with concrete block backup. However, instead of the customary shelf-angle detail, the architect elected to use a detail commonly used in the Washington, D.C. area: supporting the weight of the brick on brick headers. As required by the city's building code when this method of construction is used, the brick header must be bearing for at least two-thirds of its length; for an 8-in (203-mm) header this will amount to over 5 in (127 mm) of support.

While the method utilizing brick headers for support of a brick wall may indeed result in a secure and proper wall, that is not always the case. The bicentennial structure was one such example which ended in failure. Why?

Determining this was exactly our task when we were asked to investigate the circumstances and the causes leading to the failure. After securing the structure, we reviewed the details of construction by studying the drawings and the specifications. We analyzed stresses in typical construction details

Figure 4.61 Steel channel strongbacks were bolted to concrete columns behind brick face to prevent the collapse of the wall.

and opened exploratory probes to evaluate the existing conditions. This is a summary of the facts:

1. The design called for a 5-in (127-mm) bearing of the brick on the concrete slab, with a ½-in (13-mm) space between the face of the slab and the back face of the brick (Fig. 4.62a).

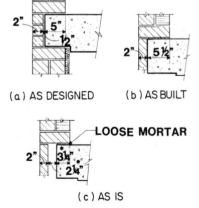

(a) AS DESIGNED   (b) AS BUILT

(c) AS IS

Figure 4.62 Details of bearing conditions at slab end. (a) As designed; (b) as built with 2-in (51-mm) soap; (c) as is (second bay from southeast corner), after shifting as a result of the shrinking of the structural frame. (*Note:* Dimensions with asterisk were measured in field.)

2. The actual construction utilized a 2-in (51-mm) brick "soap" and no space (Fig. 4.62b).

3. The as-is conditions indicated a shift of the brick in the outward direction so that the existing space between the back of the brick and the face of the concrete slab was 2¼ in (57.2 mm). The amount of bearing left was only 3¼ in (82.6 mm), leaving an overhang of 4¼ in (108.0 mm). Such a condition is obviously unstable (Fig. 4.62c).

4. Careful observation of the floor at the maximum bow indicated that the mortar bed below the main header bearing directly on the slab had broken. Moreover, the observation confirmed that, in fact, an outward shift had occurred. This left a compression member two stories high (Fig. 4.63), which failed by bowing outward.

A study of the relationship between shortening length and the corresponding bow indicated that a rather small relative shortening is required to cause a rather large horizontal shift (Fig. 4.64).

Shrinkage computations of the slab also indicated that as much as ¾ to 1 in (19 to 25.4 mm) of horizontal shrinkage may be expected to occur at each edge of the slab. The large plan dimensions of the building were the controlling factor here. This was sufficient justification for the extremely large gap observed between the brick and the face of the slab.

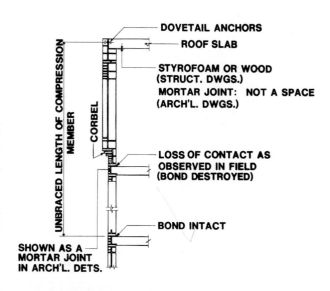

TYPICAL   SECTION

Figure 4.63 Loss of support at midheight of wall caused outward buckling and bowing.

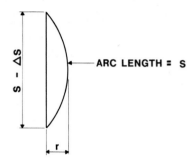

VALUES OF r FOR A PARABOLIC ARC				
arclength / Δs	21.0'	32.0'	43.0'	54.0'
0.100"	3.07"	3.79"	4.40"	4.90"
0.010"	0.97"	1.20"	1.39"	1.60"
0.005"	0.65"	0.90"	0.98"	1.10"

Figure 4.64 Geometrical relation between unbraced wall and bow r. Minute shortening caused big bows.

Correction of the hazardous condition was by the introduction of new steel shelf angles. These were installed after the brick wall was jacked with a hand jack. The jacking assured the actual transfer of load from bearing of the headers on the slab to direct bearing on the shelf angles, which, in turn, transmitted the load to the concrete frame. The repair was performed in sections for safety and stability reasons.

Alternative 1 (Fig. 4.65) was actually used. This method allowed repair

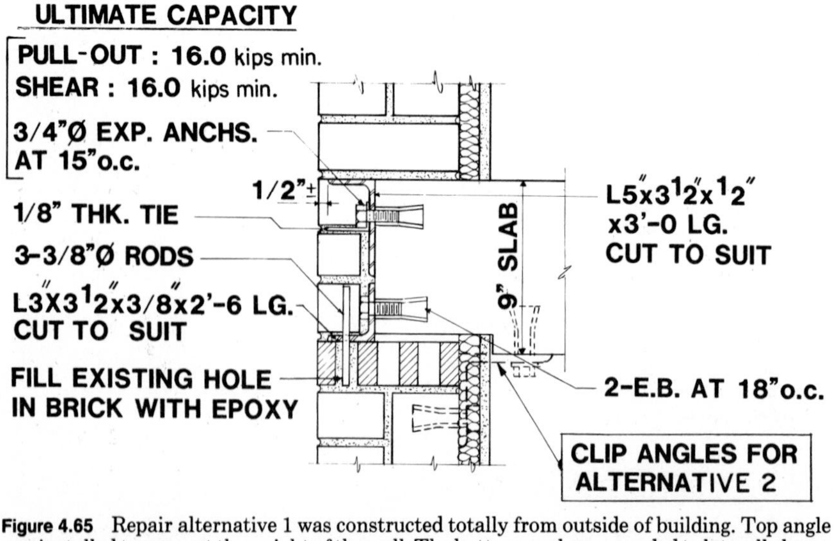

ULTIMATE CAPACITY

PULL-OUT : 16.0 kips min.
SHEAR : 16.0 kips min.

3/4"Ø EXP. ANCHS.
AT 15"o.c.

1/2"±

1/8" THK. TIE

3-3/8"Ø RODS

L3X3$\frac{1}{2}$x3/8x2'-6 LG.
CUT TO SUIT

FILL EXISTING HOLE
IN BRICK WITH EPOXY

9" SLAB

L5x3$\frac{1}{2}$x$\frac{1}{2}$"
x3'-0 LG.
CUT TO SUIT

2-E.B. AT 18"o.c.

CLIP ANGLES FOR
ALTERNATIVE 2

Figure 4.65 Repair alternative 1 was constructed totally from outside of building. Top angle was installed to support the weight of the wall. The bottom angle was needed to laterally brace the wall. Repair alternative 2 was similar to 1 except that the bottom angle was installed inside the office space (tenant interruption). Both repair alternatives utilized hand jacks to transfer the load to steel angles.

without the need to enter the occupied spaces to install the top restraining angles.

Alternative 2 provided a more positive method of support at the top but was ruled out because it necessitated work inside the office spaces.

Lessons

1. Provide horizontal soft joints in large masonry structures. Concentration of compression stresses may occur even in comparatively low structures.

2. The top of backup walls must be anchored positively to the underside of the spandrel beams to avoid movement or even collapse of the wall.

4.8.5   633 Third Avenue, Manhattan, New York

This 44-story high-rise building in Manhattan is amazingly similar to the Bicentennial Building just described. The enormous structure (Fig. 4.66), believe it or not, was designed and constructed and enclosed with a beautiful glazed green brick *without the use of a single shelf angle*. The distress was at various levels in the form of cracks, bulges, and spalls.

On opening exploratory probes, we found that the only provisions for support were brick headers bearing on the edges of the spandrel beams. The fact that many of these headers were cracked was in itself sufficient to rule them ineffective and unreliable.

In order to secure the entire facade and provide a positive method of support, a new structural system of steel shelf angles was introduced. The shelf angles were welded to the existing steel spandrel beams by means of auxiliary steel angles. The entire assembly was welded and painted in place. The new supports were added at every third-floor level to reduce the cost and the duration of the repairs. It was necessary to remove three courses of brick to install the new shelf angles (Figs. 4.67 and 4.68). To ensure proper drainage, new plastic weep holes were installed (Fig. 4.69).

Lessons

1. Exterior masonry used as the building envelope must be supported by means of steel shelf angles or bear directly on continuous extensions of the spandrel beams ("eyebrows").

2. Soft joints must be provided at masonry facades, not only for concrete structures, which are always subjected to the shrinkage/creep phenomenon, but also to allow for movements in buildings of structural-steel-framed construction.

4.8.6   Upstate mental health hospital

This structure brings with it the remarkable story of failures which is now being brought into the open. For many years I avoided mentioning this

Figure 4.66   General view of New York 44-story high-rise building constructed *without* any shelf angles.

Figure 4.67 Close-up of welded details for support of new steel shelf angle. Space below outstanding legs of shelf angle was left to create a soft joint to allow for movement. Joint was caulked later.

Figure 4.68 General view of new shelf angle installed to support three levels of existing brick facing.

Figure 4.69   New plastic tubes ("weeps") were placed to drain the cavity behind the angle.

case. The fact that the structure was owned by a public entity and was designed by a very prestigious engineering firm makes this story even more fascinating.

From the first day of our investigation it became clear that *practically every single element of this structure had failed.* The slabs were constructed as prestressed single tees, double tees, open-core slabs, and thin slabs supported on open-web steel joists. Concrete columns and girders were precast in a yard away from the site and transported and erected in place. Every one of these structural elements failed and had to be repaired.

Even more incredible was the fact that after additional steel supports were introduced to correct the problem, it was discovered that the caissons supporting the entire structure were bearing on sand rather than on rock as designed. It was finally realized that the most cost-effective correction would be to raze the structure and reconstruct properly. However, this option was not followed because of the possible embarrassment to all parties involved.

The masonry failures which will be discussed here relate to the brick bearing walls constructed to support the deep concrete beams at the living units.

These bearing walls were constructed in two varieties: an 8-in-wide (203-mm) and a 12-in-wide (305-mm) wall. The design called for bearing the concrete beams *on the edge of the walls.* A neoprene bearing pad was

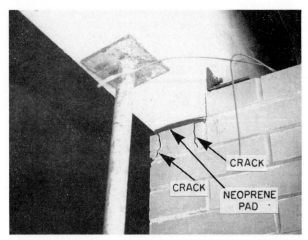

Figure 4.70 Precast-concrete beam supported on neoprene pad cracked the brick unit directly below the pad. Beam was temporarily shored.

required for support (Fig. 4.70). In some instances two girders were supported on a 45° wall as shown in Fig. 4.71.

Since hollow bricks were actually used directly below the concrete beams, rather than solid ones required by the design, they should have been filled with mortar. Unfortunately, this was not done, and the bearing pad

Figure 4.71 At corner of the living units, the precast-concrete beams were supported on 45° brick walls. Brick cracked at joints of supports.

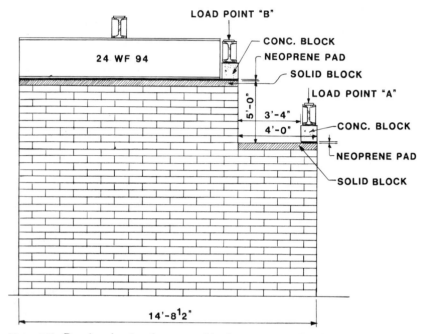

Figure 4.72 Drawing showing the proposed load test program. Load was imposed by steel beams pulled down by steel rods.

Figure 4.73 Actual test setup prior to loading.

under pressure deformed and exerted horizontal pressure on the side wall of the hollow brick unit which resulted in its fracture and failure (Fig. 4.70).

Full-scale laboratory tests were conducted at Columbia University (Figs. 4.72 and 4.73). These tests, in which we were able to precisely simulate the failure (Fig. 4.75), confirmed the conclusion as to the exact mode and cause of failure (Fig. 4.74).

Lessons

1. New, untried systems or construction methods should not be used on large-scale projects.

2. The use of many different construction systems — e.g., structural steel, cast-in-place concrete, precast concrete, open-web joists, double tees, single tees, hollow core slabs — within one project is not cost-effective. Such use may also lead to failure in the absence of a well-coordinated effort.

3. Heavy concentrated loads (such as the bearing of beams) should not be placed at the edge of masonry walls.

4. Bearing mats such as neoprene pads should not be placed directly on hollow brick or block units.

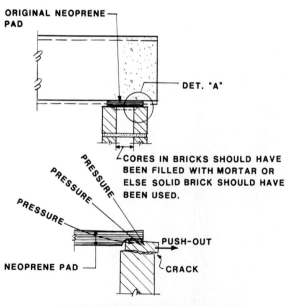

DETAIL "A"

Figure 4.74 Diagram showing exact mode of failure (neoprene "squeeze").

(a)

(b)

Figure 4.75 After edge-loading wall tests, (a) above and (b) below, edge cracks were caused by downward high pressure squeezing neoprene pads laterally.

4.8.7 Prince County, Maryland, wall collapse

A severe windstorm on April 3, 1975, left the Washington, D.C. area with a great deal of damage. In addition to extensive glass breakage, there were many blown roofs and collapsed walls.[5]

The Twin Towers of Prince County were constructed as symmetrically opposite identical high-rise structures. At the center, these towers were connected by a ground-level lobby (Fig. 4.76). A gap of only 25 ft (7.6 m) existed between the two structures (Figs. 4.76 and 4.77).

The connection of the brick wall at its top to the bottom of the roof slab was by a mortar joint (Figs. 4.78 and 4.79). A positive mechanical connection such as one afforded by a metal anchor was not called for in the design nor was it installed. Since the slab directly over the brick face wall was designed as a cantilever, no weight was actually bearing on the wall to increase the pressure on the top of the wall and transmit by bond and friction the horizontal wind reaction loads.

During the storm, the maximum wind speed measured was only 52 mi/h (83.7 km/h). However, this wind speed was sufficient to collapse the uppermost section of the brick and block wall at one of the towers (Fig. 4.80). The wall immediately below the blown-off wall was later secured with timbers and cables (Fig. 4.81).

To confirm the mode of failure, the wind action was repeated in the laboratory on a scale model. Accordingly, it was confirmed that because of the narrow space between the two towers, a Bernoulli effect was created, resulting in suction loads on the wall. These loads, acting toward the center

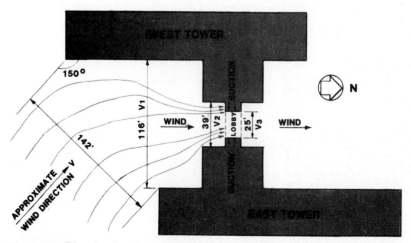

Figure 4.76 Plan showing the two towers with necking down of the space between them from 39 ft to 25 ft (11.9 m to 7.6 m).

Figure 4.77 Looking upward, one can see the narrow wind passage causing the Bernoulli suction effect.

Figure 4.78 Connection of brick wall at its top was by mortar joints which had cracked. No positive mechanical anchorage had been provided.

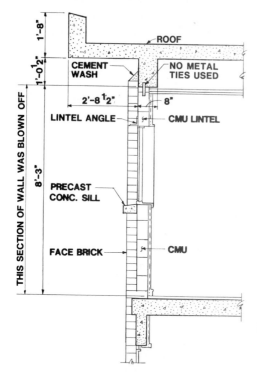

1'-8"

1'-0½"

8'-3"

THIS SECTION OF WALL WAS BLOWN OFF

ROOF

CEMENT WASH

NO METAL TIES USED

2'-8½"   8"

LINTEL ANGLE   CMU LINTEL

PRECAST CONC. SILL

FACE BRICK   CMU

Figure 4.79 Cross section of blown-off wall.

BLOWN – OFF WALL

Figure 4.80 The two towers were constructed with a mere 25-ft gap between them. The Bernoulli suction effect blew the top wall 19 stories to the ground (arrow shows missing wall).

of symmetry between the two buildings, were resisted only at the bottom of the wall by the friction afforded by the weight of the wall. At the top of the wall, no such resistance existed. Consequently, the wall collapsed, falling 19 stories on top of the connecting lobby and demolishing it (Fig. 4.82). Fortunately, no one was injured.

Lessons

1. Exterior masonry walls must be anchored at their top to the structure to avoid movement or even total collapse. This anchorage must be provided by a positive connection such as a steel anchor. A mere mortar joint is not sufficient to provide such resistance.

2. Exterior masonry walls cut by many openings such as windows or doors should be reinforced and stiffened by masonry piers or other structural elements to amply resist wind loads.

3. High-rise tower structures should not be built too close to each other, leaving such narrow spaces between them as to create a Bernoulli-type suction effect.

Figure 4.81 The remaining wall directly below the blown-off section was secured temporarily with wood beams and diagonal cables.

Figure 4.82   Debris after brick wall panel was sucked out, falling 19 stories and crashing on top of connecting lobby.

4.8.8   Bronx housing complex, New York

This well-maintained housing project experienced an unusual amount of brick-face distress during the years 1974 to 1990. A 1977 review of the data from the original drawings and specifications disclosed the following:

1. A specially manufactured "jumbo" brick unit measuring 8 × 10 × 8 in (203 × 254 × 203 mm) was used for this project. Some had a handle for lifting it into place during the bricklaying operation.

2. The brick walls served as both enclosure and as bearing walls supporting the wooden floor joists.

3. Since the construction period continued for several years, the brick used in the various buildings was supplied during 1971 and 1972 from different shipments baked at different temperatures on different dates. Similarly, the mortar was mixed at different times by various bricklayers, and its consistency varied accordingly.

4. The damage to the bricks was primarily in the form of spalling or delamination. There was also damage in the form of cracking, open or missing mortar joints, and efflorescence.

5. The spalling damage was particularly heavy at the roof parapets (Fig. 4.83), but over the years, extended downward.

6. Coping stones were not properly sealed by elastic sealer so as to cut off the path of entry of water. These joints were found to be completely open at various locations.

7. The jumbo brick units were designed with horizontal cores to serve as water drainage channels. Many of these cores were full of mortar, preventing drainage (Fig. 4.84). Solid bricks, however, had vertical cores.

The investigation conducted by my firm included review of documents, analyses, and laboratory tests. The first phase of the study was to prepare a detailed condition survey of the damage. The survey at the parapets involved tapping each brick unit to establish whether it was about to spall. After sounding of the units with a hammer attached to a long handle, an accurate record was prepared, and a statistical evaluation made.

The interesting result of these studies was that a good correlation was established between the degree of damage in the form of shaling and spalling and the brick measure of the coefficient of saturation. In those buildings where this coefficient was higher than 0.75, the degree of damage was more severe. The higher this coefficient, the more intensive the damage.

A secondary correlation related to the quality of the mortar i.e., its level of deterioration, altogether missing mortar in the joints, soft or otherwise defective mortar — related directly to the strength of the mortar. Mortar of 1500 psi (10.3 MPa) strength performed better than mortars of lower

Figure 4.83 Typical brick spall. This spalled face is seen after it fell on the roof membrane.

Figure 4.84 Horizontal drainage cores were found to be full of mortar.

strength. Workmanship typically was also below acceptable standards with mortar joints varying extensively in width (Fig. 4.85). This could only partially be explained because of the large size of the brick units.

In general, with most clays, bricks which have been baked in the kiln for longer periods of time have a darker color, and tend to resist freeze-thaw thermal cycles better than underbaked units of light color. At the Bronx complex, we found this correlation to exist. Fig. 4.86 shows the darker units intact adjacent to spalled lighter-colored units. The underbaked brick was the culprit in all cases.

Corrective actions

1. All spalled brick units were removed by cutting and were replaced in the late 1970s with equivalent units.

2. All open mortar joints were tuck-pointed.

3. Joints below the coping stones were raked out and replaced with caulked joints, using elastic sealants.

Figure 4.85 Open joints in stone coping serve as entry to water. Note poor workmanship of mortar joints varying from narrow joints (arrow) to very wide joints (arrow).

Figure 4.86 Inside face of brick parapet shows spalls (brick units 38, 39, and 40). Note brick units 36 and 37, which are much darker and have *not* spalled.

The owner was fully advised that additional damage was to be expected and a long-range repair program had to be anticipated, with proper budgets. In later years, in fact, the damage continued unabated at an alarming rate, even after additional repairs were carried out on a continuous basis. A total recladding is under consideration. The brick continued to spall in spite of the fact that the joints were pointed and surface sealer was applied. The defective brick was determined to be by far the major factor leading to this failure. Safety bridges were installed to protect the public from falling brick.

Lesson. Brick for exterior use should have a low coefficient of saturation, which minimizes deterioration by freeze-thaw cycles. The quality of the brick and its resistance to deterioration should be checked by laboratory tests.

4.8.9 Manhattan Towers, New York City

Crushing of exposed faces of concrete slabs in a flat-plate high-rise apartment complex (Fig. 4.87) was accompanied by brick masonry spalling at the contact with the slabs. After a big sliver of concrete fell to the sidewalk (Fig. 4.88), a full-scale investigation was ordered.[*],[†] The pattern of failure was determined by review of field records and photographs. The exposed "eyebrow" after the spalling is shown in Fig. 4.89. Damage was random as to location and pattern, and failures were scattered throughout the structure. Damage was somewhat heavier at lower floors, but not predominantly so.

Material testing at the laboratory established the modulus of elasticity of the brick to be $E = 6.0 \times 10^6$ psi (42×10^3 MPa) — a very high value for face brick.

Computer analyses with assumed creep and shrinkage were conducted for a value of the modulus of elasticity of brick which had been obtained from laboratory tests. The computer model is shown in Fig. 4.90.

Stress-relief tests were then conducted to establish the existing level of compressive stress in the brick. These tests were performed by first attaching electric strain gauges to the face of the brick and the concrete, and later establishing the relief in stress as a result of cutting the brick and mortar joints (Fig. 4.91). The results of the field tests and computer analysis are shown in Fig. 4.92. Both the theoretical analyses and the field relief tests showed good correlation and amazingly high values of compressive stress

---

* Dov Kaminetzky, "Anatomy of Failures of Roofs and Other Structures," American Society of Civil Engineers Convention, May 11–15, 1981, New York, N.Y., Reprint No. 81-167.

† Dov Kaminetzky, "Application of Polymers for Rehabilitation of High-Rise Concrete Structures," International Symposium, India Institute of Technology, Bombay, India, pp. 75–88.

Figure 4.87 General view of the exposed "eyebrow" slabs in this award-winning New York high-rise building.

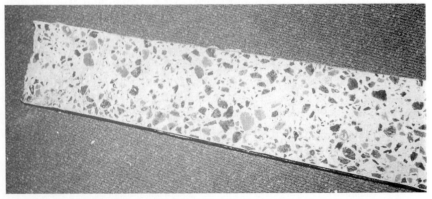

Figure 4.88 Large sliver of spalled concrete from "eyebrow" fell to the sidewalk after separation from exposed slab.

Figure 4.89 Exposed "eyebrow" of slab after spalling.

in the brick, with an average level of 2000 psi (14 MPa) and a maximum value of 3300 psi (23 MPa). The fact that the mortar joint directly below the concrete slab was only partially filled intensified the stresses and spalling at the face of brick and concrete (Fig. 4.93).

The revealing meaning of these tests and the analytical studies is that a large portion of the load was transferred from the concrete columns to the masonry. This phenomenon is apparent in many high-rise concrete structures where soft joints have not been provided. In effect, the brick and block masonry in such cases carries a big portion of the dead load. This fact has been generally known for some time, but the magnitude of the load carried by the masonry and the compressive stresses in the brick, at 2000 psi (14 MPa), were unexpected.

The following were the findings of this investigation:

1. A major part of the dead load of the structure (including floor slabs and walls) was being carried by the exterior masonry.

2. The effects of creep on stresses in the masonry was considerable even at the upper floors and increased at the lower levels (trapezoidal load distribution, rather than the previously assumed triangular one).

3. The effect of introducing soft joints at alternate floors was not sufficient for relieving stresses throughout the structure. There was stress relief only at the soft joints, while stresses were still locked at the hard joints. The only effective method of stress relief was to introduce *soft joints at every floor level.*

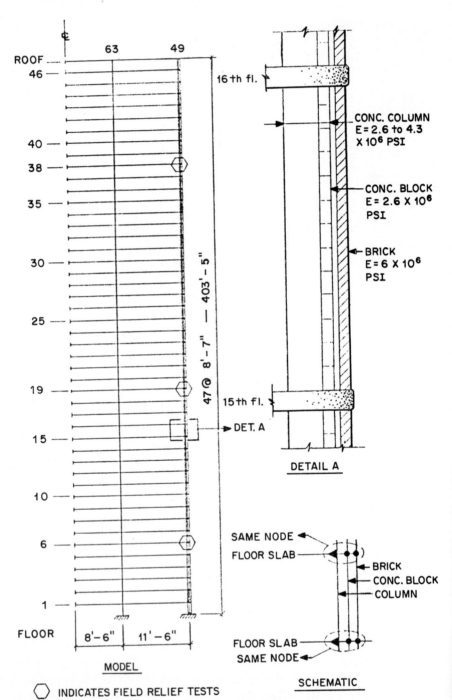

Figure 4.90   A computer model was used for stress analysis.

Figure 4.91 The setup of the stress relief accomplished by breaking the brick can be seen in this photograph.

Corrective actions. To correct the overstressed condition at this structure and to repair the already existing damage, the first step taken was to cut new soft joints at each level. This was accomplished either by saw cutting at the relatively thick mortar joints or by removing one course of brick (at the very thin mortar joints) and replacing it with a shallower course of brick (Fig. 4.94).

Once the stresses causing the damage were removed, the repair operation was started, and proved effective after its completion.

The repair of spalled sections of concrete was accomplished by using polymer mortar, which proved most effective for this purpose (Figs. 4.95, 4.96, and 4.97).

Lessons

1. The most effective means to avoid compression stresses in an exterior brick facade is to introduce soft joints at every level.

2. Soft joints are required not only at brick faces utilizing the shelf angle method of support, but also the so-called "eyebrows" (exposed concrete spandrel beams).

3. The magnitude of compression stresses in a masonry facade may be reliably measured by instrumentation.

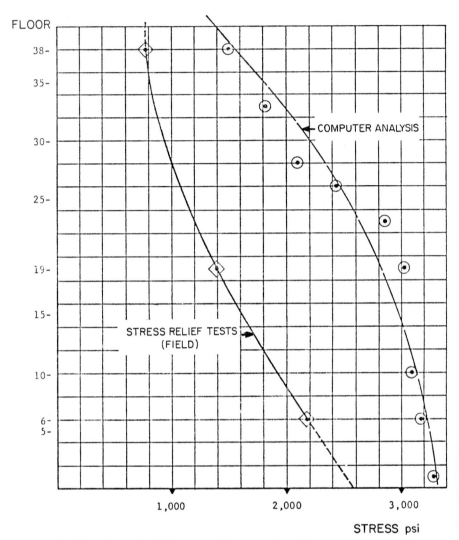

Figure 4.92   Chart shows that field stress-relief tests confirmed the computer stress analysis.

4. The effects of shrinkage and creep of high-rise concrete structures
   increase at the lower levels of the building, but do exist also at the higher
   stories and cannot be disregarded.

4.8.10   Pilot House, Florida

The need for soft joints in brick masonry[6] is now generally accepted in the
construction industry. However, the need for similar joints in stucco facades in not quite so well recognized.

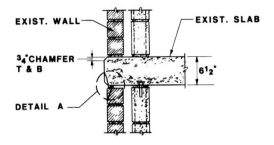

SECTION

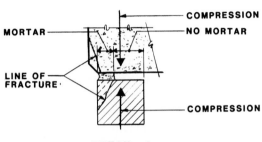

DETAIL A

Figure 4.93 Failure mechanism at hard mortar joints. Top mortar joint was partially filled with mortar at the front face. Soft joints will eliminate this problem.

The Pilot House, a high-rise condominium project, had many construction problems from its inception. One such problem was the extensive cracking of its stucco facade (Fig. 4.98).

The structural frame of this building consisted of steel girders and columns. At the short ends of the structure, steel wind trusses were placed.

Cutout probes [12 × 12 in (305 × 305 mm)] were removed from the building face. These indicated that the stucco was applied in two layers on diamond-shaped steel mesh. Consistently, the cracks in the stucco coin-

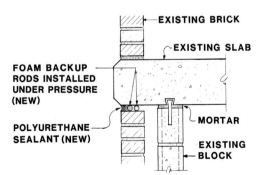

Figure 4.94 Detail of cutting new horizontal soft expansion joint below the slabs and sealant installation.

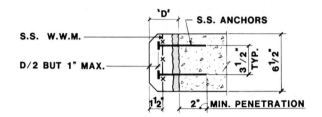

CROSS SECTION

Figure 4.95 Detail of repairs to the spalled "eyebrow" slabs using stainless steel anchors. (*Note:* Omit W. W. M. where *D* is less than 1 in.)

cided with the lap in the mesh. Because this lap was inadequate [only a ½-in-long (12.7-mm) lap instead of the required lap of 1 in (25.4 mm) minimum], the tensile stresses could not be transmitted through the joints, and they cracked (Fig. 4.99). Prior attempts to seal the cracks in the stucco failed because the structural integrity of the joint was not restored and the tensile stresses were not eliminated. Cutting movement joints through the stucco was the solution.

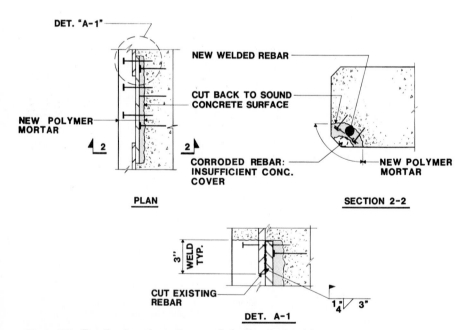

Figure 4.96 Details of repairs to the corroded rebars in the slabs.

Figure 4.97   "Eyebrow" repairs showing polymer mortar troweled on spall.

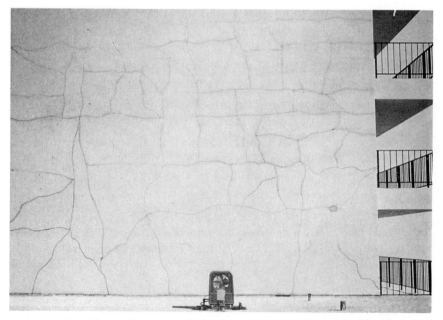

Figure 4.98   Cracking at the stucco walls developed throughout the entire end walls.

Figure 4.99   Cracked stucco extended through section without adequate lapping of steel mesh.

Lessons

1. Expansion movement joints must be provided in stucco enclosures in order to avoid cracking.

2. Exterior stucco walls must be completely separated and isolated from the structural frame, so as not to fail after attracting high wind loads.

3. Steel-mesh reinforcing within stucco walls must be sufficiently lapped at splices to avoid cracks and separations.

4. Structural frames must be reinforced by rigid shear walls or trusses to reduce excessive movements resulting from wind loads.

4.8.11   Hofstra University dormitory building

Water penetration into the occupied rooms of this college dormitory launched a comprehensive investigation to determine the source of leaks and to correct the problem. Once the list of leak complaints was made available, the adjacent walls were probed from fixed scaffold platforms (Fig. 4.100).

The cavity spaces adjacent to the spandrels were uncovered, and found to be full of mortar (Fig. 4.101). The flashing membrane was observed to be torn at several locations (Fig. 4.102). In addition, large volumes of trapped water were found in a number of places, indicating a total failure of the

Figure 4.100 After continuous leaks and many unsuccessful repairs, this dormitory wall was probed from scaffolds. Large quantities of trapped water flowed from the probes.

Figure 4.101 The cavity was found to be full of mortar droppings.

Figure 4.102   The membrane flashing contained many holes and cuts, permitting water to travel inside to occupied rooms.

drainage system of the cavity. This led to accumulation of water in the spandrel beam "container," which overflowed, with the water spilling into the rooms.

Corrective actions.   At locations of water stains, the cavities were opened and all mortar droppings were removed. New weep holes were drilled in all walls near (above and below) the recorded leaks; great care was exercised not to penetrate and tear the existing flashing membrane. This operation had to be repeated several times until all the leaks were eliminated. The only other approach to repair was to completely open the wall cavities on the periphery of the building and rebuild the adjacent masonry. This solution was not followed because of the obvious high costs involved.

Lessons
1. The cavity between the exterior brick and interior block wythe must be constantly kept free of mortar droppings, which bridge the cavity and plug the weep holes.
2. The flashing membrane must be protected against tears during construction. Where holes and tears are found, they must be repaired by sealing with proper mastic. Where budget permits, use metal flashing.

4.8.12 Brittany Apartments, Arlington, Virginia

The main expansion joints of this project failed (Fig. 4.103). Because of the large movements generated in such joints, the joint filler material (sealant) was incapable of accommodating the required elongation. Such wide joints must be bonded properly to resist the separation tensions at the interface (contact between the joint sealant and the brick).

In addition to the expansion joints which were not operating properly, distress in the form of diagonal brick cracks developed as a result of floor deflections of the long floor cantilevers on each side of the expansion joints (Fig. 4.103).

Lessons

1. Very wide expansion joints must be designed to accommodate large movements, which are associated with high tension loads at the surfaces bonding the sealant to the brick. Because common sealants are incapable of developing such high tensions, the following solutions are proposed:

   a. Use more frequent expansion joints with narrower widths.

   b. Use precompressed joint material.

   c. Use mechanical assemblies with moving metal cover plates.

Figure 4.103 This very wide expansion joint failed and separated from the brick (arrow). Probe is seen on right at a diagonal crack caused by vertical floor drop.

2. Avoid *long* cantilevered slabs at expansion joints, which are always the cause of large vertical slab deflections.

4.8.13 Cornell University, Ithaca, New York

Shortly after completion of the agricultural-school wing of this university building, it became apparent that its beautiful brick face was not properly anchored to the block backup. This caused movements of the brick units, resulting in shifting and cracking.

A review of the contract documents indicated that masonry anchors were, indeed, called for but were not installed. Soft joints, on the other hand, were not incorporated in the original design, and their omission was a factor in the distress, since there was no relief for the vertical compression stresses.

Corrective actions. The repair program included the introduction of new soft joints and the placement of new stainless steel anchor pins to tie the face brick to the backup. The pins were placed by first removing a brick, installing a hooked pin in mortar into the block, and rebricking the brick to close the space. The newly cut soft joints and the holes for the new pins may be seen in Figs. 4.104 and 4.105.

Figure 4.104 New horizontal soft joints (arrows) were cut into this newly completed structure to reduce the high vertical compression stresses.

Figure 4.105 Because of lack of adequate masonry ties, new anchors were installed in a regular pattern. One brick was removed for each new anchor.

Lessons

1. Quality control programs must be established to verify that adequate numbers of anchors are installed in masonry facades.

2. Horizontal soft joints must be incorporated in design and construction of exterior masonry to minimize distress.

4.8.14   57th Street condominium high-rise, New York City

This 45-story apartment cooperative has had more than its share of problems. The distress ranged from masonry brick cracks, spalled brick, and ripples in the brick to water leaks, spalled stone facings, and broken concrete balcony corners. Many of the building corners and piers exhibited long vertical brick cracks (Figs. 4.106 and 4.107). Cracks and bulges also appeared over windows (Fig. 4.108). In July 1975, an entire brick pier separated and fell to the sidewalk. This was followed by the issuance of a violation by the New York City Building Department.

After many attempts at repairs (four engineering firms and four separate repair programs were involved, in turn), we were retained to study the continuing problem of brick distress.

Figure 4.106 Vertical cracks developed at the corners. Many were only surface-sealed and later reopened.

Figure 4.107 Some of the cracks extended for several stories (arrow).

Figure 4.108 Cracks over windows were prevalent.

Our first important observation was that, to our amazement, the fundamental design of this tall high-rise structure was based on a concrete frame concept without any shear walls. Obviously, a very slender and flexible structure resulted, with corresponding large movements and deformations. One look at the vent sleeves below the windows will suffice: the diagonal cracks emanating from the corners of these openings tell the story (Fig. 4.109). The exterior brick walls became de facto shear walls, attracting with their relatively high stiffness the acting wind shears. As a result, cracks and spalls developed throughout the structure.

The entire library of construction defects was revealed during the initial probes and later during the repair operations:

1. Inoperative soft joints (they were, in fact, hard joints).

2. Excessive overhangs of the brick. In some instances the brick units had *absolutely no bearing* on the shelf angles (Figs. 4.110 and 4.111).

3. Missing shelf angles.

4. Missing shelf angle anchor bolts.

5. Missing adjustable inserts for shelf angle anchor bolts.

6. Missing dovetail anchors (Fig. 4.112).

7. Missing dovetail anchor slots.

(a)

(b)

Figure 4.109   Diagonal cracks (arrows) caused by wind loading developed at vent openings are shown in (a) elevation and (b) close up.

Figure 4.110 Excessive overhang of brick units is shown in this photograph. This defect should have been apparent to the masonry inspector.

8. Short shelf angles around the corners (Fig. 4.113 shows the crack directly in line where shelf angles are missing).

9. Excessive shims [sometimes up to 2 in (51 mm)] and short and loose shims (Figs. 4.114 and 4.115).

10. Missing weep holes.

Figure 4.111 Entire brick face overhung the shelf and was bearing only on the brick below.

Figure 4.112 Only two dovetail anchors (arrows) were found when this pier was exposed.

Figure 4.113 Probe shows missing shelf angles at cracked corner.

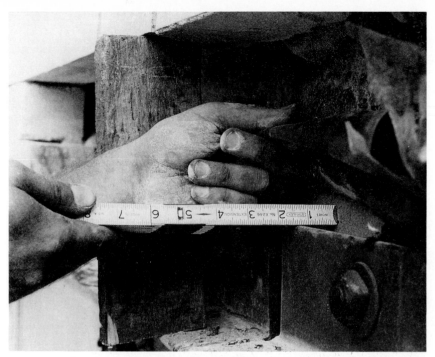

Figure 4.114 Gaps as wide as 1 in (25 mm) or more between the face of the spandrel and the back of the shelf angle were found to be full of shims.

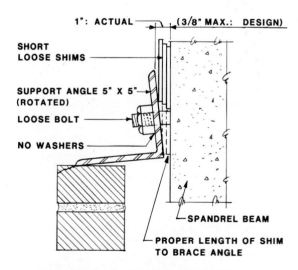

SKETCH: FIELD OBSERVATIONS OF DEFECTIVE BRICK SUPPORT ANGLE

Figure 4.115 Sketch of short shims incapable of preventing rotation of shelf. Where short and loose shims are used, the unbraced bottom of the angle rotates, causing masonry distress.

11. Unpainted shelf angles.

12. Shelf angles attached by friction clips (Fig. 4.116).

13. Missing or torn flashing.

14. Unlapped or short flashing.

And many more!

In addition to the above listed problems, this project utilized glazed brick, which has an unenviable history of failure in severe environments.

An intensive repair program was launched. Among the most important features of this program were:

1. Introduction of new soft joints (Fig. 4.117).

2. Repair of existing soft joints.

3. Pinning of the entire facade with new stainless steel pins.

4. Replacement of spalled brick with new brick units.

5. Repair of wide brick cracks [wider than ¼ inch (6.4 mm)] by "toothing-in" new brick units.

6. Caulking narrow cracks with elastic sealants.

7. Installation of "piggyback" steel angles where the brick overhang was greater than 2 in (51 mm) (Fig. 4.117).

8. Attaching galvanized Gripstay steel slotted channels (Fig. 4.118).

Figure 4.116 Improperly used shims were installed in front of the shelf angles rather than behind. Only friction was left to prevent failure.

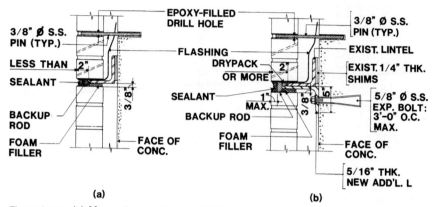

3/8" Ø S.S. PIN (TYP.)

LESS THAN

SEALANT

BACKUP ROD

FOAM FILLER

EPOXY-FILLED DRILL HOLE

2"

3/8"

FACE OF CONC.

(a)

FLASHING

DRYPACK OR MORE

SEALANT

BACKUP ROD

FOAM FILLER

2"

1" MAX.

3/8"

5"

3/8" Ø S.S. PIN (TYP.)

EXIST. LINTEL

EXIST. 1/4" THK. SHIMS

5/8" Ø S.S. EXP. BOLT: 3'-0" O.C. MAX.

FACE OF CONC.

5/16" THK. NEW ADD'L. L

(b)

Figure 4.117 (a) New soft-joint details. (b) Where excessive brick overhangs were found, piggyback angles were installed.

Failure of the proper execution of the repair was discovered later when "dry" (no-epoxy) holes and pins were found (Fig. 4.119).

The option of repair by complete stripping and brick replacement was recommended but was ruled out by the owners because the cost of such extensive rework would have been prohibitive.

Figure 4.118 Galvanized steel slotted Gripstay channels were installed in lieu of missing dovetail slots.

Figure 4.119 Defects in the repair were discovered when new stainless steel anchors were found to be "dry" (no epoxy).

Lessons

1. Exterior masonry construction requires close attention to details during both design and construction phases because of the many components involved.

2. Where either the structural frame or the masonry facing is out of tolerance, use either shelf angles with wider horizontal legs or weld backplates to avoid loose shims.

3. Wind resistance of high-rise structures must be provided by shear walls or shear trusses to avoid distressing the exterior enclosure.

4. Repair of distressed structures must be performed under extreme care and with an increased level of inspection.

4.8.15   Gimbel's Department Store, Paramus, New Jersey

A white brick wall which served as facade for this department store for over 25 years fell to the ground one day in January 1984, leaving the entire backup wall exposed (Fig. 4.120).

As soon as we arrived at the site, we called for temporary shoring of the remaining questionable wall (Fig. 4.121). Probes were initiated to determine the existing conditions and to attempt to establish the proximate cause of the collapse.

The probes yielded the following information:

1. Adjustable anchor slots had pulled out from the concrete beam encasement (Fig. 4.122).

2. Paper was embedded in the mortar joints (Fig. 4.123).

3. Masonry ties were either short or broken off (Fig. 4.124).

Figure 4.120 Without warning, this entire brick face panel collapsed to the ground, exposing the block wall behind.

Figure 4.121 Temporary shores were immediately installed at one of the remaining bays.

Figure 4.122   Adjustable insert slots supporting the relieving angle pulled out of the concrete and failed.

Figure 4.123   Paper was found in mortar joints, placed there in an apparent attempt to prevent loose mortar from dropping into the brick cores.

Figure 4.124   Metal ties were found to have broken. Some were just too short.

To attempt to establish the existing level of vertical compression within the brick, a stress-relief test was conducted with the use of mechanical gauge inserts which were installed at fixed intervals. After a brick unit was removed (Fig. 4.125), the resulting drop in stress was established by strain measurement and stress/strain computations. These showed a considerable level of stress [250 psi (1.7 MPa)]. This level of stress was sufficiently high in our judgment to initiate a buckling mode. What triggered the collapse at the particular day was not determined. A review of the temperatures at the nearest weather station did not indicate any unusually high temperatures for the date.

Corrective actions.   Because of the great uncertainty of the stability of those walls within the complex that showed definite bulges, an accurate instrument survey was performed, resulting in the total removal and replacement of certain panels which were considered to be in danger of imminent collapse.

Lessons
1. Paper sheets should never be placed in masonry joints in order to save mortar by preventing the mortar from filling the cores.
2. Masonry metal ties must be placed so that they are well-embedded within the brick face [a maximum of 1 in (25.4 mm) from the face].

Figure 4.125 Mechanical gauge inserts (arrows) were placed in a vertical line. One brick was removed in order to measure the existing strains.

3. Damaging compression stresses may develop even in low-rise structures, a fact that confirms the need for soft joints.

4.8.16 Morgan Avenue wall collapse

During a windstorm, a section of the exterior wall of an industrial building collapsed and fell to the ground in Long Island City, New York (Fig. 4.126). The wall was constructed of concrete-block units and was designed to be braced by the structural-steel frame.

To secure the remaining portions of the walls, the contractor installed temporary timber bracing (Fig. 4.127). Upon arrival we determined that this bracing was wholly inadequate to support the walls. It must be perfectly clear that a brace with an unbraced length ratio of L/D (where L is the total unbraced length and D is the minimum lateral dimension) of more than 50 is totally useless.

A review of the actual construction indicated that the block walls were not at all tied to the steel column (Fig. 4.128), a design omission.

What was even more shocking was that we found on examination that the mortar joints included paper (Fig. 4.129) and wood strips (Fig. 4.130), which prevented any bond whatsoever between the block and the mortar.

Figure 4.126 The entire block wall of this industrial building under construction collapsed to the ground in a windstorm. It was not braced at its top. Joints broke with block left completely clean of mortar (arrow).

Figure 4.127 Temporary shores were placed by the contractor. They are obviously totally inadequate.

Figure 4.128 The design failed to require ties to brace the wall at the steel columns.

Figure 4.129 Paper found in the mortar joints also prevented any mortar-to-block bond.

Figure 4.130 Thin wood strips were placed in the mortar joints, resulting in practically no bond between mortar and block.

Corrective actions. In lieu of total removal of the existing block walls, a course of action which would have caused serious occupancy delays, economic hardships, and lengthy and costly lawsuits, the following steps were recommended and taken:

1. Additional block piers were added to the existing walls for adequate stability.

2. Additional metal ties were installed to secure the block walls at the steel columns.

3. Additional anchorage to the block wall was provided at the steel joists at the roof.

4. The collapsed wall was replaced with a new wall using standard mortar joints.

Lessons

1. Masonry walls receiving their lateral support by the roof structure must be properly shored by adequate temporary bracings.

2. The design of exterior masonry walls should include permanent metal ties for securing the walls to the structure.

3. Foreign inclusions such as sheets of paper or wood should never be placed within masonry mortar joints.

4.8.17   Moonachie, New Jersey, wall collapse

During the course of construction, in May 1965, a 12-ft-high (3.7-m) masonry block wall collapsed (Figs. 4.131 and 4.132), injuring 14 workers. The metal scaffolds used during construction fell to the ground. A close-up of the wall after failure is seen in Fig. 4.133. The wall prior to the collapse is shown in Fig. 4.134.

The attorneys representing the masons alleged that additional temporary bracings should have been placed at the top of the wall to give the wall sufficient strength to withstand wind pressures. Strong winds were reported on the day of the accident.

The most important questions that had to be answered were: Did the wind blow down the wall, which fell carrying the scaffolds with it? Or did the overloaded scaffold legs sink in the soft ground, tumbling over and pulling the almost completed wall along?

Our review of the facts revealed that:

1. Parts of the scaffold still standing were tied by twisted wires to the wall and the scaffold legs sat without base plates on loose, short wood planks without bedding under the planks.

Figure 4.131 Concrete-block wall failed during construction. It was not laterally braced.

Figure 4.132   Top view from the remaining scaffolding shows total devastation with no mortar strength in the joints.

Figure 4.133   Close-up view shows joint reinforcement in the mortar joints (see arrows) which could not and did not prevent this failure.

Figure 4.134 Wall under construction just prior to collapse shows the scaffolding alongside the wall.

2. The top of the scaffold fell farthest away from the debris.

3. Parts of the upper scaffold frame were left hanging in the air still attached to the masonry.

4. The scaffold frame at the failed section was four tiers high; the rest was only three tiers.

These observations indicated that the most probable failure mode was that of the scaffold pulling the wall over, with the pull being transmitted through the twisted wires. The overturning force was generated by heavy construction material which was loaded on the scaffold, coupled with the insufficient resistance of the soil to the pressure on the scaffold legs.

Our analysis of the stability of the wall and its strength to resist overturning caused by lateral wind pressures showed that the unbraced wall was able to resist winds in excess of the 30-mi/h (48-km/h) winds reported at the time at the nearby airport. The fact that an adjacent section of wall which had been completed just prior to the collapse, and was in fact higher than the collapsed section, did not fail led us to conclude that wind was not the overriding factor causing this collapse.

Lessons

1. Masonry walls must be temporarily braced at the top to prevent their overturning from foundation instability and lateral loads such as wind. These braces should be left in place until the permanent roof structure has been installed.

2. Construction scaffolds and other formwork should be built on mudsills adequate to properly distribute the loads to the soil.

4.8.18 630 Third Avenue, New York City

Several bows in the exterior brick wall of this New York City office building alerted the owners of the structure to investigate.*

Our firm's probes immediately revealed that "soaps" (thin units of brick, usually cut down from standard-size brick) were used by the masonry contractor in order to clear the concrete spandrel beams, which were cast out-of-tolerance. This situation caused the relatively thin section of the brick veneer to buckle under the weight of the brick above (Fig. 4.135). The soft joints were ineffective.

In fact, the failure mode was such that the mortar joints at the flashing lines became effective hinges around which the rotation actually occurred (Fig. 4.136).

Corrective actions. Since it was established by a thorough probing program that "soaps" were used throughout the south wall enclosing the stairway, this wall was completely removed and replaced with a new wall placed further out, leaving sufficient space for the installation of standard brick units adjacent to the spandrel beams. This course of action was determined after it was concluded that pinning would be neither effective nor reliable.

---

* Dov Kaminetzky, "The Three Orphans," *Concrete International: Design and Construction,* American Concrete Institute, Detroit, Mich., June 1988, pp. 47–50.

Figure 4.135 Section of this high-rise brick face buckled and had to be removed for public safety.

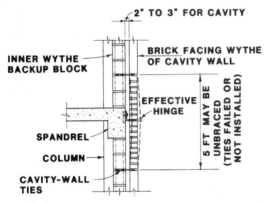

Figure 4.136 Buckling mechanism of brick face of cavity wall under compression. Note "effective hinge" which developed in the absence of ties and may be seen in Fig. 4.135.

Lessons

1. The use of "soaps" in masonry construction should be avoided.

2. Sufficient clearance must be provided in the design to accommodate standard construction tolerances.

4.8.19 Manhattan meat market

One morning, in the fall of 1981, an old building located on the west side of Manhattan collapsed, with bricks flying onto the sidewalk. One of the occupants of the building was trapped in the ruins and died. The building was at the final stage of reconstruction, after the entire structural portion of the work had already been completed.

The structure came to rest in the familiar V-shape funnel, which pointed to the center bearing wall as being at the core of the collapse (Fig. 4.137).

Among the allegations made during the litigation which followed the collapse was that beam pockets were not filled in properly with masonry and mortar. These were pockets cut into the existing main bearing wall at the center of the structure. The pockets accommodated steel beams supporting the new concrete floor for a fish store. The unfilled pockets were similar to the pockets at the north wall which were, in fact, filled in (Fig. 4.138).

The strength of the old masonry mortar was also an issue in the litigation. Since this mortar was a lime mortar which was commonly used at the time of the original construction, estimated to have been in 1882, it was agreed by most consultants investigating the collapse that the bond strength of the 100-year-old mortar was probably nil.

The actual trigger for the collapse was never found, since no construction activity was going on at the time of the failure. Theories involving possible vibration from heavy street traffic or weakening of the structure

Figure 4.137   This old timber-framed building collapsed in the classic V-shaped funnel after renovation, killing one of its tenants.

Figure 4.138   Pockets for steel beams were blocked with brick at bearing wall.

during prior blasting in the adjacent building lot could not be substantiated by investigators.

Lessons

1. Old structures must be carefully examined prior to rehabilitation. Deteriorated mortar must be replaced *prior to* any other work.

2. Pockets cut in old walls for beam supports must be properly filled with masonry, concrete, or mortar to reinforce the wall.

4.8.20 Maryland naval facility

This is an ideal example of *how not to solve construction problems.* When it was discovered that the precast-concrete panels being erected for this project were cracked, all construction work stopped. (See also Sec. 2.15.10 and Figs. 2.108 to 2.112.)

Arguments started, and during 9 months of charges and countercharges the job was at a standstill. Temporary shoring was placed meanwhile to support the cracked panels.

The possible causes for the cracking which were considered were:

1. Design error

2. Erection stresses

3. Temperature stresses

4. Stresses due to roof loads

Because the precast panels were of irregular shape (Z-shaped), the design, stress distribution, manufacture, handling, and transportation were complex.

A finite-element analysis was undertaken and indicated high stresses at an unexpected location at the edge of the panel slightly above the break line. The stresses were sufficiently high under the weight of the panel itself so that when additional stresses due to temperature rise, wind loads, or effect of roof loading were considered, the modulus of rupture (tensile strength) of the concrete was exceeded. The stresses calculated by a computer model proved the inadequacy of the design; the panels were locked into position and therefore were overstressed and cracked. A typical panel is shown in the horizontal position in its lifting frame in Fig. 2.108.

While the job was stopped, the costs of overhead and equipment soared, and a relatively small repair problem turned into an enormous claim.

Corrective actions. The precast panels were repaired by injecting epoxy into the open cracks and later prestressed by using steel brackets attached to the back of the units.

Lessons
1. Precast wall elements of unusual shapes should be avoided where possible. When such shapes are dictated by aesthetics, the panels must be separated from the structure and properly reinforced for all possible loads, including erection loads and temperature variations.
2. Construction problems must be solved and rectified expeditiously so as to minimize enormous indirect costs and claims by the various trades.

4.8.21 Gimbel's 87th Street

In 1979 a passerby was injured when a piece of stone covering fell to the ground outside a Gimbel's department store in Manhattan. During the investigation which followed, the stone covering was found to be bowed excessively. It was stacked, with metal anchors tying it to the block backup. The stone facing consisted of thin sheets of Carrara marble which had been brought from Italy.

The design required each pair of stacked panels of marble to be supported by a steel shelf angle attached to the concrete beams. In between vertical columns of marble, alternating columns of dark slate sheets were installed. Figure 4.139 shows a general view of the building facade. Figures 4.140 and 4.141 show the support and anchorage details.

Figure 4.139 View of the building showing missing stone panels.

Figure 4.140 View from bottom looking up at steel shelf angle support after removal of stone unit below.

Figure 4.141 View from the top showing thin stone unit [1¼ in (32 mm) thick] and its bent anchor and dowel.

We started the investigation after the second panel fell onto the sidewalk. The gaps in the marble face, where panels fell or had been removed, can be seen in Fig. 4.139. A review of the entire facade disclosed that many of the marble slabs, which were 1¼ in (32 mm) thick, were bowed severely. Marble units removed from the building face retained this bow indefinitely, and therefore were considered to have failed. Selected marble slabs were taken to the New Jersey Institute of Technology for testing, and their tensile strength, modulus of elasticity, and expansion properties were established. The units were tested in compression machines in the bending mode (Fig. 4.142) to total failure by breakage.

As a result of this bowing, the vertical sealed joints between the units failed and separated, tearing the joint sealant (Fig. 4.21). The usual overshimming was found here, too (Fig. 4.143), and added to the bowing of the marble units.

One of the interesting facts yielded by the laboratory tests was that the exterior portions of the marble, which were exposed to weather for several years, differed substantially from the interior sections of the stone, which were not directly exposed. This, in fact, changed the marble to a nonhomogeneous material which acted as a bimetal shoe. When the stone was exposed to high temperatures, it bent into a curve which increased as the marble units were restrained.

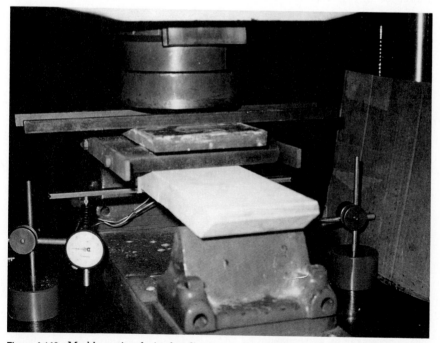

Figure 4.142 Marble section during bending test in the compression machine in the laboratory *prior to* failure.

Figure 4.143 Excessive shimming of the shelf angles was discovered after removal of marble units. This resulted from violation of permissible frame tolerances.

The mathematical relationships used in the temperature analysis are indicated in Fig. 4.144. Some of the properties established by the testing were as follows:

1. *Modulus of elasticity.* 5.26 × 10⁶ to 13.33 × 10⁶ psi (36 to 92 × 10³ MPa)

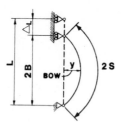

y, IN	B, IN	Δ_L = (L-2B,IN)	ΔT
0.10	25.4997	0.0006	1.3
0.20	25.4990	0.0020	5.3
0.30	25.4976	0.0048	12.5
0.40	25.4958	0.0084	22.0
0.50	25.4935	0.0130	34.5
0.60	25.4906	0.0188	49.8

GEOMETRY OF SHORTENING

Figure 4.144 The bows resulting from temperature differentials are shown. Note that when panels are restrained, relatively large bows may develop at rather low temperature differentials. Original $L = 50$ in; $\Delta_L = \alpha \Delta_T L$, where α = temperature coefficient = 7.4×10^{-6}, T = temperature, in degrees Fahrenheit; y, B, Δ_L, in inches.

2. *Temperature coefficient.* 6.0 to 7.4 $\times$ 10^{-6} °F (3.3 to 4.1 $\times$ 10^{-6} °C)

3. *Compressive strength.* 5970 to 10,760 psi (41 to 74 MPa)

4. *Modulus of rupture.* 432 to 882 psi (3.0 to 6.1 MPa)

Corrective actions. It was concluded that the existing marble slabs had failed and could not be safely supported. There was no alternative but to remove the entire marble face, while the slate slabs were retained.

New aluminum angles were attached to the block backup to support the new white aluminum panels and secure the retained dark slate units (Figs. 4.145 and 4.23 on p. 303).

Figure 4.145   Total marble removal was inevitable because the marble had failed and buckled and it was highly risky to leave it in place. Vertical aluminum angles were expansion-bolted into block backup to secure the dark slate panels.

Lessons

1. Avoid the use of stacked stone panels, especially for thin units [2 in (51 mm) thick and less].

2. Natural-stone units cut from materials which are affected by exposure to the environment (lose strength or modulus of elasticity) should not be used as exterior walls.

4.8.22 New Jersey Medical Headquarters

The headquarters for this medical research company consisted of a main tower structure (Fig. 4.146) and a low-level one-story extension. The masonry enclosure of the structures was white Carrara marble.

Several years after completion, distress in the form of bowed, shifted, and buckled units was first noted (Fig. 4.147). In addition, the sealant within the joints, especially in the long horizontal section of the building, was squeezed out as a result of thermal expansion of the units (Fig. 4.148).

Figure 4.146 High-rise marble faced tower prior to stone removal.

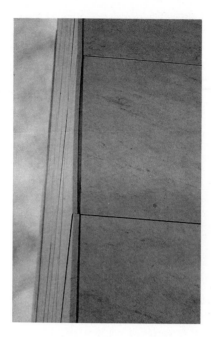

Figure 4.147   Buckled units and failed joints are clearly visible.

During the extensive investigation, it was determined that the actual construction of the facade utilized a supplementary steel support system on which the marble units were set. Unfortunately, this system was not installed as intended. The marble units ended up stacked one on top of the other, instead of being supported independently with definite joints between the units (Figs. 4.149, 4.31, and 4.32 on p. 310). Figure 4.31 (on p.

Figure 4.148   Squeezed and failed joints resulting from horizontal expansion of marble units.

Figure 4.149 Marble units were first stacked, thus compressing soft tape, and only then secured, negating the effectiveness of the soft joint. Shelf angles should be secured one by one as setting of stone proceeds.

Figure 4.150 Close-up of test apparatus showing extensometers and electrical strain gauges.

309) shows the design, and Fig. 4.32 (on p. 310) shows the actual conditions as installed in construction with the units stacked, with no soft gap.

The results received from the laboratory tests (Figs. 4.150 and 4.151) were similar to those obtained from the Gimbel's 87th Street case. These tests confirmed that there was a reduction over time of the flexural strength of the marble units from 1400 to 1200 psi (9.7 to 8.3 MPa). This conclusion was reached after comparing samples of exposed marble taken from the exterior of the structure with unexposed samples taken from the interior of the building.

More significant was the reduction over time of the modulus of elasticity from 3.5×10^6 to 1.7×10^6 psi (24×10^3 to 12×10^3 MPa). This, in effect, meant that the marble stone "softened" and deflected more rapidly when subjected to loads.

Corrective actions.   The only avenue for correction was total removal of the marble. Replacement was by new white marble sheets, which were

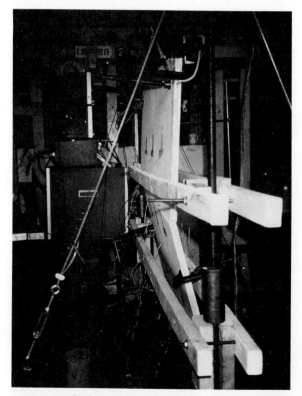

Figure 4.151   Broken marble panel after test failure. Load was applied in testing machine.

installed around the entire building (Fig. 4.152). Three horizontal soft joints were also introduced.

Lessons

1. Stone units designed to be supported on auxiliary metal supports must be properly installed so that adequate soft joints are provided to accommodate future movements.

2. Avoid the use of stacked stone panels, especially for thin units [2 in (51 mm) thick and less].

3. Natural-stone units cut from materials which are affected by exposure to the environment (lose strength or modulus of elasticity) should not be used as exterior walls.

4.8.23 Eastern courthouse granite facade

The first indications of trouble at this granite-enclosed court building were black shadow lines which appeared at several of the dark-colored panels. It was evident that shifting of these units had occurred. The cause of the shifting was unknown; however, it was decided to secure some of the displaced units. Heavy black roundheaded bolts were installed (Fig. 4.153), and an in-depth investigation was initiated.

The entire building facade was surveyed and scaffold drops were hung.

Figure 4.152 General view of reclad building facade.

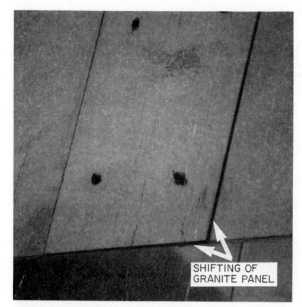

SHIFTING OF
GRANITE PANEL

Figure 4.153 Loose granite panels were secured temporarily
using exposed bolts.

Stone panels were removed and the following details were checked:

1. Bearing of the stone on the shelf angles
2. Existence, condition, and number of anchors
3. Straightness of stone panels
4. Condition of mortar and/or sealant within the joints
5. Existence and condition of mortar spots behind the granite
6. Level of compression stresses within the stone units

The stresses were determined by stress-relief and temperature measurements. The data were collected and recorded in a comprehensive report (Figs. 4.154 and 4.155). A prior report had been prepared by another engineering firm. That report consisted of similar raw data; however, its conclusions were completely different.

The intensive studies showed that while the existing stresses were not at all significant, there were many construction and design difficulties and defects which required repair:

1. Soft joints were not indicated in the design and were not installed.
2. A substantial number of stone anchors were missing, torn, or otherwise defective.

Figure 4.154   A stress-relief test was conducted to establish level of existing stresses. Mortar joint is cut.

Figure 4.155   Surface temperature is measured with hand-held instrument.

3. Anchors actually installed were made from two-piece units connected together by mortar globs.

4. Several granite panels had shifted.

5. The bearing of the granite panels on the steel shelf angles was inadequate at various locations.

Corrective actions.   To correct the above-listed defects, an intensive repair program was launched. This program included an extensive pinning of the panels to supplement the existing anchors. The stainless pins were set in epoxy and their heads were recessed in the granite face and covered with thin granite disks (Figs. 4.156, 4.157, and 4.158). All broken panels were to be replaced. A monitoring program to detect shifting of units was effected. Continuous readings of the absolute and relative location of the panels were taken.

Lessons

1. Do not use mortar to transfer stresses from one metal anchor to another.

2. Avoid the use of stacked stone panels, especially for thin units [2 in (51 mm) thick and less].

4.8.24   The facade of the Marriott Hotel

This 4-year-old structure experienced water penetration problems quite early after its completion. The facade of the building was an exposed aggregate-type stucco finish with metal-stud backup.

Reports by other consultants concluded that the main cause for the water on the interior surfaces of the guest rooms was condensation. After measurements of relative humidity and related temperatures were made both inside the wall and on its faces, we determined that the main problem *was not condensation*. The exact path by which the water entered the rooms had yet to be determined. For this purpose we used the box test (Fig. 4.159), building the box according to ASTM specification E-514.[2] The totally enclosed box contained nozzles through which water was sprayed under pressure to simulate a wind of 62.5 mi/h (100 km/h). As the self-circulating pump collected the water and resprayed it on the wall, all losses could be determined. The interior of the hotel room was exposed to inspection, so it was also possible to see the entry of water at close range.

The box test led us to the unquestionable determination that the water was penetrating through cracks and pinholes in the stucco finish, and, in addition, through gaps in the window construction. Later, wall samples measuring 24 × 24 in (610 × 610 mm) were cut from the actual wall and sent to the laboratory for permeability tests. These tests demonstrated

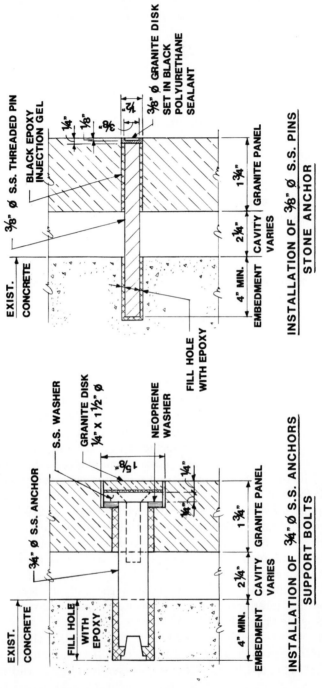

EXIST. CONCRETE

3/8" ⌀ S.S. THREADED PIN

BLACK EPOXY INJECTION GEL

1/4"
1/8"
3/8"
1/2"

3/8" ⌀ GRANITE DISK SET IN BLACK POLYURETHANE SEALANT

FILL HOLE WITH EPOXY

4" MIN.	2 1/4"	1 3/4"
EMBEDMENT	CAVITY	GRANITE PANEL
	VARIES	

INSTALLATION OF 3/8" ⌀ S.S. PINS STONE ANCHOR

EXIST. CONCRETE

S.S. WASHER

GRANITE DISK 1/4" x 1 1/2" ⌀

1 5/8"

NEOPRENE WASHER

3/4" ⌀ S.S. ANCHOR

FILL HOLE WITH EPOXY

1/4"
3/8"

4" MIN.	2 1/4"	1 3/4"
EMBEDMENT	CAVITY	GRANITE PANEL
	VARIES	

INSTALLATION OF 3/4" ⌀ S.S. ANCHORS SUPPORT BOLTS

Figure 4.156   Repair by stainless steel pins.

402

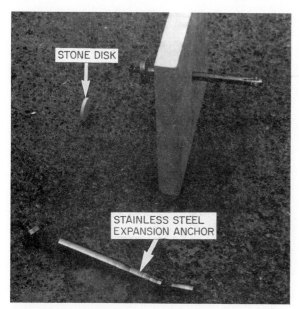

Figure 4.157   Stainless steel repair pins, ¾ in (19 mm) in diameter, are shown with their stainless metal heads, washers, and stone disk used to cover the bolt.

Figure 4.158   Circular repair plug is obvious at close range but invisible at normal distances.

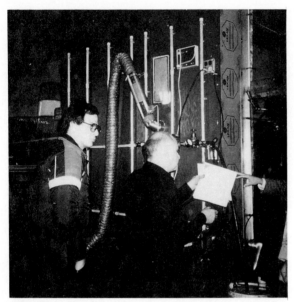

Figure 4.159   Test box (shown from rear) utilizes water under pressure to simulate 62.5 mi/h (100 km/h) wind pressure. Water loss is measured.

that a small pinhole is capable of admitting unexpectedly large quantities of water. It was also demonstrated that a *good surface sealer can reduce the amount of water permeating through stucco, but not reduce water flow through large pinholes.*

Corrective actions.   The proposed correction method included the repair of a big section of wall by the application of a surface sealer and repair of cracks and pinholes on a one-by-one basis using an exact leak survey. The owner, however, elected to completely reclad the building.

Lessons

1. Thin stucco-type masonry must be used in conjunction with flashing systems to prevent water penetration.

2. Even a very small hole or a very narrow crack form paths sufficient for conveying large quantities of water.

3. Surface sealers are not effective in permanently sealing cracks in stucco-type enclosures.

4.8.25   Minnesota utility headquarters

Several years after completion of construction, the owners of a prestigious midwestern utility company headquarters became aware of constant shifting of the interior face of the block walls at various levels. Soon afterward, cracks appeared at the exterior brick. The concern for the safety of the facade grew up to the point that steel strongbacks were placed on many elevations to safeguard the building.

The probes that followed revealed a whole catalog of horrendous irregularities and defects. The stability of the facade itself was so questionable as to force the consultants investigating the structure to recommend replacement of the entire face. We were called in to render a second opinion, and to conduct an in-depth study into the possibility of repair rather than total replacement. Our conclusion was a confirmation that repairs would not be an acceptable solution.

In addition to certain design deficiencies such as the lack of movement joints, the construction defects included:

1. Block backup walls built on top of styrofoam pads (Fig. 4.160).

2. Brick masonry built on top of mortar joints laid on top of wood sheets (Fig. 4.161).

3. Shelf angles with great eccentricities welded to spandrel plates with the aid of welded nuts (Fig. 4.162). Some were welded with flat horizontal steel bars (Fig. 4.163).*

---

* Dov Kaminetzky, "Investigating Masonry Failures," *The Construction Specifier*, Construction Specification Institute (CSI), Alexandria, VA, January 1989, pp. 42–47.

Figure 4.160 Soft styrofoam pads were found below block walls, contrary to standard requirements for stability of walls.

Figure 4.161 Thin wood sheets were found in mortar joints, eliminating the bond between mortar and brick.

4. Block backup walls built on top of a collection of masonry debris of broken bricks placed loosely *without* mortar (Fig. 4.164).

5. Shelf angles welded only at their top, with the bottom of the leg unwelded and separated (Fig. 4.165).

6. Unauthorized cutting of flanges of steel beams to provide clearance for brick face (Fig. 4.166).

Lessons

1. Even proper and well-designed facades may fail if adequate attention, care, and inspection are not provided during construction.

2. Where the backup spandrel beams are out of tolerance, stiff brackets designed by engineers should be used to rectify the resulting conditions.

Figure 4.162 This unbelievable makeshift of welded nuts, above, and worthless welding, below, intended to connect the shelf to the spandrel plate, was doomed to fail.

Figure 4.163 Bars improperly installed in a flat position are totally ineffective and offer no strength in supporting the shelf angle.

Figure 4.164 This incredible jumble of broken brick pieces was found loose and without mortar.

Figure 4.165 This steel shelf angle for support of brick face was welded in a skewed position with only one weld at the top.

Figure 4.166 The flange of this steel beam was cut (arrow) in the field without authorization. The cut dangerously weakened its structural strength.

References

1. "Standard Specification for Mortar for Unit Masonry," ASTM Standard C 270-86b, American Society for Testing and Materials, Philadelphia, 1986.
2. "Standard Test Method for Water Permeance of Masonry," ASTM Standard E 514-74, American Society for Testing and Materials, Philadelphia, 1974.
3. "Standard Specification for Building Brick (Solid Masonry Units made from Clay or Shale)," ASTM Standard C 62-87, American Society for Testing and Materials, Philadelphia, 1987.
4. "Penetration of Moisture into Masonry Walls," *Brick Engineer's Digest,* Brick Manufacturers' Association of N.Y., Inc., October 1936.
5. "Investigation of Wind Damage in the Metropolitan Washington, D.C. Area, April 3-4, 1975," *Technical Note no. 909,* National Bureau of Standards, Washington, D.C.
6. D. Kaminetzky, "Preventing Cracks in Masonry Walls," *Architectural Record,* November 1964, p. 210.
7. ATM Standard E-514, American Society for Testing and Materials, Philadelphia.

5

Foundations

The foundation is the structural element providing support for the various loads acting on a structure. It is the link between the structure and its ultimate support, which is the soil itself. The actual transfer of load may be by direct bearing on soil or rock, or by intermediary elements such as piles or caissons.

When we speak about foundation failures, we often refer to both the failure of the structural elements of the foundation, such as footings or piles, and the failure of the soil itself. While the first type of failure may be the result of overloads on the foundation or of its understrength, the second type results from overconfidence in the test borings or other subsurface information, or loss of bearing value because of adjacent work. Foundation failure resulting from failure of the footing itself is a rare occurrence. Once such case was the fracture of the posttensioned concrete foundation mat in Washington, D.C., which was caused by a design error (see Sec. 2.16.1).

The foundation is usually an artificial structural element, while the soil is a material found in its natural state, disturbed or undisturbed by humans. Artificially made soils are also used in construction, being mechanically compacted soils or soils made totally from nonnatural materials.

The construction of a foundation introduces new conditions into the soil. This happens because the subgrade is exposed and, therefore, unloaded, because changes of the internal friction of the soil are effected by blasting, or because the soil is densified by pile intrusion.

Ground conditions often vary considerably from one location to another within the confines of a single construction site. Sometimes the variation is so great that, when two borings do show identical soil layers, the validity of all borings will be questioned. Rock exposures may vary from fine gray syenite to hard seamy limestones, granites, and schists of various hardnesses. These vary to such an extent that they are ringing hard or so soft that a pipe pile will penetrate over 10 ft (3.0 m) before indicating a 30-ton (267-kN) resistance. In some areas even seams of serpentine and asbestos are found.

On the other hand, granular materials will range from hardpans to fine, uniformly sized, running sands. Soils also often contain clay deposits and organic silts. And, of course, there are the exposed and buried mud deposits with organic layers intermixed in a series of lenses to considerable depths. Because natural soils are often so varied and inconsistent, difficulties in foundation design and construction are expected and are encountered, often with associated trouble.

Actual *water level* in rock excavations sometimes has no relation to the level indicated by borings or to conditions next door. A typical example was the completed work for the New York Lincoln Center underground parking facilities and mechanical plant, where groundwater was first encountered at quite a high level. This water later disappeared after some anchorage drill holes were completed. We later discovered water almost 20 ft (6.1 m) lower than the groundwater encountered in Philharmonic Hall, adjacent to it.

Bedrock surfaces are often very erratic and irregular. During the excavation for the Sixth Avenue subway in New York City during the 1930s, it was found that a strip only 100 ft (30 m) wide was 40 ft (12 m) lower on one side than on the other.

During the construction of the new Bellevue Hospital in Manhattan in 1960, the main hospital building was supported on piles as long as 100 ft (30 m). The adjacent parking facility, which was located no farther than 50 ft (15 m) away, was founded on concrete piers bearing directly on rock.

It is not uncommon to find large vertical mudholes or wide gaps filled with silt within a rock structure.

With such erratic rock strata, rock blasting has a serious effect on the behavior of rock. Rock seams must be found, traced along their length, and cleaned, packed, and tied across with grouted dowels and rock anchors. Rock as a supporting subgrade can be made safe and sound, but only with careful planning and special work.

Granular soils are found in different varieties. There are some very good and some not so good granular soils. Some dense, consolidated, sandy glacial gravel deposits can safely sustain 10 tons/ft² (957 kPa), while uniformly fine-grained and very loose silty sands are treacherous and extremely difficult to control. Glacial deposits also vary. Sometimes old sand fills are taken to be natural deposits. Standard borings are of little value to distinguish the artificially made from the natural geological sand layers. Only actual load tests are effective indicators, which often will show surprising looseness of these fills.

Layers of varved *silts and clays* cover many rock troughs and old shore-lines. These layers are quite stable when left alone, but become liquid when disturbed. Dewatering must be done slowly, to permit the seams to drain out, otherwise "boils" and collapse of braced sheeting excavation will re-sult. Pile driving also generates local liquid conditions in which the mate-rial flows readily. The plastic flow of shorelines continuously pulls river structures outward. Even structures supported on piles carried to rock are affected and rotate with time.

Buried layers of *peat* are sometimes interspersed with silt or sand de-posits and are found at great depths. Where piles receive their support at soil layers overlying peat deposits, serious settlements are common. Also, where pumping occurs in areas where footings are bearing above peat layers, settlements are frequent.

Pumping water out of weak soils is known to cause consolidation of these soils and result in settlement damage to existing structures. Several pre-cautionary methods are presently being employed to avoid this damage. Watertight sheeting enclosures with deep cutoffs and internal pumping has been often found to be an adequate solution.

It is well-recognized that all loads must be so transferred to the underly-ing soils that the resulting settlements can be tolerated by the structure without distress. At the same time, stability must be maintained over the life of the structure.

The 10 most common categories of foundation failures are:

1. Undermining of safe support
2. Load transfer failure
3. Lateral movement
4. Unequal support
5. Drag-down and heave
6. Design error
7. Construction error
8. Flotation and water-level change
9. Vibration effects
10. Earthquake effects

It is obvious that a foundation failure is a serious event, since it may trigger the collapse of the entire structure. This is critically so because most structures are systems in which the support of the upper levels is always dependent on the structural integrity of the lower elements.

Ironically, some foundation failures have been economical successes. One such example is the Leaning Tower of Pisa in Italy (Fig. 5.1). A close

Figure 5.1   Famous Leaning Tower of Pisa is an example of a foundation failure that became an economic success.

look at this tower will reveal that the tower started to lean during its construction. The reason for this conclusion is a slight change in the slope of the tower, near its midpoint. It is an indication that the builders attempted to correct the foundation settling problem (Fig. 5.2).

5.1   Undermining of Safe Support

Robert Frost in his poem "Mending Wall" describes the troubles with stone walls in New England and concludes, "Before I built a wall I'd ask to know what I was walling in or walling out."[1] To paraphrase this warning: before a foundation is designed, one should know *what loads are to be carried and what soils are expected to carry the loads.*

The need for a thorough soil investigation prior to the undertaking of a construction project cannot be overemphasized. In addition to the careful

Figure 5.2 Variable slope of the Tower of Pisa is an indication that the tower started to tilt during construction and that the upper levels of the tower were adjusted by the original builders.

study of the soil strata directly below the proposed structure, existing adjacent structures must be reviewed with care. The need for temporary supports must be evaluated. A well-designed bracing and shoring system is often needed to prevent a lateral shift. A permanent support structure such as underpinning should be installed where the new construction will undermine an existing present support system. Where these provisions are ignored or entirely omitted, serious distress follows (Fig. 5.3) and sometimes tragic consequences result (Fig. 5.4). Another example is the total collapse of a five-story building on 34th Street in New York City that occurred when the vertical support was lost as a result of loss of lateral restraint. This happened when the excavation for a new high-rise building came too close to the foundation of the existing old building. Fortunately, the collapsed building was evacuated a short time earlier because of unsanitary conditions. As a result of the collapse, the rubble filled the cellar completely and piled up almost to the second-floor level.

Figure 5.3 This three-story brownstone, damaged by adjacent excavation, had not been underpinned.

Figure 5.4 The excavation of a trench adjacent to the eastern wall of this five-story wall-bearing building caused the total collapse of the building shown in aerial view. The collapse caused a halt to midtown traffic, and even the subway trains had to stop. The owner of the building died in the collapse.

Excavations for new sewer trenches adjacent to existing buildings have frequently caused undermining of footings, resulting in distress and even total collapse. In the President Street collapse in Brooklyn, and other similar collapses, several occupants of existing buildings were killed or injured when their buildings were undermined. When these excavations were close to the existing footings, and within the influence line (Fig. 5.5), the footings were undermined, often with disastrous results.

5.1.1 New York suburb sewers

Construction for a new sewer main project in Brooklyn required the excavation of deep cuts adjacent to existing old residential buildings. The sewer lines were designed to run too close to existing footings. In addition to the fact that the new cuts were within the influence lines of existing footings, the contractor used vibratory pile drivers to install the piles. There was no monitoring of the vibration levels during pile driving. As a result, many nearby buildings settled, cracked, tilted, and shifted laterally. After the

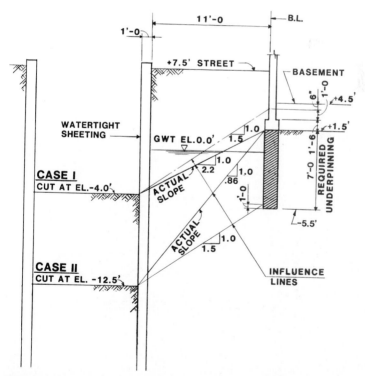

Figure 5.5   When excavations are dug close to existing footings and within the influence line, footings can be undermined, often with disastrous results, unless they are underpinned. In this example, there are poor soil conditions and water is present. Use soil influence line of 1 : 1.5 for watertight conditions. Case I: no underpinning required; Case II: underpinning required.

damage occurred, diagonal braces were added to provide lateral support (Fig. 5.6).

One building collapsed, killing one of the tenants (Fig. 5.7). The investigation revealed that the building was constructed with timber footings, a construction method common at the turn of century. The footings themselves were found to be in good condition but had been disturbed by the sewer activity. Once the excavation was within the influence line, vertical support was lost, causing footing settlements. The high-grade steel bracing system (Fig. 5.8) could not save this building because soil support was lost *below* the footing level.

To restore the integrity of the vertical support, it was initially decided to underpin the buildings. The bid prices obtained for underpinning were so high, however, that it was more cost-effective to buy the damaged buildings and demolish them.

Lessons

1. Careful *preconstruction* study to determine the need for protection and underpinning of adjacent structures is required. This study should include review of existing plans, taking sufficient soil borings, and evaluating the condition and strength of existing buildings and other nearby structures.

Figure 5.6 Diagonal braces were added for lateral support of adjacent buildings during excavation for a sewer. These braces are obviously too flexible to offer meaningful support.

Figure 5.7 This is all that remained after the collapse of a building which resulted in loss of life.

Figure 5.8 Top of excavation cut was well braced. This was not sufficient to save the building in Fig. 5.7, since some bracing was installed after movement had already taken place.

419

2. Excavations for sewers or other utilities should be moved away from existing buildings and so located that the excavation cut is outside footing influence lines. Where this is not possible proper underpinning and support should be provided *prior* to excavation.

3. Use of vibratory equipment for pile driving should be avoided in loose sands and similar soils.

4. Vibration readings, using seismographs, should be taken as often as needed to establish the suitability and proper energy of the driving hammers.

5.1.2   Electronic plant, upstate New York

This new plant was under construction when the entire building started to sink vertically and slide laterally toward the low adjacent valley. Instrument survey readings confirmed these visual observations. The footing movements, as usual in these cases, were accompanied by distress in the building in the form of cracking of slabs and concrete grade beams (Fig. 5.9). The main question that had to be answered was: What was the effect of large movements — of the order of 4 in (102 mm) — on the structural-steel frame and its bolted connections?

To answer this question we had to determine the level of stress in the loaded and deformed steel beams and the actual loads in the connection

Figure 5.9 Deep grade beams were severely cracked.

Figure 5.10 Investigation of building distress established that the excavation within the foundation influence line led to loss of vertical support, causing footing settlement. In this photograph, a stress-relief test is being conducted to measure stresses in bolts.

bolts. For that purpose we used an innovative stress-relief method (Fig. 5.10). This method, which our firm pioneered, proved that the level of stress actually present in the steel members and connections was low and acceptable.

This procedure requires the attachment of electrical strain gauges to the steel flanges of the beams. Through each strain gauge, a hole is then drilled directly into the steel. The reduction of stress at the hole is measured by the strain gauge connected to a Wheatstone bridge. The relieved stress thus measured gives the approximate stress level then existing in the loaded steel beam.

The only structural corrective measure that was executed was to enlarge the existing footings, so as to enable the structure to support the additional live loads without considerable settlement. The method of repair is shown in Figs. 5.11 and 5.12.

Lessons

1. High footings adjacent to severe soil slopes should be designed by using low allowable bearing pressures so as to reduce possible settlements.

2. In extreme cases, the use of retaining walls should be considered.

3. The actual stress level in a deformed structure is the true yardstick for evaluating the adequacy of a distressed structure.

4. The stress-relief method proved to be an effective tool in strength evaluation of distressed structures.

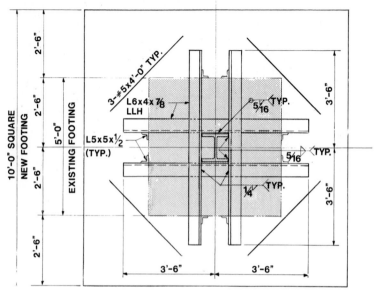

Figure 5.11 The existing small footing was increased in size in order to support additional live loads.

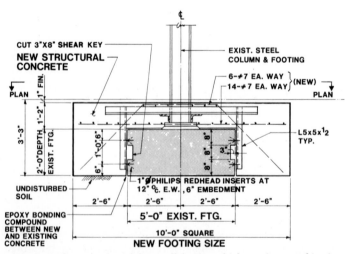

Figure 5.12 Cross section of the small footing which was increased in size. New concrete was anchored with steel expansion bolts.

5.1.3   East-side hospital, New York City

During excavation for a new high-rise hospital structure, an alarm went out to everybody connected with the project. It was discovered that, as a result of rock removal by the foundation contractor, the adjacent high-rise apartment building to the west was precariously sitting on a "sliver" of rock (Fig. 5.13). Construction work came to a halt, and serious consideration was given to evacuating the fully occupied apartment building.

Probes into the wall revealed the following conditions (Fig. 5.14):

1. The width of the rock sliver, of unknown strength, was approximately 14 in (355 mm). Incredibly, it was the only structural element supporting the high-rise building.

2. Even more remarkably, the footings called for in the original design for the existing apartment building were not built at all.

3. The easterly wall of the building cellar was not a standard concrete cellar wall but was actually a rock faced with stucco finish.

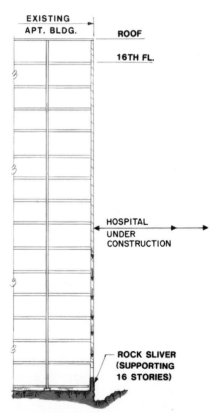

EXISTING
APT. BLDG.

ROOF

16TH FL.

HOSPITAL
UNDER
CONSTRUCTION

ROCK SLIVER
(SUPPORTING
16 STORIES)

Figure 5.13   During excavation for a high-rise hospital structure, it was discovered that the adjacent apartment building was dangerously sitting on a rock sliver, as a result of adjacent rock excavation.

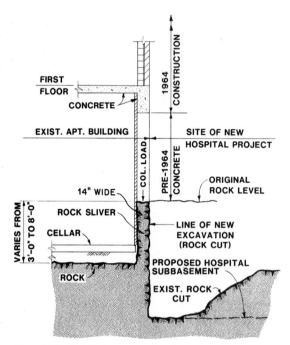

Figure 5.14 Detailed cross section showing geometry of rock sliver.

4. The plans that had been filed with the Department of Buildings did not show the as-built conditions.

An emergency plan for underpinning was immediately launched. This plan included new concrete underpinning piers (braced laterally) cut into the rock sliver and bearing on the rock below the existing cellar level (Fig. 5.15). The plan was executed perfectly, and the alarm was rescinded.

As usual, the responsibility for the unexpected additional cost was in dispute. The questions asked by everybody were:

1. Was the original 1964 construction, which deviated from the design, bearing the apartment building columns on the neighbor's rock, legal?
2. Did it follow accepted good construction practice?
3. What responsibility, if any, does the structural engineer of record have for the contractor's variation from the plans?

Concerning the first question, our research into the building code showed that the code was completely silent on this issue. Regardless, we found no requirement forbidding the practice.

Our opinion was that it is poor construction practice for a building owner to utilize a neighbor's rock for bearing for the owner's building, especially

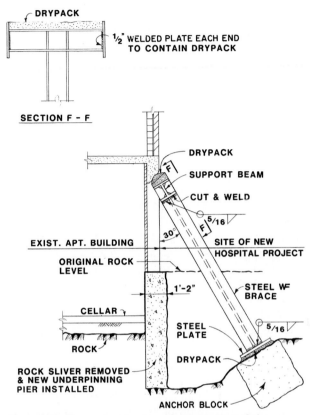

Figure 5.15 Emergency plan for underpinning included new concrete piers cut into rock sliver and bearing on rock below existing cellar level. Diagonal braces were installed to brace the existing building.

since the owner has no control over nor any right to limit the neighbor's construction activity.

The engineer of record had no "controlled inspection" responsibility for this project, nor was he required to perform such an inspection of the construction according to the then pertinent code.

According to many building codes, when neighboring construction goes *below* the bottom of an existing footing, it is the responsibility of the neighbor doing the new construction to underpin the existing structure provided the new excavations are more than 10 ft (3.0 m) below curb level.

The New York City building code, for example, states:[2]

> When an excavation is carried to depth more than 10 ft [3.0 m] below the legally established curb level, the person who causes such excavation to be made shall at all times and at his own expense, preserve and protect from injury any adjoining structures . . .

Lessons

1. Do not use the rock outside your property for support of your structure, regardless of the apparent immediate savings.

2. You may support your structure only on the rock totally within your own side of the property line.

3. Rock slivers should not be relied on for foundation support.

5.1.4 Manhattan hospital complex, New York City

When the structures adjacent to a deep excavation began cracking, it slowly became clear that something was wrong. In this example, an excavation extending across the entire site was progressing toward a total depth of 45 ft (14 m) below grade. The existing structures contained only single-level basements. Initially, it was expected that much of the excavation would be into sound rock, which was identified as gneiss. The groundwater level within the buildings was controlled by a sump pump located below an existing basement slab.

During the design phase, it was obvious that the adjacent structures would have to be underpinned. The New York City building code requires that underpinning be designed by a professional engineer. This requirement was met, and a design using underpinning pits with posttensioning cables to resist lateral earth pressures was devised.

As excavation for the pits was proceeding, movement was detected in the building directly east of the site. This movement was recorded by the telltales (movement gauges) which were properly installed for monitoring such movements. When water was found several feet below the pits, attempts were made to dewater the soil surrounding the pit excavation. But, as recorded by the inspection engineer, " . . . this proved futile because of the impermeable nature of the soil existing at this location."

Dewatering by well points was then tried. This also proved ineffective because the silts in the soil clogged the well-point screens. A third alternative, using jackpiles, was then considered. This method of pushing pipe segments into the ground uses the building weight as a reaction for the jacking forces. In this project, unfortunately, it readily became apparent that the footings of the existing building, which consisted of decomposed rubble, did not offer sufficient resistance to the jacking loads. Even an attempt to use a steel beam to jack against failed, and the scheme was then abandoned. While these efforts were going on, a further adverse movement was detected in the adjacent building. At this time the safety of the structure was questioned, and temporary rakers (sloped braces) were installed. For all practical purposes, construction operations came to a stop.

The two buildings on the east exhibited serious cracking throughout (Figs. 5.16 and 5.17). The one nearest the excavation settled, causing the

Figure 5.16   The facade of the building adjacent to the excavation exhibited distress in the form of cracks, indicating that something was seriously wrong.

Figure 5.17   Footing movements were accompanied by building distress such as cracking of partitions and settlement of stairs.

entire building to pivot about the foundation walls. This pivoting effect, in turn, caused the entire party wall to move westward, leaving a 6-in (152-mm) gap at the top (Fig. 5.18). This movement caused enormous distress on the interior of the building. Floors settled, ceilings and walls cracked, and even the elevator got stuck as its shaft deformed and would not permit free movement of the cab. At the fourth floor there was actually a separation between the wood joists and the party wall, causing the floor to settle by as much as 1⅛ in (28 mm) at one point.

The foundation contractor, realizing the danger, immediately installed vertical shores under the west end of the wood joists, from the first floor all the way up to the roof. It was at this point that we were retained and recommended that the contractor take weekly movement readings with surveying instruments, and continue doing so until the settlement stopped. He was also instructed not to proceed with any restoration work until all settlement had stopped.

The New York Times printed the following article regarding this project:[3]

> George N. and Barbara K. don't live here any more. Their former apartment, in a brownstone in the Grammercy Park area, once looked like the home of many a professional couple — antique furniture, luxuriant plants and exposed wooden beams, lots of beams.

Figure 5.18   A 6-in gap opened at the top of the building as one building pivoted toward the excavation.

. . . Their apartment is one of the occasional casualties of excavation work, the bulldozing, blasting and drilling that create the cavernous holes meant to hold a new building's foundation.

Though New Yorkers frequently sue or arrange out-of-court settlements with excavators (claims range from apartment damage to sexual inadequacy caused by noise), George and Barbara just decided not to renew their lease.

"The whole experience was very unnerving," Mr. N. recalls. "I was really concerned how much physical damage would be done to my home by the end of a day. *It started out as a hair-line crack and got bigger every day.* [Italics mine.] I saw the baseboard move further away from its original place on the floor. Then we had to take down a shelf in the closet because it was no longer wide enough. *The wall had moved four inches [102 mm] toward New Jersey."* [Italics mine.]

. . . A side of the ceiling was now without support and, just as Mr. N. had feared, it soon gave way. Fortunately, their apartment was saved by beams, which had been installed by workmen a few days earlier. [*Author's note:* The beams were the vertical shores referred to above.]

. . . Dr. F., a physician, and her husband, resident-owners of the building next to the excavation, never contemplated selling their property.

Although the roar of the machinery can reduce telephone conversations to the F. residence to frustrating shouts, Dr. F. says that somehow her practice at home (psychiatry) "has not been disturbed."

. . . Mr. R., project engineer for the excavation company, says that the unknown factor in that site was an unexpectedly high underground water elevation. "We had what you call test borings. . . . As it turned out, these test borings were greatly different from what we encountered."

When all efforts to support the building failed, the *method of last resort* was used: very costly soil solidification by freezing the soil. The saturated soil proved to be an advantage for the freezing process. This process, with all its pipes and equipment, took about 3 weeks to install (Fig. 5.19). It proved to be a total success; practically all movements within the adjacent ailing structure stopped. Once the soil under the building was consolidated, the original conventional underpinning design was implemented.

Just prior to freezing the soil, a cross-lot bracing system was installed (Fig. 5.20). The purpose of this system was to eliminate any further movement in the building on the east and to stabilize it during the subsequent construction operations. The giant cross-lot braces, which spanned the entire lot, permitted construction to proceed without interruption.

After completion of the underpinning, the adjacent buildings were restored to their original condition.

Lessons

1. Accurate preconstruction subsoil study is essential to the success of foundation work. Inadequate test data can cause serious delays, cost overruns, damage, and even collapses. In this case, if accurate informa-

Figure 5.19 The soil directly below the undermined building was solidified by freezing.

Figure 5.20 Giant cross-lot steel bracing system by itself could not save the building because soil support was lost *below* footing level. Freezing of the soil was necessary.

tion had been obtained, design changes and soil solidification could have been planned prior to start of construction.

2. Adjacent structures must be carefully monitored during underpinning operations, so that, if and when movements are detected, changes in methods or procedures can be implemented.

5.2   Load-Transfer Failure

A well-designed and constructed rigid-frame structure will tolerate even substantial foundation movements. When an assembly of walls, floors, frame, and partitions is rigidly connected, the system will adequately adjust itself to differential foundation movements. The load transfer is made by frame action through the support offered by the foundation. When this interconnected rigidity is absent, the load transfer will be through a single support so that the load will go directly to the soil vertically. Where this single support in the soil is missing, the structure will fail, unless the structural rigidity will transfer it horizontally to other footings and then, in turn, to the soil.

Where the adjacent rigidity is present but lacks sufficient strength, the adjacent structure will fail. Figure 5.21 is a diagrammatic description of this action. Four people are carrying a log, its weight uniformly distributed. When A steps into a ditch (an inadequate foundation) A's portion of the load is suddenly transferred to B, who may be unable to support the additional load.

A typical example is a 16-story warehouse building in lower Manhattan, supported by four continuous pile caps and timber piles. Pumping operations at a neighboring 20-ft-deep (6-m) excavation for a depressed roadway

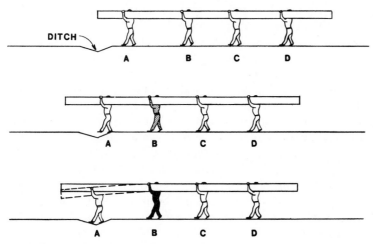

Figure 5.21   Diagram of four people carrying a log.

lowered the water table and caused one line of piles to settle and pull the pile cap down. This released the exterior support so that the load transfer to the interior line of columns overloaded their steel beam grillages. The 24-in (610-mm) beams collapsed down to a 12-in (305-mm) depth, with consequent 12-in (305-mm) settlement of every floor in the building. The floor beams resting on steel girders between columns had only soft support, so they took the new trough shape without failure.

The entire building was shored and new shallow grillages were installed to replace the failed ones. Instead of the entire structure being jacked, which would have had definite risks, especially for an old structure, each floor was releveled and the building was put back into use.

5.3 Lateral Movement

It is well-known that 1 in (25 mm) of lateral movement of a foundation causes more damage than 1 in of vertical settlement. Lateral movements are caused by either the elimination of existing lateral resistances or the addition of active lateral pressures and loads. The changes in the active pressures and passive resistances as a result of variable water conditions must be considered. Saturation of the soil often increases the active pressures and reduces the passive resistances.

Many collapses have occurred as a result of demolition of adjacent buildings. In these instances the backfilling of adjoining cellar space generated lateral soil pressures on the old cellar partitions. These walls, in turn, failed because they were not designed and constructed as retaining walls.

Lateral flow of soil under buildings is also known to cause collapses of buildings. In São Paulo, Brazil, such lateral soil flow caused the total collapse of a 24-story office building in 1943.

A similar case occurred in 1957 in Rio de Janeiro, Brazil, where an 11-story residential building collapsed completely after the excavation of an adjoining wall removed the lateral restraint of the building piers. The attempted underpinning, which was started too late, could not save the rotating structure.

There also are numerous cases of wall failures from broken drains alongside the footings, with washout of the soil during heavy storms.

Change in pressure intensity against walls often causes failure, especially in the unreinforced concrete basement walls for residential buildings. These walls are unfortunately seldom investigated for high soil pressures. Surcharging soil on land adjacent to structures often causes large lateral pressures. Mounds of debris from demolitions are frequently piled adjacent to basement walls which had not been designed to resist such loads. These bowed basement walls often cave in and cause the total collapse of the structure.

The stability of retaining walls is often similarly affected, resulting in failure by overturning, either because of inadequate base width, or because of high soil pressure caused by defective drainage or adjacent load surcharge.

Failure of buried sanitary structures is also common, especially when they are emptied for cleaning or repairs.

5.3.1 New Jersey postal facility— Cracked pile caps

Shortly after completion of the construction of this mailbag repair shop, the entire structure started to shift sideways, causing many of the piles to bend and the pile caps to crack (Fig. 5.22). In addition, an interior block wall parallel to the direction of the shift also cracked (Fig. 5.23).

Our firm was called to investigate and determine the structural adequacy of the entire facility. What was immediately evident was the very heavy vertical gravity loads imposed on the piles. Figure 5.24 shows heavy bales of paper and cloth sitting directly on the already shifted and distressed piles.

The area directly east of the plant was found to be surcharged with mounds of heavy garbage (Fig. 5.25). It quickly became clear that the soft clay soil directly underneath this extreme garbage weight was surcharging the soil, causing lateral pressures on the long, unbraced steel piles. Pre-

Figure 5.22 These H piles shifted laterally and bent severely, cracking the concrete pile caps at their top.

Figure 5.23   The stepped diagonal cracks in these block walls show both lateral shift and downward settlement.

Figure 5.24   To make matters worse, the damaged piles were heavily loaded with stacked bags and bales.

Figure 5.25 The cause for the lateral soil shift was identified as the high surcharge on the right (arrow).

vious attempts to stop the bending of the piles by welding steel cross-ties had been unsuccessful, and instead of stopping the lateral shift, they only transferred the lateral loads to the adjacent piles, which eventually failed, too. The additional lateral cross-ties may be seen in Fig. 5.26.

We immediately recommended that the owners reduce the actual live loads on the piles, by moving the heavy bales to unaffected areas. A complete underpinning program involving drilling through the existing slabs and jackpiling down through the soil was then recommended. The owners procrastinated for a while, but finally a program similar to one proposed by our firm was implemented several years later.

Lesson. The placement of high surcharge adjacent to structures generates high lateral loads that can cause shifting of piles and bowing of foundation walls, and should be avoided.

5.3.2 Collapse of Hartford, Connecticut foundation walls

A large housing development on sloping terrain had concrete foundation walls enclosing the basements. The walls were unreinforced, as is common for low-rise residential buildings. Rather than use controlled granular fill, the contractor used clay backfill during a dry period of weather. This clay

Figure 5.26 Structural-steel cross-ties were welded in place to stabilize the piles and prevent collapse.

backfill shrank away from the concrete walls, leaving narrow gaps along the full height of the walls. After a heavy rainstorm, the runoff filled these gaps, causing very high hydrostatic lateral pressures. Some of the walls which were not yet braced at the top toppled over, and some bent, cracked, and completely failed (Figs. 5.27 and 5.28).

Lessons

1. Foundation walls must be braced at top and bottom prior to any back-filling.

2. Only well-controlled and drainable, porous backfill should be placed against foundations and other retaining walls, to avoid shrinking soils and subsequent development of high lateral soil and water pressures.

5.3.3 57th Street garden wall collapse

A 125-ft-long (38-m) garden wall, 10 ft (3.0 m) high, supporting a rear patio, collapsed without warning during heavy rain, leaving the concrete slab and its garden furniture hanging (Fig. 5.29). The superintendent told us that he was inside the building when he heard a tremendous noise.

Upon inspection we found that the garden wall consisted of a double wythe of 4-in (102-mm) brick walls without any headers, durawall, or ties

Figure 5.27 Foundation wall cracked, above, and buckled, below, during construction as a result of high lateral hydrostatic pressure caused by clay backfill after heavy rainstorm.

Figure 5.28   These foundation walls later totally collapsed under the high lateral pressure.

Figure 5.29   The rear garden wall supporting this patio collapsed under heavy lateral pressure caused by saturated soil.

connecting the two walls together. A wall of this construction could not have possibly sustained the lateral pressure of a 5-ft (1.5-m) height of saturated ashes and cinders. There was an allegation that the load imposed by hanging two telephone cables from the rear wall had caused the wall to collapse. Certainly the additional weight of the cables could not have seriously affected the amount of lateral pressure.

Lessons

1. Unbonded brick walls cannot be used as retaining walls, especially where water pressures are likely to build up.
2. For stability, garden walls must either have wide bases, be equipped with "dead men" (heavy anchors in soil), or be massive to resist lateral loads.

5.4 Unequal Support

A basic rule of engineering is that there is no load transfer without deformation. When loads are transferred to the soil through the foundation, this transfer is associated with a deformation of the soil. In other words, all footings settle when they are loaded. The amount of settlement is equal for different footings when soil resistances are identical and load distributions are equal. When they are not, differential settlements occur and load transfer or tipping of the structure will result. The portion of the structure founded on the weaker soil will tip away. Where the framework is not continuous, the brittle masonry enclosure will crack (usually in a diagonal pattern) during the shear transfer. There are a great number of such structures, with foundations partly on rock and partly on soil, partly on rock and partly on friction piles, partly on stiff soil and partly on soft soil. There are no doubt many structures with footings bearing on different soils with different and unequal soil bearing resistances. The results are invariably distress and severe cracking (Fig. 5.30).

All these soil support deficiencies may be corrected and are often so rectified. However, these corrections are usually quite costly and involve underpinning the weaker soil or the weaker support. The use of additional piles and the installation of jackpiles are some of the most commonly used and most successful methods. The famous Transcona grain elevator in Manitoba, Canada (Fig. 5.31), showed unequal settlement and tipped. This structure consisting of 65 concrete bins, 93 ft (28 m) high on a concrete mat 77 × 175 ft (24 × 53 m), settled 12 in (305 mm), then tipped 27° when about 85 percent full of grain. The structure weighed 20,000 tons (178 MN) and contained at the time grain weighing 22,000 tons (195 MN). When repair work was begun in 1914, the massive structure, which was 34 ft (10 m) out of level, was jacked back to a vertical position and underpinned to rock — a remarkable engineering feat.

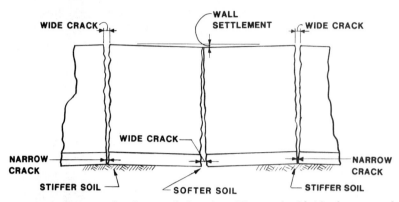

Figure 5.30 Structures bearing on soil of varying stiffnesses should either be separated by expansion joints or else designed for differential settlement. Cracks of varying widths indicate rotation resulting from vertical movement or settlement of wall.

Sometimes, other methods of soil stabilization by cement or chemical injection are used. Another correction procedure is the addition of a subsurface enclosure (usually a tight sheet pile cell) which increases the bearing value of the soil. This was done on a grain elevator in Portland, Oregon in 1919, where 2600 pinch piles were added to enclose the foundations after the structure showed progressive unequal settlement and was tipping.

Figure 5.31 This well-known grain elevator in Canada tipped as one unit and then was jacked back to a vertical position.

Not so fortunate were the owners of a smaller grain elevator in Fargo, North Dakota, which tilted in 1955 when it was only 1 year old. This elevator, which contained 20 bins 120 ft (37 m) high, was supported on a concrete mat measuring 52 × 216 ft (16 × 66 m). As the mat tilted, the bins crumbled and the structure was wrecked and had to be abandoned.

There are many classical examples of buildings tilting as a result of unequal settlements. The Tower of Pisa is probably the best known and researched example and is still in use, as is the Palace of Fine Arts in Mexico City (Fig. 5.32) with its settlements measured in meters. The Guadalupe Shrine and Cathedral, also in Mexico City (Fig. 5.33), is so distressed, cracked, and tilted that it makes visiting tourists nervous.

Buildings partially on earth and partially on rock have too often been designed on the incorrect assumption that the mere use of the local building code's allowable bearing values will assure equal settlement. Where such variable bearing conditions exist, a separation joint must be provided so that each section of the building can then act as an individual separate structure. These buildings, especially when they utilize bearing walls, often crack almost immediately where such a joint is missing. Thus, nature provides the omitted joint (Fig. 5.34). Underpinning and other repair costs together with the additional economic loss caused by delayed or vacated

Figure 5.32 Palace of Fine Arts in Mexico City, which has continuously settled for years.

Figure 5.33 The Guadalupe Shrine, which has cracked and rotated, has been attracting tourists for years.

use far exceed the cost of properly designed and constructed separated foundations.

We have also investigated several instances where soil shrinkage due to local dewatering triggered differential settlements even though the original construction was proper and followed faultless design.

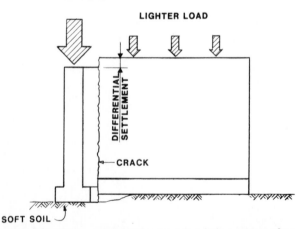

Figure 5.34 Where an expansion joint is missing, nature often creates it by vertical cracking. Differential settlement will occur when either the loads are of different intensities or the soil-bearing resistance varies.

5.5   Drag-Down and Heave

As a footing is loaded, the supporting soil reacts by yielding and compressing to provide resistance. The actual compressing of the soil takes place rapidly in the case of granular soils but much more slowly for clays. Once this compression occurs, the structure remains stable, since the foundation no longer settles. This stability depends on fringe areas as well as the soil directly below the footing, or, in the case of piles, on the soil near the pile tip. If the soil below the footing is removed or disturbed, settlement or lateral movement is induced.

Similarly, if the fringe area is removed, the soil reaction pattern must change. Even if sufficient resistance still remains without measurable additional settlement, the center of the resistance is changed, thus affecting the stability of the footing.

When the fringe area is loaded by a new structure, causing new compression in the affected soil volume, the old area can be required to carry some of the new load by internal shear strength. In such a case, there will be an additional unexpected new settlement of the previously stable building. In the event the new building is not separated from the existing construction, the settlement caused by the new load will cause a partial load transfer to the existing wall, with possible overloading of the previously stable footing. In plastic soils, these new settlements often are accompanied by upward movements and heaves (Fig. 5.35) some distance away. Since the liquid in the soils cannot change volume, every new settlement must produce an equal-volume heave.

When an entire structure settles at a slow rate for a long period of time, these settlements may be acceptable as long as they are *uniform*. *Large differential settlements* will result in damage.

Disregard of such elementary considerations is often the cause of much litigation stemming from cases where new foundation construction in

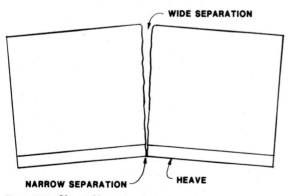

Figure 5.35   Upward heaving always is accompanied by cracks with wider separation at the top.

urban areas suddenly places neighboring existing construction in danger of serious damage or even complete failure.

Under several building codes (the New York City building code is one example), if new construction requires excavation more than 10 ft (3.0 m) in depth adjacent to an existing foundation, all precautions to avoid damage to the existing conditions are the obligation of the new project. If the new excavation depth is 10 ft (3.0 m) or less, the obligation for protection falls on the owner of the existing building should its foundations be higher than the new footings.

In medium- to well-compacted sands, pile driving demands heavy power, and the vibration impulses can shake adjacent structures to the point of serious damage.

Drag-down from soil shrinkage, when water tables recede or the soil is desiccated by tree and shrub growth, with consequent differential settlements, is a phenomenon observed in many localities. The failure of a theater wall in London in 1942 was correlated with the growth of a line of poplar trees. In Kansas City, Missouri, 65 percent of the homes in one residential area were found to be affected by soil desiccation from vegetation growth in foundation plantings.

Piles embedded in soil layers that will consolidate from dewatering or from load surcharge are subject to overloading when the greater soil density increases surface friction and the new soil loads "hang" on the piles. This phenomenon is sometimes referred to as *negative friction*. The added load may cause considerable increase in settlement and may even pull the pile out of the pile cap or pull the pile cap free of the column or wall. Dewatering operations are commonly caused by new construction or pumping by water supply companies.

In some instances, as consolidation of organic clay strata takes place, piles continue to sink with additional settlement of the structure. This was the case at the Queens Apartments in New York.

5.5.1 Queens Apartments, New York

Settlements and masonry cracks developed in many levels of the buildings of this project (Fig. 5.36). The cracks appeared in both interior partitions and the exterior brick walls. These structures, located in Flushing, New York, were constructed over marshland and were supported by timber piles.

An earlier subsurface investigation of this project indicated that it was likely that at least some of the piles in the three-pile group supporting the distressed corner of the building had broken. We could not determine this fact without excavation or probes. The breakage may have occurred when the piles were driven or subsequently when the loads resulting from consolidation of the organic clay stratum were imposed on them.

Figure 5.36   The settlement of timber piles driven into a marsh caused the settlement of and cracks in this building.

In an effort to assess the potential for additional settlements and their likely magnitude, a laboratory and a field program were performed. In the laboratory, the rate of settlement due to a secondary compression of samples of the organic clay was measured. In the field, a sensitive settlement plate (Fig. 5.37) was monitored for movement every 2 weeks for about 6 months by a depth caliper accurate to 0.001 in (0.025 mm).

The area selected for installation of the plate was the interior corner directly outside the laundry room. This area was chosen for its close proximity to settlement cracks; it also provided sufficient overhead clearance for drilling. A 1-in-diameter (25-mm) heavy-duty pipe was driven into the lower sand stratum to provide a fixed reference point. A 2-in-diameter (51-mm) outer casing protected this pipe from drag-down effects from the overlying fill and clay.

Analysis of laboratory data has suggested that settlement of organic clay, as a result of the fill above it, can be expected to be about 1 in (25 mm) every 10 years. The field measurements from the settlement plate indicated even greater settlement.

Figure 5.38 presents the data obtained from the settlement plate device, with settlement rates of up to about ¼ in (6.4 mm) per year. Thus, the field data suggested settlements about twice those indicated by the laboratory data.

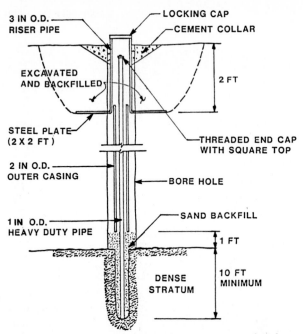

Figure 5.37 Field instrumentation for measurement of settlement of the soil surface steel plate settlements were measured.

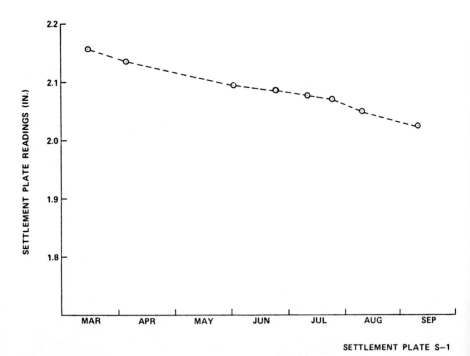

SETTLEMENT PLATE S–1

Figure 5.38 Chart showing soil settlements as measured by downward plate movement over a 7-month period.

Lessons
1. Do not guess that a structure that has settled will not continue to do so. Instrument the foundations and the soil to assess the potential future settlements.
2. Placing a structure on piles bearing in marshland always involves risks of settlements.

5.5.2 Edgewater, New Jersey high-rise

The piles supporting the first-floor slabs of this complex settled and pulled the slab downward. This caused settlements and cracks of the slabs and tilting and cracking of partitions (Fig. 5.39). The first signs of distress were sticking doors and windows.

An in-depth structural investigation was carried out and revealed that while the main supports for the columns were bearing on steel piles driven to resistance in rock, the piles supporting the slabs were bearing only on wooden friction piles. The pattern of slab deflection was confirmed by instrument survey and the corresponding contour map (Fig. 5.40). Probes below the slab showed that the backfill had settled approximately 3 in (76 mm) (Fig. 5.41). The negative friction effect of the downward settlement of the backfill and consolidation of the soil pulled the piles down. Consequently, the concrete slab attached at the top of the piles was dragged down.

The repair required cutting into existing slabs and adding new rein-

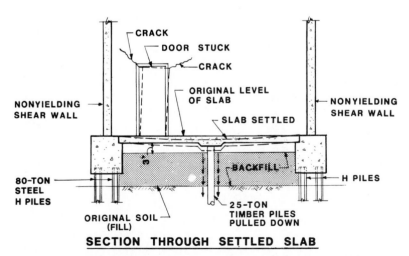

SECTION THROUGH SETTLED SLAB

Figure 5.39 Slabs supported on timber piles were dragged down by negative friction and settlement. Main walls supported on steel piles driven to refusal did not settle, which exacerbated the distress. Partitions cracked and doors tilted and stuck as a result of differential slab deflections.

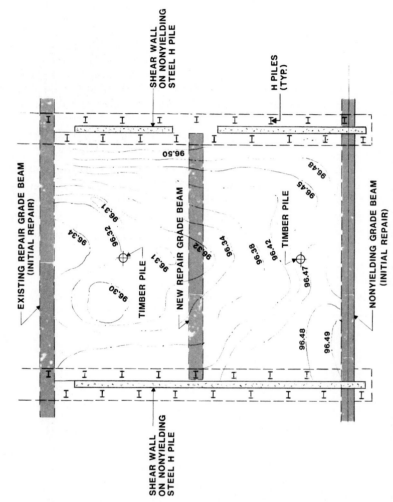

Figure 5.40 Slab contours show survey of slab elevations with settlements of up to 2½ in (64 mm), the result of pile drag-down.

Figure 5.41 Excavation probe showed that the soil backfill settled, leaving a 3-in (76-mm) gap below the slab.

Figure 5.42 New concrete beams were cast in slots cut into existing slabs. The new beams shifted the slab load to the steel piles.

forced-concrete beams which transferred the slab loads into the reliable column piles (Fig. 5.42).

Lesson. When piles are driven into compressible soils, negative-friction effects should be taken into consideration. Additional soil surcharge will increase the drag-down forces.

Heaving of footing contact surface is known to neutralize soil bearing by frost action. We have investigated several cases where during extremely cold winter, frost penetrated below slabs on grade and cracked buried water and sprinkler lines, causing ice buildup which heaved the structure above and even sheared some of the supporting columns.

5.6 Design Error

Unfortunately, many foundations are designed with insufficient prior subsurface investigation or with disregard of the true soil conditions. This results in inadequate support for the structure, often requiring expensive corrective work later.

Following standard structural design procedures for foundations will result in an adequate factor of safety; most errors are in judgment, leading to assumptions or decisions which are not consistent with the actual behavior of the soil later.

A common design error, usually made in an effort to save initial construction costs, is to provide pile support for the walls and the roof, while the main floor is placed on more or less compacted sand over fill, rubbish, peat, organic silt, and other compressible layers. We have seen many industrial plants where the roof was supported on piles while the floor, supporting expensive equipment, was placed over soil fill. It seems more logical to support expensive equipment on piles and let the roof settle. Heavy factory machinery generally requires installation on precisely level floors, and even minimal differential settlement can become cause for concern. Even in warehouse use, floor deformations become objectionable, making efficient use of forklifts impractical.

It is obvious that placing floor slabs on soil fills is a false economy. True cost efficiency is measured by the total of the initial and in-life service costs.

5.6.1 Alaska pier collapse

This unusual failure was caused by a design error resulting from inadequate preinvestigation study of local data and conditions. The formation of huge ice blocks atop the pier pilings was totally unexpected; the pier designers had not taken into account loads of such magnitude being imposed on the structure. The total collapse of the pier occurred at low tide when the total weight of the ice was suddenly suspended from the sloped piles (Fig. 5.43).

There was a unique condition that contributed to this collapse: the extremely large variation between low and high tide, which was approximately 20 ft (6.1 m). The blocks of ice that had formed during the winter suddenly lost their support when the thaw came in the spring. The enormous weight of the ice was shifted to the piles, which bent, causing fracture of the concrete pile caps. The thawing of the ice at the contact surface with the piles also caused the ice blocks to slide down diagonally, while wedged between the piles (Fig. 5.44). The bending stresses at the ends caused by the wedging forces (see force and moment diagrams shown in Figs. 5.45 and 5.46) severely cracked the concrete pile caps (Fig. 5.47), leading to the eventual collapse of the pier.*

Our files show that the partially completed pier was supported on pile bents spaced 20 ft (6.1 m) on center. The bents consisted of both vertical and diagonal piles. Twenty-inch (508-mm) octagonal prestressed vertical piles and 20-in-diameter (508-mm) batter steel piles were arranged in pairs forming quasi-A frames. The length of the piles was between 70 and 80 ft (21 and 24 m). The piles were driven into 15 ft (4.6 m) of silt, then 55 ft (16.8 m) of silty gravel and sand underlain by layers of clay. By midwinter, huge blocks of ice, approximately *20 ft (6.1 m) thick*, had formed on the piles. The blocks, weighing 50 tons (445 kN) or more, slid down the sloped piles, damaging the piles and pile caps.

---

* Dov Kaminetzky, "Problems and Risks Related to Marine Projects," American Society of Civil Engineers, Met Section, New York, N.Y., March 1, 1977, pp. 40–60.

Figure 5.43   General view of the pier with piles crusted with ice.

Figure 5.44 As the temperature rose, ice started to melt and big blocks became wedged between vertical and sloped piles.

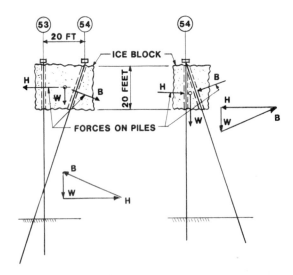

LATERAL FORCES ON BATTER PILES
AND VERTICAL PILES

Figure 5.45 Loading diagrams of wedging action of ice blocks.

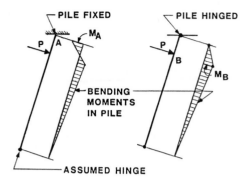

ACTUAL CONDITION   DESIGN ASSUMPTION

LATERAL FORCES ON BATTER PILES
DUE TO LOAD OF MELTING ICE

Figure 5.46 Moment diagrams of wedging action of ice blocks, showing high moments which cracked the pile caps.

Figure 5.47 The concrete pile caps, which were too shallow to start with, cracked and failed under the heavy ice wedging forces and bending moments.

The recorded telephone conversation between the design engineers on the project went like this:

D.: We've got great big problems up here . . . I went down the dock . . . a good half of our pile caps . . . are cracked.
They're cracked bad. They've got an inch and a half [38 mm] crack in some of these. We know it's because of the tidal action and the ice.

A.: It looks like . . . the weight of the ice is causing such a moment up there that it is cracking the heck out of it. We're going to have to take the panels off, go in there and remove the caps and redesign the whole thing. . . .

D.: We've got a good connection up there . . . it's very, very, rigid, but it's bending that's doing it. Would have saved us I think, if we'd had a lot more of those stirrups in there, but there are only no. 5 bars [16 mm diameter] at about 18 inches [457 mm] on center and it's just not enough.

While this failure was considered by many investigators to be caused principally by a design error, it was also established that damage was caused to the vertical piles during installation with overjetting and pounding with the diesel hammer. Insufficient penetration of the batter piles was also confirmed in several cases.

Lessons

1. The effect of ice formation and resulting forces must be considered in the design of structures in northern exposures.

2. Piles subjected to lateral loading and bending must be sufficiently embedded into the concrete pile caps.

3. The consequences of extreme variations of tidal rise and fall should always be taken into consideration in design of marine structures.

5.7 Construction Errors

There are two common sources of construction errors:

1. Temporary protection measures during the construction phase

2. Foundation work itself

Most foundation failures involve the first source of error, relating to temporary shoring and bracings, and temporary cofferdams for lateral protection and pumping operations. Because of the temporary nature of these structures, it is common that factors of safety are kept to a minimum for economic reasons.

Many construction errors on foundation work also are a result of the same deficiencies that exist in superstructure concrete work, such as improper concrete quality and improper rebar placement. Where deficient material is provided, in general there is absolutely no difference between superstructure and foundation work. The exceptions are foundation elements such as piles and caissons where "blind work" is involved. Cavities within cast-in-place piles are often found where quality control was lacking. The omission of verification methods *early* in the construction progress is partially responsible for the enormous losses caused by such errors.

The unique cases involve the placement of foundation elements such as footings, piers, or piles on inadequate bearing soil strata. Examples are footings bearing on soft soils, and piles driven "short" so as to hang high in improper and insufficient bearing soil.

In 1958, a newly constructed 11-story building in Rio de Janeiro, Brazil, toppled and overturned. It was reported that the driven piles were about 20 ft (6.1 m) shorter than piles used for adjacent structures.

Initially the settlement of the building was considered to be "natural." After several days of attempts to save the leaning building, the building was evacuated. Hours after the evacuation, the structure fell flat in just 20 seconds.

5.7.1 Lower East Side collapses

On April 4, 1984, two structures, located at Delancey Street in lower Manhattan, collapsed during construction (Fig. 5.48) about 1 hour after a 3½-in

Figure 5.48 General view of the structure immediately after the collapse. Rescue teams are looking for missing workers.

(89-mm) concrete floor slab had been poured. Two workers were killed by the collapse, including the general superintendent for construction.

On arrival on the scene a day later, I found that the debris was still being removed from the site. There was severe damage to the two adjoining structures, both of which were in precarious condition. One structure was subsequently demolished and the other was braced and shored.

Our investigation found several improper construction procedures, including inadequate field supervision. We also identified a number of design deficiencies.

On exposure after the collapse, several of the footings were found to have been cast at less than half (46 percent) the size required by the design drawings. Out of three interior footings, two footings had been cast eccentrically to the centerline location [one 7½ in (191 mm) and the other by as much as 11½ in (292 mm)]. The eccentrically positioned footings were visibly rotated and pitched from one end to the other by as much as 5 in (127 mm). The general layout of the building may be seen in Fig. 5.49.

The structural analysis indicated that the bearing capacity of these footings was considerably reduced. Using the actual loads computed to be present at the time of collapse, we estimated that even for the optimum possible bearing capacity of these footings, they would have been loaded at 25 percent over ultimate capacity, *and had to fail.* See Figs. 5.50 and 5.51 for footing diagrams and load table.

Another serious deficiency we uncovered related to the method of temporary support provided for the central bearing walls. These walls were subjected to a twisting rotational couple due to the weight of the two unequal bearing walls (Fig. 5.52). The fact that the second floor was missing (it had been previously been removed) compounded this effect by the lack of lateral support. The method of temporary support used by the contractor had a considerable inherent potential for instability. Thus, when the footings settled vertically, dislodging wedges and shims, the resulting movements initiated the total collapse.

One of the striking omissions in this failure was the lack of a construction procedure, which is essential to the successful execution of a difficult foundation reconstruction of an old structure.

The design deficiencies included main girders which were calculated to be stressed between 33 to 39 ksi (228 to 269 MPa) when subjected to full dead and live loads. Even the review of the design of the footings indicated overstresses in the range of 27 to 44 percent.

Needless to say, the lack of shop drawings did not help matters.

Lessons

1. Eccentric loading of footings should be avoided at all costs, unless provided for in the design by oversized footings.

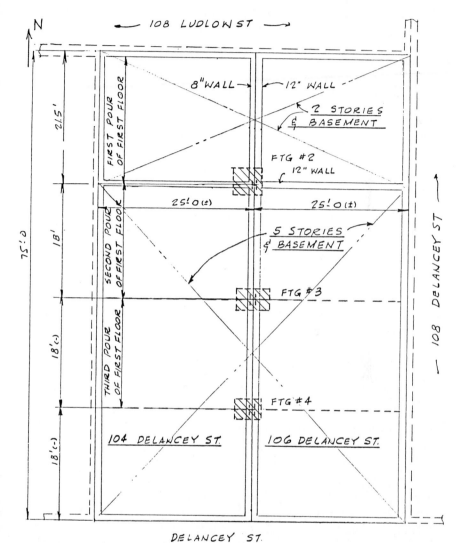

N

108 LUDLOW ST →

108 DELANCEY ST. →

8" WALL →   ← 12" WALL

2 STORIES
& BASEMENT

FTG #2
12" WALL

25'-0 (±)   25'-0 (±)

5 STORIES
& BASEMENT

FTG #3

FTG #4

104 DELANCEY ST.   106 DELANCEY ST.

21.5'

FIRST POUR OF FIRST FLOOR

18'

SECOND POUR OF FIRST FLOOR

18'(-)

THIRD POUR OF FIRST FLOOR

18'(-)

75'-0

DELANCEY ST.

Figure 5.49 General layout of the collapsed building, showing the three critical footings (hatched areas). 108 Ludlow was later demolished. 108 Delancey was underpinned and saved.

2. A construction procedure outlining step by step the method of temporary support, underpinning, shoring, and bracings is a must in construction of new foundations for old and deteriorated structures.

3. Lateral stability and eccentric effects on critical bearing walls must be considered and accounted for.

4. There is no substitute for good control of a project, including the preparation of shop drawings and thorough field inspection.

FOOTING SIZE AND AREA			
FOOTING NO.	AS DESIGNED	AS CONSTRUCTED	REMARKS
FOOTING #2	5'-6"X 8'-6"=46.75sf	4'-8"X4'-7"= 21.4 sf	FOOTING CAST ECCENTRIC TO SUPPORTED COLUMN
FOOTING #3 (TOP AREA)	5'-6"X7'-0" = 38.5sf	5'-9"X3'-6" = 20.1 sf	FOOTING CONCENTRIC WITH SUPPORTED COLUMN BUT
FOOTING #3 (BOTTOM AREA)	5'-6"X7'-0"= 38.5sf	3'-9"X 3'-6" = 13.1sf	SIDE SLOPES, RESULTING IN AN ECCENTRIC BEARING SURFACE
FOOTING #4	5'-6"X7'-0" = 38.5sf	4'-2"X 3'-4"=13.9sf	FOOTING CAST ECCENTRIC TO SUPPORTED COLUMN

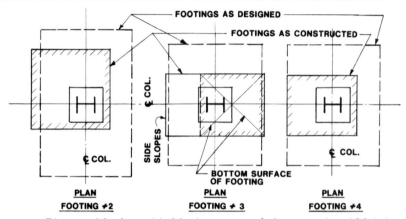

Figure 5.50   Diagram of the three critical footings, measured after excavation of debris from the collapse. The incredible violation of the design intent and standard construction practice is glaring.

5.7.2   Apple-juice factory — Drilled-in piers

When I first saw this project, which had come to a halt, both owners and contractors were aware of the serious problem they were facing. The exposed sides of several concrete piers could be seen to contain concrete contaminated with large quantities of brown mud (Fig. 5.53). Windsor probe tests had already been taken, and these confirmed the nonuniformity of the strength of the concrete. It was absolutely clear to everybody that expensive corrective work was required. The nature and extent of this work and the causes for the defects were sure to be in contention.

The foundation was being constructed to support an apple-juice plant located in Fleetwood, Pennsylvania. The foundation system consisted of concrete drilled-in piers (caissons), which were cast in place into big holes augered into the ground. These piers ranged in size from 30 to 42 in (0.8 to 1.1 m) in diameter, with bells 42 to 90 in (1.1 to 2.3 m) in diameter, and were 15 to 55 ft (5 to 17 m) long. They were designed to bear on bedrock with a bearing value of 12 tons/ft² (1149 kPa). Unfortunately, *all the piers had been completed* at the time the defects were discovered.

The design of drilled piers with bells in clayey or silty sand is feasible, provided the soil in which the bells are formed is cohesive. That condition

1*	2*	3†	4†	5	6	7	8
ALLOWABLE LOADS		DESIGN LOADS		DES.LD. ÷ ALLOW.LD.		DES.LD. ÷ ALLOW. LD.	
N.Y.C. BLDG. CODE	ONE-THIRD OF ULTIMATE	WITH 1-12" & 1-8" WALLS	1-12" WALL ONLY	3 ÷ 1	3 ÷ 2	4 ÷ 1	4 ÷ 2
FOOTING #2	128^K	67^K	336^K	315^K	2.63 (163% OVER)	5.03 (403% OVER)	2.46 (146% OVER)	4.70 (370% OVER)
FOOTING #3 (FULL AREA)	121^K	216^K	358^K	313^K	2.95 (195% OVER)	1.66 (66% OVER)	2.59 (159% OVER)	1.45 (45% OVER)
FOOTING #3 (BOTTOM AREA)	79^K	272^K	358^K	313^K	4.53 (353% OVER)	1.32 (32% OVER)	3.96 (296% OVER)	1.15 (15% OVER)
FOOTING #4	83^K	114^K	358^K	313^K	4.30 (330% OVER)	3.14 (214% OVER)	3.77 (277% OVER)	2.74 (174% OVER)

	9	10	11**	12	13
	ESTIMATED LOAD AT TIME OF COLLAPSE		ULTIMATE LOAD	COLLAPSE LD. ÷ ULTIMATE LD.	
	WITH 1-12" & 1-8" WALLS	WITH 1-12" WALL ONLY	(PER WOODWARD-CLYDE)	9 ÷ 11	10 ÷ 11
FOOTING #2	251^K	217^K	200^K	1.26 (26% OVER)	1.09 (9.0% OVER)
FOOTING #3 (FULL AREA)	210^K	142^K	648^K	0.32 (32%)	0.22 (22%)
FOOTING #3 (BOTTOM AREA)	210^K	142^K	167^K	1.26 (26% OVER)	0.85 (85%)
FOOTING #4	210^K	142^K	342^K	0.61 (61%)	0.42 (42%)

ALLOWABLE BEARING DETERMINED BY WOODWARD-CLYDE CONSULTANTS INC.
*BASED ON ACTUAL FOOTING SIZES; N.Y.C. BLDG. CODE ALLOWABLE, 3 TONS PER SQ.FT.; EFFECT OF ECCENTRICITY NOT INCLUDED.
**EFFECT OF ECCENTRICITY CONSIDERED.
†DESIGN LOADS USED BY OUR FIRM.

LIVE LOADS
1ST FLOOR: 100 PSF (RETAIL)
2ND FLOOR: 75 PSF (RETAIL)
3RD, 4TH, 5TH FLOORS: 100 PSF (LIGHT STORAGE)
ROOF: 30 PSF

Figure 5.51   Load table showing that the footings *as constructed* had to fail.

may be realized where the bells are above the groundwater level. Where the shafts are flooded, as the case often was at this construction site, the success of belling is highly questionable. Especially doubtful is the proper installation of the bells and maintaining them intact without the sides caving in and collapsing (Fig. 5.54). The failed caissons confirmed this opinion. The presence of groundwater at several of the caissons should have been expected from the 23 borings taken at the site prior to construction.

What was more stunning was the interference of the design engineer of record with the contractor's methods and operations. When it became clear that it was impossible to pump the holes dry, the engineer directed the subcontractor to use larger pumps. (The groundwater infiltration was greater than the capacities of the pumps by more than 1000 gal/min (3785 L/min). When the contractor did not comply, the engineer *bought large pumps with his own funds* and delivered them to the contractor.

Thus the engineer of record committed three serious errors:

A first major technical error. It should be common knowledge that it is

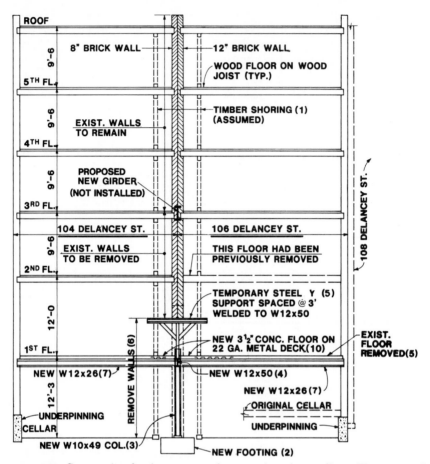

ROOF

8" BRICK WALL ——— ———— 12" BRICK WALL

9'-6
5TH FL.

WOOD FLOOR ON WOOD
JOIST (TYP.)

9'-6   EXIST. WALLS
        TO REMAIN
4TH FL.

TIMBER SHORING (1)
(ASSUMED)

9'-6   PROPOSED
        NEW GIRDER
        (NOT INSTALLED)
3RD FL.

104 DELANCEY ST.   106 DELANCEY ST.

9'-6   EXIST. WALLS   THIS FLOOR HAD BEEN
        TO BE REMOVED   PREVIOUSLY REMOVED
2ND FL.

108 DELANCEY ST.

TEMPORARY STEEL Y (5)
SUPPORT SPACED @ 3'
WELDED TO W12x50

12'-0

1ST FL.

REMOVE WALLS (6)

NEW 3½" CONC. FLOOR ON
22 GA. METAL DECK (10)

EXIST.
FLOOR
REMOVED (5)

NEW W12x26 (7)

NEW W12x50 (4)

NEW W12x26 (7)

12'-3

ORIGINAL CELLAR

UNDERPINNING
CELLAR

UNDERPINNING

NEW W10x49 COL. (3)

NEW FOOTING (2)

Figure 5.52 Cross section showing sequence of construction prior to collapse. The center wall was subjected to twisting moments as a result of the missing second floor on one side. Numbers shown thus (1) describe contractor's sequence of operations.

improper to cast concrete into holes while pumping at the same time. Such pumping only serves to disturb the wet concrete and mix it with soil, resulting in contaminated and defective concrete as in Fig. 5.54 (a properly filled caisson is shown in Fig. 5.55).

A second technical error. There was no way for the engineer's inspector who was lowered into the shaft to verify in the flooded shafts the adequacy of the bearing rock and whether the rock was level or sloped (Fig. 5.56).

A policy error. An engineer should never interfere with the contractor's methods and means, and definitely should not supply the contractor with the tools to perform the engineer's instructions.

Figure 5.53   Several drilled concrete piers were excavated and, to everybody's surprise, exhibited brown sections of mud-contaminated concrete.

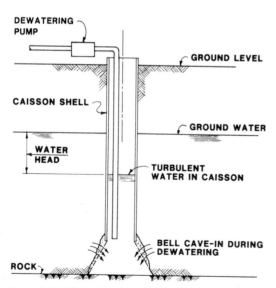

Figure 5.54   Water head created during pumping inside caisson shell created turbulence, causing cave-in of the caisson bell.

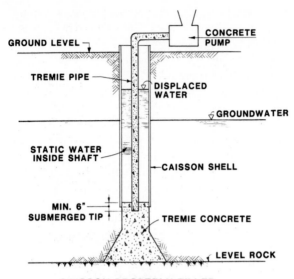

CAISSON PROPERLY FILLED

Figure 5.55 Keeping static water inside the caisson shell while pumping tremie concrete with a submerged tip is the best way to achieve quality concrete within the caisson.

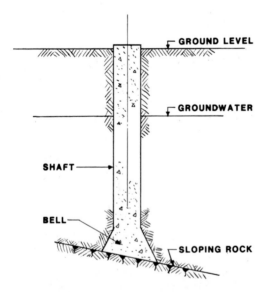

IMPROPERLY SEATED CAISSON

Figure 5.56 The water-flooded shafts were not inspected visually, which could be done only by divers. The resulting construction on sloping rock is a serious structural defect.

Moreover, the subcontractor for the drilled piers should not have knowingly followed the erroneous instructions, or should have known as a supposed expert in the field that they were flawed.

An experienced caisson contractor would also have known to continuously keep the bottom of the tremie (underwater concrete) pipe inserted into the fresh concrete (Fig. 5.55). Improper tremie placement is shown in Fig. 5.57.

While the selected foundation system for this structure was totally inappropriate (a system of H piles or even a continuous concrete mat should have been used), the construction methods failed to produce a reliable foundation. Not only was the concrete defective in many instances, but even the quality and levelness of the rock bearing strata could not be confirmed.

The sad ending to this story was that the foundation system as constructed had to be completely abandoned and lengthy litigation ensued.

Lessons

1. Designers should carefully check the feasibility of excavating bells in soil prior to concrete casting, especially in water.

2. Tremie placing of concrete should always be in still (static) water. Pumping water out of the shell during concrete placement operations should not be attempted.

3. The installation of drilled concrete piers should be performed only by experienced contractors.

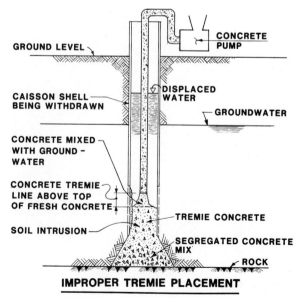

IMPROPER TREMIE PLACEMENT

Figure 5.57 Dropping tremie concrete is a sure way to obtain defective concrete.

4. Full-depth concrete test cores should be extracted from the *first* five piers, to verify the quality and strength of the concrete. The *early detection* of deficiencies is essential.

5. Engineers should not direct contractor field operations, should not supply them with tools to perform the work, and should not interfere with their methods and means unless emergency safety is involved.

5.7.3 The North River project — Caisson deficiencies

Manhattan's enormous North River water pollution control project* was constructed on a 30-acre (121,400-m²) concrete platform supported on 2500 drilled caissons (Fig. 5.58). The caissons extend to the Hudson River bottom and are socketed into the rock underlying the lower sand layers. They are from 80 to 250 ft (24 to 76 m) long and are constructed of steel pipe shells in diameters varying from 36 to 42 in (0.9 to 1.1 m). For economic reasons, the caissons were designed to carry very heavy loads of up to a maximum of 5000 tons (45 MN) each.

During the construction of the platform, two major problems arose:

---

* Dov Kaminetzky, "Foundation Problems in New York Metropolitan Area — North River Project," American Society of Civil Engineers, Met Section, New York, N.Y., November 1987.

Figure 5.58 Aerial view of the 2500 installed caissons in the Hudson River. Manhattan's upper west side and Henry Hudson Parkway are in upper right of photograph.

1. Defective concrete was found to have been cast in the caisson shells.

2. Caissons shifted laterally at the top and bent.

These problems came to light after a substantial number of the caissons had been completed.

A full-scale study was launched, and it was determined that the problem of the *defective concrete* was caused by improper tremie concrete placement. It became clear early during the installation of the caissons that about one-quarter of the caissons had low-strength concrete and discontinuities of several varieties: uncemented aggregates, unhydrated cement, and sand layers intermixed with the concrete.

After being driven into the river, the steel shells were seated in the rock. Sockets were then drilled in the rock, and the shells were emptied of river mud and sand. The contractor made several attempts at casting the caissons in the dry after sealing the bottom and pumping the water out of the shell. When these attempts failed, a tremie method (casting concrete in water) was used. This method, which unfortunately did not utilize pressure pumping of concrete, also failed, a fact that was brought to light only later, when the accidental lateral shift of the caissons caused concern, necessitating full-scale in-place load tests. The first of the load tests (described later) resulted in failure of the tested caisson, which sank vertically, a failure which confirmed the concrete defects within the caissons.

The concrete subsequently was repaired by new high-strength [75 ksi (517 MPa)] rebar bundles installed in vertical holes drilled in the caissons. These holes were then pressure-grouted with high-strength grout.

The *shifting of the tops* was determined to be the result of one-sided soil dredging of a deep trench. The creation of the trench caused unbalanced lateral soil pressure on the caissons, resulting in horizontal movement and excessive bowing. This horizontal movement of the tops exceeded 2 ft (0.61 m) in many cases. The caissons were later forced back into position by a pulling program involving the application of high lateral loads coupled with jetting of the caissons.

It was only after the lateral shift of several hundred caissons that the concrete problem was discovered, when it was decided to core-drill a few of the caissons. The coring revealed many deficiencies within the caissons, where some "concrete" near the bottom of the caissons was nothing but sand (Fig. 5.59). Before any remedial measures were considered and proposed, it was decided to load-test one of the affected caissons. In order to determine the load capacity of the shifted and bent caissons, a full-scale load test was ordered. The tested caisson was 36 in (0.9 m) in diameter, 98 ft (30 m) long, socketed 5 ft (1.5 m) into rock, with a design capacity of 680 tons (6.1 MN). The top of this caisson had been displaced laterally as much as 2.35 ft (0.71 m) after it was installed.

The vertical load was applied to the test caisson by means of four hydraulic jacks; the jacks reacted against a loading frame held in place by four

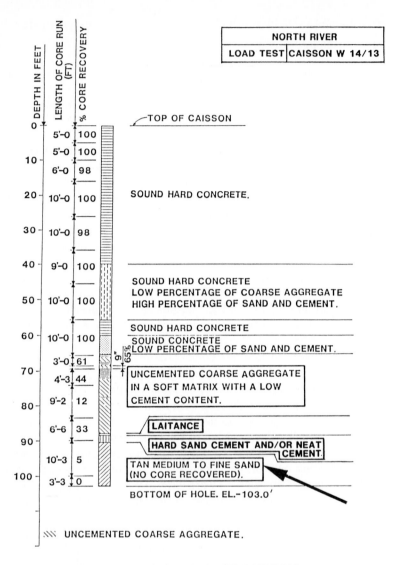

CORING LOG – LOAD TEST CAISSON

Figure 5.59 Investigatory core drilled 100 ft deep (30 m) into the caisson revealed defective concrete, uncemented aggregate, and fine sand instead of concrete at bottom of caisson (arrow).

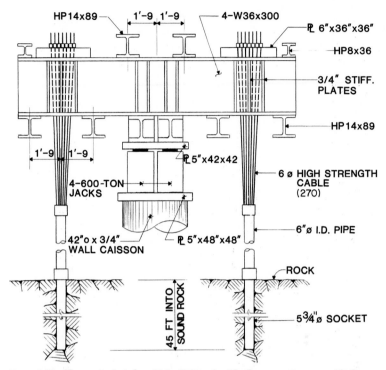

Figure 5.60 Heavy steel girders 36-in (0.91-m)-wide-flange sections provided reaction to four jacks test-loading the questioned caisson.

vertical cable anchorages grouted 45 ft (13.7 m) into sound rock (Fig. 5.60). Four 0.001-in (0.025-mm) dial extensometer gages were mounted on an independent common frame to measure vertical displacements at the top of the caisson. The top of the test caisson was braced to adjacent caissons to prevent lateral translation during the test. The braces were pin-connected to eliminate undesired transferral of the vertical load through the bracing system. The load was applied in two phases.

During the first phase, a load of 850 tons (7.5 MN) was applied. The load was cycled in small increments. All loads were maintained until the vertical displacements were stabilized. Only the 765- and 850-ton (6.8- and 7.5-MN) loads were maintained for 15 and 18 h, respectively. Later the load was *cycled 5 times*. The vertical displacement under a load of 850 tons (7.5 MN) (125 percent of the design load) after 17 h was 2.6 in (66 mm) with a permanent set of 2.2 in (56 mm).

During the second phase, the load was increased and cycled further in increments to a maximum of 1360 tons (12 MN). This maximum load was held for 60 h and then recycled again 10 times. To everybody's shock and disbelief, the caisson failed to support the load and sank. An additional deflection of 3.6 in (92 mm) was measured for a total vertical displacement of 5.8 in (147 mm). The total permanent set was 4.5 in (114 mm) (Fig. 5.61).

The excessive vertical displacement and permanent set were determined to have been caused by either one or a combination of two factors:

1. The existence of uncemented zones within the lower 40 ft (12.2 m) of the caisson resulted in the transfer of the entire caisson load to the steel shell, which was subjected to a stress of 50 ksi (345 MPa).

2. Lacking sound concrete in the socket itself (there was actually uncemented sand in the socket), the cutting edge of the shell was punched further into the socket.

It was also evident that it was not possible to determine whether the lateral shift of the caisson was, in fact, a significant factor causing the vertical displacement of the caisson and its eventual failure. It was for this reason that an additional caisson was tested two months later. This later test proved that the lateral displacement of 2.5 ft (0.76 m) at the top of the caisson had no discernible effect on the elastic behavior and load-carrying capacity of the structurally sound portion of the caisson. Therefore, all caissons with displacements of less than 2.5 ft (0.76 m) were incorporated into the structure. The unbalanced lateral loads resulting from the lateral shift were subsequently balanced by batter caissons.

A second defective caisson was then tested similarly. This caisson was also cored with an NX-size double-tube-core barrel prior to load testing. What we found was that the upper 80 ft (25 m) contained generally sound concrete, unlike the first tested caisson. To measure the vertical displacement at the bottom of the caisson, measuring devices were inserted into 1-in-diameter (25-mm) pipes installed in holes drilled in the caisson. The devices were ½-in-diameter (13-mm) smooth steel rod telltales lowered into the pipes. This caisson held the load for the required 60 h, with a total displacement of only 0.35 in (9 mm). This led us to conclude that the caissons were repairable and could be incorporated into the structure without any reduction of their load capacity. We also determined that the presence of the powdery cement layer was apparently the result of the concreting procedures employed, whereby up to three bags of cement were deposited to absorb the water which was not removed by pumping out of the shell.

Remedial work.   The caissons were repaired by bridging the discontinuities with grouted rebars. Two 10-in (254-mm) holes were drilled within the concrete for the full depth of the caissons, including the socket, into rock. After each hole was cleaned of loose concrete, rebar bundles, consisting of 12 no. 11 high-strength [75-ksi (517-MPa) yield] bars tied together around pipe spacers, were lowered into the holes and grouted (Figs. 5.62 and 5.63).

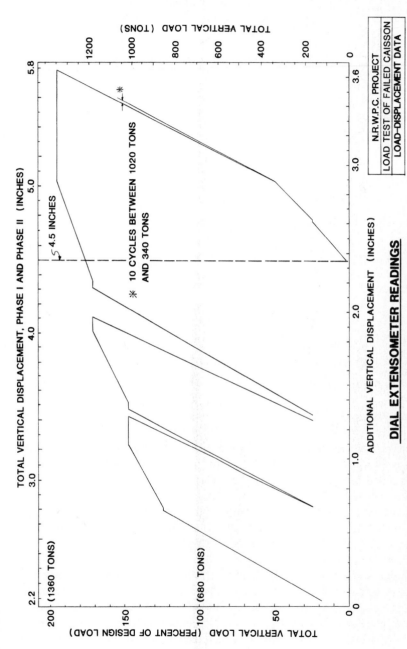

Figure 5.61 Load-displacement chart shows that load which was cycled 10 times ended with total vertical displacement of 4.5 in (114 mm).

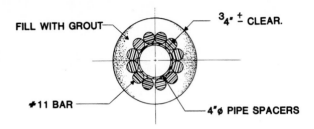

FILL WITH GROUT

$3_4"$ $\pm$ CLEAR.

#11 BAR

4"φ PIPE SPACERS

10" φ HOLE WITH 12-#11 REBARS

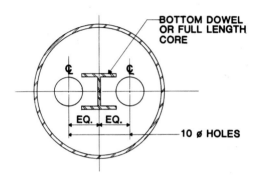

BOTTOM DOWEL
OR FULL LENGTH
CORE

₵   ₵

EQ. EQ.

10 φ HOLES

PLAN OF CAISSON

Figure 5.62   Caissons were repaired by using two 10-in (254-mm)-diameter holes drilled into concrete. The holes were then filled with grouted bundles of heavy bar reinforcement.

Lateral shifting of top of caissons.   As soon as it was discovered that a large group of caissons had shifted laterally after installation, an immediate investigation was launched. Survey teams using laser-based instruments took readings of 180 individual caissons to establish their locations accurately. The tops of the caissons were surveyed for a year. (See typical movement chart in Fig. 5.64.) Lateral displacements of as much as 2.5 ft (0.76 m) were measured. Temperature variations were found (by analytical computation) to account for only a small portion of the movement. Examination of the horizontal steel 8HP36 staylath members (Fig. 5.65) and their connections showed them to be in good condition and without any signs of distress. Analytical computations established that very high loads indeed would be needed to cause such a shift. It was also determined that a one-sided soil pressure is capable of generating such high loads.

The actual behavior of the group of caissons, which were tied together by means of steel bracings, followed the pattern of lateral shift resulting from an unbalanced lateral load. It was soon discovered that such unbalance did, in fact, occur as a result of a sizable dredging operation performed to remove existing boulders which were obstructing caisson driving opera-

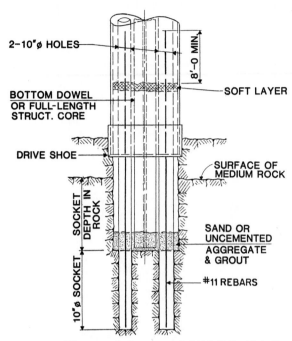

2-10"⌀ HOLES

8'-0 MIN.

BOTTOM DOWEL
OR FULL-LENGTH
STRUCT. CORE

SOFT LAYER

DRIVE SHOE

SURFACE OF
MEDIUM ROCK

SOCKET
DEPTH IN
ROCK

SAND OR
UNCEMENTED
AGGREGATE
& GROUT

10"⌀ SOCKET

#11 REBARS

Figure 5.63   The grouted rebars extended 12 ft (3.7 m) into the rock.

tions. This determination was quite conclusive, since it was established that, while most caissons moved in the westerly direction (toward the dredged trench, which was excavated parallel to the shoreline), there were several caissons located west of the trench which, indeed, moved toward the trench in the easterly direction.

At this time we had to determine whether the movements in the southeast section of the foundation system were continuing and/or could be expected to continue after the completion of the concrete deck supported by the caissons. It was also required to assess the effect of such possible movements on the superstructure.

To answer these questions, we released a number of caissons from their staylath bracings and installed inclinometers in drilled holes within the caissons. The inclinometers, which are devices that measure *tilt* or the angular deviation from the vertical, were the slope-indicator type.

The results of the surveys (Fig. 5.66) indicated that, except for the initial adjustment of the casing after installation, no movement was occurring. It was, therefore, concluded that the westward movements of the caissons and/or soil in the southeast area of the project had ceased.

The main concern at this juncture was to establish whether the struc-

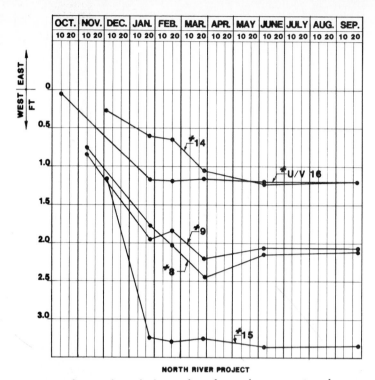

Figure 5.64 Survey of top of caissons showed westerly movements, as large as 2.5 ft (0.76 m), until March. No significant movement was recorded afterward, indicating stabilization.

Figure 5.65 The tops of the caissons were braced with 8-in (0.20-m) steel staylath members in all directions.

472

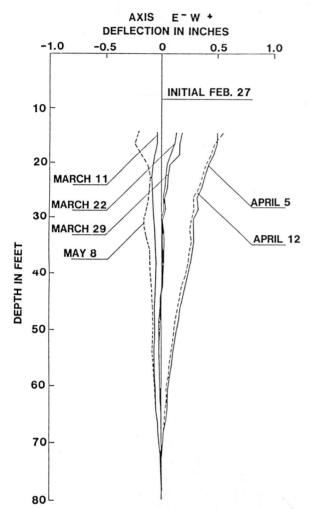

AXIS   E ⁻ W ⁺
DEFLECTION IN INCHES

Figure 5.66 Inclinometer reading also showed no significant
movement of caissons after March.

tural integrity of the caissons was preserved. The stress analysis that
followed did confirm this fact, and we then looked for a method of moving
the displaced caissons back to their original vertical positions.

Remedial work. Pulling procedures, using "come-alongs" to move the
caissons laterally, were developed. The loads (measured by dynamome-
ters) were applied in small increments, so as not to damage the caissons.
Jetting of the river bottom was utilized to assist in pulling the caissons to
the desired locations.

It should be pointed out that the enormous construction difficulties encountered in this project were effectively resolved by the excellent cooperation and combined efforts of the owner and both the design and the construction teams.

Lessons

1. Tremie casting of concrete into long caissons should be performed under pressure by the use of pumps.

2. The bottom of the tremie pipe must at all times be embedded into the fresh concrete.

3. The quality and strength of concrete in caissons, and the reliability of the tremie concrete procedures, *must be confirmed by core drilling at the beginning of the operation.*

4. Cutting deep trenches adjacent to caissons or similar pile foundations will result in a lateral shift and should be avoided.

5. Good cooperation among the various parties involved in a construction failure is essential to a speedy, successful, and litigation-free resolution.

Figure 5.67 A close inspection of the brickwork coursing showed that the initial cracking occurred during construction and prior to masonry brick placement. See additional brick course left of crack (arrow).

5.7.4 Marble Hill's Missing Pile

When cracks in the brick wall of this one-story structure started to widen, it became apparent that a serious condition existed below the concrete grade beam supporting the wall. After the soil adjacent to the grade beam was removed by excavation, a long diagonal crack was uncovered in the beam.

The crack previously had been patched crudely with mortar, which itself was badly cracked and served no purpose whatsoever. There had been no attempt to find the culprit!

The failure was a classical shear crack with a vertical shift on both sides of the crack. A close look at this brick wall told the story of a settlement that had taken place *prior* to construction of the brick wall. There was an additional course of brick on the left side of the crack (Fig. 5.67). The solution to this mystery was found when, after further excavation, it was discovered that a pile designed to support the grade beam was missing. It was clearly a construction error. The attempt to patch the crack obviously was not adequate to correct the serious omission.

Lesson. When cracking occurs, determine the cause of the distress. Do not guess! Only when the true causes are found can an efficient, long-term correction be designed and executed.

5.8 Flotation and Water-Level Change

Except in well-consolidated granular soils, a change in water content will modify the dimensions and structure of the supporting soil, whether from flooding or from dewatering. Many cases are recorded where pumping by occupants for cooling water, or by water companies to increase the capacity of their water supply, results in receding of ground levels which, in turn, causes settlements with severe damage.

Pumping from adjacent construction excavations has also affected the stability of existing spread footings and even caused drag-down on short piles. This fact is in part responsible for the insertion of the requirement of recharging of groundwater levels in some foundation contracts. The effectiveness of these recharging pools has been and continues to be a touchy subject among foundation designers and geotechnical engineers.

Construction of new dams has also been found to be responsible for lowering of river levels, thereby causing severe damage and cracking of adjacent structures.

Clay heaves from oversaturation must also be expected, and, in such soils, the structures must either be designed to tolerate upward displacement or else the supporting soil must be protected against flooding. Many parts of the world have trouble from heaving bentonitic clays, for which a perfect solution is yet to be found. Where water infiltration into the soil is effectively prevented, and there is no change in water level, the soil volume should stay stable.

5.8.1 Brooklyn hospital — Settlement of delivery room

This three-story hospital had performed perfectly for many years after its construction. Suddenly, masonry walls started to crack, for no apparent reason (Fig. 5.68). The most severely affected wing of the building contained both the newborn delivery room and the general surgery room, whose continuous operation could not be stopped. I was actually measuring cracks inside the delivery room (dressed in scrubs) while a newborn was delivered. It was then that I found it necessary to temporarily support this wing, accomplished by using a diagonal timber bracing system (Fig. 5.69).

Once the structure was secured, we started the investigation of the sudden cracking and distress. To our amazement we found that the water supply company operating in the area had recently abandoned a great number of wells. As a result, the groundwater level had risen sharply in the

Figure 5.68 Wide crack suddenly and inexplicably developed at hospital's corner, directly below the surgery room and infant delivery room.

Figure 5.69 Heavy diagonal timber braces were installed against vertical steel strongbacks and concrete pads cast in the ground to prevent additional settlements. These bracings were added as a precaution.

entire area. The immediate result was water penetration through foundation walls of the hospital, which had not been waterproofed to such a high level. To solve the new problem of water infiltration into basement areas, pumps had then been installed by the hospital management. However, an unforeseen side effect was that the continuous pumping removed most of the fines in the soil directly below the footings supporting the affected wing, thereby causing the settlements and cracks.

There was no easy solution short of underpinning the building. This was done with jackpiles, which were pushed down into the ground, with the weight of the building serving as a counterweight (Fig. 5.70). The steel pipes, which had been pretested for the required load, were then filled with concrete and the jacks removed.

Lessons

1. Lowering or raising of groundwater level affects the bearing capacity of soils.

2. Pumping for new excavations may cause settlements in buildings. Therefore, water level readings should be monitored and protective measures taken.

Figure 5.70   Hydraulic jacks were installed and used to push new pipe piles against stiffened steel beams, picking up the building corner. The weight of the building served as counterweight.

5.8.2   Industrial building, Hackensack, New Jersey

This one-story factory and office building was located in the bottom of a geological lake, atop deep glacial deposits of silts and clays in successive layers, forming what is known technically as *varved silt deposit*. In this very loose material, areas of soft shoreland become marshland. Considerable areas of such marsh land have been filled in to form usable development property. The existence of the soft underlying material should signal potential difficulties and must be taken into account in all construction operations. This type of soil is very susceptible to drainage and is associated with a considerable volume change.

The local sewer authority had issued a contract for the construction of a new sewer. A deep trench was excavated using steel "soldier beams," 10 ft (3.0 m) on center, and timber sheeting. The excavation was braced at two levels. As the excavation approached the factory, various signs of damage started to appear in the form of cave-ins and vertical and diagonal cracking of walls and partitions (Figs. 5.71 and 5.72).

As we reviewed the damage, it was evident that the undermining was caused by the drainage operations in the sewer trench causing a shrinkage of the subsoil under the building's foundation. This was followed by the lateral movement of soil from under the building toward the deep hole at the sewer trench. The nearest edge of the sewer excavation was approxi-

Figure 5.71 Excavation for a new sewer and its associated drainage operations caused shrinkage of the soil supporting the foundation of this warehouse. Separation crack extended the total height of the wall.

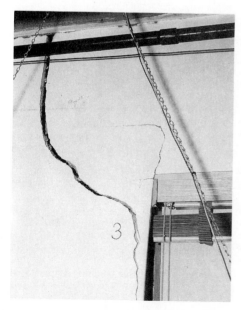

Figure 5.72 Diagonal cracks developed in interior partitions.

mately 220 ft (67 m) from the building. Yet the effect of the drainage of the sewer excavation extended to great distances on each side of the sewer, and soil settlement was even visible in the parking area as well as along the walls of the building itself.

Fortunately, the main structure of the building was supported on piles which had been carried into materials deep enough and dense enough not to be affected by this drainage. There was, however, still the effect of negative friction on the piles, with the additional soil weight adding to the loads which were already present.

Remedial work. Once the pumping operations ceased, no additional damage was recorded. Slabs which had settled by as much as 3 to 4 in (76 to 102 mm) had to be removed and rebuilt.

Slabs which settled to a lesser degree, leaving a hollow space between the underside of the slab and the ground, were restored by pressure grouting. Wall cracks were repaired with ordinary cement mortar and refinished.

Lessons
1. The groundwater level outside a deep excavation should be monitored with piezometer pipes installed at the adjacent structures which may be

Figure 5.73 Severe cracks developed in the exterior brick of this building, supported on timber piles. Lowering of water level by pumping at adjacent railroad contributed to pile settlements.

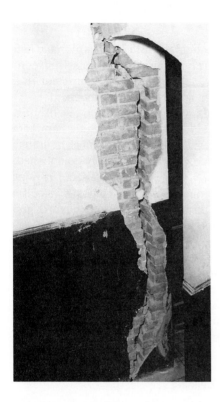

Figure 5.74 The cracks developed fully through the wall thickness to the building interior.

affected. Criteria for the maximum permissible water-level drop should be set.

2. In the event of a drop exceeding the criteria and/or the initiation of damage, the pumping methods and procedures should be reviewed to minimize the damage.

3. The services of competent geotechnical and structural engineers should be solicited.

5.8.3   West 41st Street structure — Wood piles

Damage to the exterior brick masonry of this building was caused by settlement of timber piles resulting from lowering of groundwater level at an adjacent railroad construction site. Probes below the building showed that the timber piles were rotted and could not accommodate the additional loads generated by the drop of the water level. The damage to exterior and interior masonry is shown in Figs. 5.73 and 5.74.

Lessons.   See lessons from the previous case.

5.9   Vibration Effects

Earth masses which are not fully consolidated will change volume when exposed to vibration impulses. The vibration source can be blasting; construction equipment, especially pile drivers; mechanical equipment in a completed building; or even traffic on rough or potholed pavement.

We have investigated many cases of damage resulting from pile driving. Heavy damage can be caused, in particular, when either impact hammers or vibratory hammers are used. In several cases, entire rows of buildings have had to be condemned as unsafe and were subsequently demolished.

Blasting operations must be carefully programmed, to avoid serious damage to adjacent structures. A criterion for maximum particle velocity, which is a measure of the intensity of vibration, has to be established, and constant readings by seismographs have to be taken. Damage from blasting is often potentially so severe that the size of explosive charges has to be limited to keep the vibrations to tolerable levels. Factors such as the condition, rigidity, material brittleness, and age of nearby structures must be taken into consideration. Where adjacent structures are founded on rock, the seam structure of the rock itself also must be taken into account.

A complete study of vibration transmission and correlation between intensity, wavelength, and possible damage to various types of structures is given in Ref. 4.

5.9.1   Lower Manhattan Federal Office Building

After heavy impact hammers were used to install the piles for this prestigious building, the adjacent structures were seriously shaken. Street pave-

Figure 5.75 Street pavements settled (arrow) as much as 16 in (406 mm) as a result of pile driving in the adjacent construction site.

ments settled (Fig. 5.75) as much as 16 in (406 mm) after only one-fourth the piles had been driven. The installation of steel-sheet piling was then attempted in an effort to protect the streets against further subsidence. This attempt was successful in preventing additional settlement and damage to sewer and water lines.

Damage became even heavier after the contractor shifted to vibratory hammers. High-rise buildings as tall as 19 stories were also damaged. The damage was so severe that eventually all buildings within a radius of 400 ft (122 m) were condemned as unsafe and had to be demolished.

In similar construction some 1000 ft (305 m) away, when first signs of trouble became evident, the adjacent buildings were underpinned on piles down to the tip level of the new building piles. From then on, the consolidation of sand layers as a result of pile driving had no ill effect, since sand layers were no longer supporting building loads.

Lessons

1. The use of vibratory pile-driving equipment must be avoided where possible. It may be used only after careful evaluation of potential damage.

2. Underpinning of adjacent structures must be seriously considered for support, so as to minimize damage from pile-driving operations.

3. Where possible, the use of augered piles should be considered.

5.10 Earthquake Effects

Foundations in earthquake-affected zones must be designed to tolerate the expected shock spectra provided by nature. In the past two decades there have been more than 20 major earthquakes resulting in severe damage. These were widely scattered from Mexico, Alaska, and California to Japan, New Zealand, and the Soviet Union. The effect on foundations is usually less severe than on the superstructures, especially in earthquakes of short duration.

Where earthquake shocks continue for long periods, the subsoil structure is substantially altered and a great deal of additional damage is caused as a result of the failure of the foundations. In the Anchorage, Alaska, earthquake of 1964, many houses in residential areas were rendered useless by the large consolidation and lateral flow of the underlying clays. These sensitive clays underlying the sand-gravel beds lowered buildings as much as 14 ft (4.2 m) vertically, causing total failure of the structures.

The Anchorage Cinema, however, settled 14 ft (4.2 m) as one rigid body, so that the canopy over the entrance ended up at sidewalk level (Fig. 5.76). Remarkably, there was no structural damage, not even a broken window pane. With no possible entry, however, the building was no longer usable, and it was later demolished.

Figure 5.76 After earthquake in Anchorage, Alaska, movie theater settled 14 ft (4.2 m), remarkably as one rigid body, with canopy ending up at sidewalk level.

Some vibration effects associated with earthquakes are sufficient to liquefy saturated fine sands and cause lack of support. Some of these soils require more than a single shock; at least 1 to 2 minutes' shaking is required. The need for total separation of the foundation at expansion joints (two separate footings rather than one combined footing for twin columns) is still being studied.

5.11 Conclusion

It should be understood that most structures have some tolerance to unequal settlement, but when the support is stressed beyond the elastic limit, ultimate failure is unavoidable without the immediate strengthening of the foundation. Statistically, the number of completed structures with foundation failures is small. However, foundation failures have the unique nature of often affecting entire buildings and seriously impairing adjacent structures, generating a succession of lengthy and expensive litigations.

References

1. Robert Frost, *North of Boston,* Holt, New York, 1979.
2. "The Building Code of the City of New York," Sec (26-1903. 1(b)(1), Department of General Services, New York, September 1, 1988.
3. A. Koral, "When the Earth Opens and Walls Move," *The New York Times,* July 3, 1977.
4. F. J. Crandell, "Ground Vibration Due to Blasting," Liberty Mutual Insurance Co.

6

Corrosion

It is well-recognized that the great majority of failures occur during construction and within the first 2 years of the life of the structure. Deterioration of structures, on the other hand, occurs during their later service life and involves a continuous process, contributing to loss of strength and structural integrity. Deterioration by corrosion of metals is a major factor contributing to failures of structural-steel, reinforced-concrete, and pre-stressed-concrete structures. Unprotected steel exposed to the environment oxidizes with loss of a portion of the base metal (Fig. 6.1), which turns into iron oxide. When steel is used in concrete elements as tensile reinforcement, the loss of even localized cross-sectional area, resulting in a corresponding increase in stress level, may have catastrophic results on the overall strength and survival of the corroded member.

The first symptoms of steel corrosion are the so-called rust stains which appear at the surface. We often see these stains where iron oxide in solution migrates to the surface of building materials such as concrete, stone, brick, and stucco. Once noticed, the alarm is sounded. Where the degradation process goes unnoticed for a long period, the damage is often irreversible, and a decision has to be made whether to repair the member or to replace it.

However, when corrodible metals are protected from the environment — and are not in contact with water, air, and chemicals — they will last for a long time.

Because failure from corrosion is often a very complicated process, there are many and varied methods of protection. Regardless of whether the protection is given by a layer of impervious coating or by cathodic protection, the continuous maintenance of the structure is an absolute requirement for success. There also must be a clear understanding of the corrosion process itself before design safeguards and/or a method of protection can

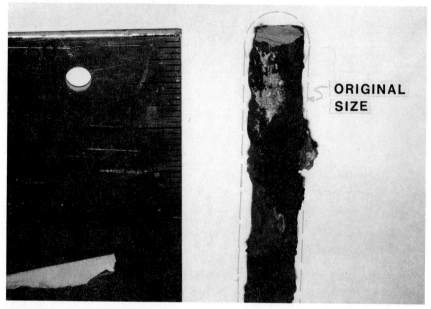

ORIGINAL SIZE

Figure 6.1 Steel rebars, when exposed to air, oxidize—losing cross-sectional area and strength.

be selected. Familiarization with stress and strain, material properties, and the effects of chemicals and temperature are necessary for an understanding of the corrosion mechanism.

Where a structure performs properly over time under constant load and environmental conditions, it can be expected to continue to do so in future service. Failure may occur when loads increase dramatically or where there is a significant reduction in strength. The steady corrosion of metal elements may be the cause of such reduction in strength.

Iron, the principal element in steel, in the presence of oxygen and moisture, tends to revert to its stable state, iron oxide (or rust). For steel to corrode, in any environment, positive iron ions must leave the surface of the steel. In air, these iron ions then combine with negative oxygen ions. Since these iron ions are, in fact, positively charged particles, their ability to leave the parent metal requires that the material surrounding the steel be a conductor. Water containing dissolved salt is a far better conductor than water alone. This electrolytic solution, or electrolyte, exists in salt-contaminated concrete, greatly enhancing the removal of iron ions—and ultimate corrosion of the steel.

At high pH, concrete provides such a favorable environment for steel embedments that the tendency of steel to corrode is negligible; the steel is "passive" (Fig. 6.2).

Cured concrete contains large amounts of lime which render the con-

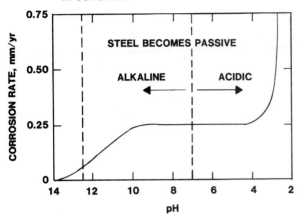

Figure 6.2 Relationship of corrosion rate to pH values. Note that, at high pH, steel becomes passive. (*G. Litvan.*)

crete strongly alkaline, with a pH generally between 12 and 13. Thus, *steel fully embedded in concrete normally does not rust.* However, if the lime in the concrete is leached out or neutralized by carbonation (the transformation of the lime into calcium carbonate through reaction with carbon dioxide), the passive state of the steel is terminated and corrosion can occur.

Nevertheless, steel corrosion will take place even in highly alkaline media if chloride ions are present. The exact mechanism through which chloride ions promote corrosion and break down the passivity of the steel is not fully understood. It is known, however, that chlorides are not consumed in the process; they act as catalysts and their harmful effect, once begun, does not diminish with time.

Research shows that a concentration of 1 to 2 lb chloride ion per cubic yard (5.8 to 11.6 N/m³) of concrete [approximately 250 to 500 parts per million (ppm)] is all that is required to initiate this destructive process. There is disagreement among corrosion researchers as to the limit (referred to as *chloride corrosion threshold*) which should be set for chloride content. There are three ways to quantify chloride content: (1) total, (2) acid-soluble, and (3) water-soluble. The most common method in use is the acid-soluble. Often chloride content, which is the amount of chloride ions present in concrete, is expressed as percent by weight of cement.

The latest ACI-318 code (1989, Article 4.3.1)[1] allows a maximum water-soluble chloride content of 0.06 percent for prestressed concrete, 0.15 percent for reinforced concrete exposed to chloride in service, 1.00 percent for reinforced concrete protected from moisture in service, and 0.30 percent for all other concrete. ACI-222R-85[2] recommends a maximum chloride content of 0.08 percent for prestressed concrete and 0.20 percent for reinforced concrete.

An additional and very important effect of chlorides is that their presence in *varying* amounts within the concrete greatly increases the electrical conductivity of the concrete, thus enhancing additional corrosion.

Another factor which must not be overlooked is the water/cement ratio of the concrete. The lowest possible water/cement ratio should be used, considering the practicality of casting. For example, concrete with a water/cement ratio of 0.4 has been found to better resist deicing-salt penetration than concretes with water/cement ratios of 0.5 or 0.6. And last, but not least, is the concrete cover which protects rebars from the exterior environment. Good, dense cover is extremely important in resisting rebar corrosion in concrete. Even in severe environments, adequate cover means a much longer time for the chlorides to reach and attack the rebars.

Uncoated steel in concrete gradually develops a thin film of oxide. This film of initial rust serves as a protective barrier between the steel and oxygen-bearing moisture that penetrates through cracks or pores in the concrete. As long as this barrier remains intact, further corrosion of the steel is retarded. This stability is upset, however, by the presence of negative chloride ions from deicing salts. Either sodium chloride or calcium chloride deicing agents ionize in water and release chloride ions, which lower the pH of the concrete and eventually cause destruction of the oxide film on the steel, allowing corrosion to proceed.

Interestingly, the chloride is not used up, but instead increases its effect as its concentration increases.

The iron oxide which is produced as a result of the exposure of the steel to oxygen-bearing moisture has a much larger volume than the parent metal. Therefore, as a steel reinforcing bar rusts, its volume greatly increases, producing internal stresses in the concrete. In a reinforced-concrete slab where the top reinforcing bars are in layers just below the surface, for example, this pressure first splits the concrete above the rebars away from the concrete below the bars — toward the path of less resistance. Delamination ensues, which initially may or may not be visible from the surface. It is just a matter of time, however. The rebars closest to the surface will be affected first. At this stage, the concrete slab ceases to be a coherent structure, and the bond between the concrete and the steel rebars (which is critical for any reinforced-concrete structure) is destroyed. (See Fig. 6.3 for the four stages of the corrosion process of a reinforcing bar.)

Continued internal pressure will eventually crack the concrete through to the surface, creating a spall, which is a complete separation. The mechanical action of the traffic pounding from above and the uplift caused by freezing and expanding water from below will soon carry away the spalled concrete, forming the all-too-familiar pothole.

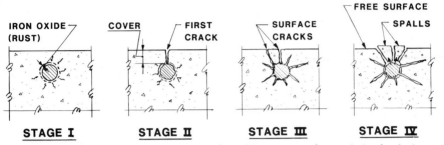

STAGE I   **STAGE II**   **STAGE III**   **STAGE IV**

Figure 6.3   Mechanism of rebar corrosion and resulting concrete damage in its developing stages.

Other types of corrosion modes include galvanic corrosion and stress corrosion, the latter process being less common.

Galvanic corrosion is an electrochemical process that can be best described by a model of a galvanic or voltaic cell (Fig. 6.4). The cell (battery) consists of a negative electrode or anode (a readily oxidizable substance, such as iron), a positive electrode or cathode (an oxidizing substance or receiver of electrons, such as copper), and an electrolyte, or ionically conducting solution (in this example, water with dissolved copper sulfate), separating the anode and cathode. An electric current of iron ions flows between the two electrodes as iron dissolves in the electrolyte, liberating electrons, which migrate to the cathode. The more noble metal (copper) forms the cathode, while the less noble metal (iron) becomes the sacrificial anode.

The battery analogy is often used to describe galvanic corrosion because two different (dissimilar) metals and a conducting liquid solution are needed for the process to take place.

Galvanic cells may operate at the surface of the metal, between dissimilar metals in electrical contact, or between areas of unequal electrolytic concentration. Where a corrosion process has already damaged an existing protective film, the rate of corrosion can be expected to increase. This rate will progress faster in the presence of accelerators such as chlorides.

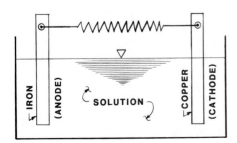

Figure 6.4   Model of a galvanic cell with two electrodes and an electrolyte.

The following are the more common commercial metals and alloys, in order of electrochemical activity:

1. Zinc
2. Aluminum
3. Steel (an alloy of iron and other elements)
4. Iron (cast and wrought)
5. Nickel
6. Tin
7. Lead
8. Brass (an alloy of copper and zinc)
9. Copper
10. Bronze (an alloy of copper and tin)
11. Stainless steel (nickel- and chromium-rich steel)
12. Gold

There are many other alloys of the these metals that occupy various positions in the series.

When any two metals in this list are in contact, with an electrolyte present, the less active metal or alloy (with a lower number) is corroded. The electrolyte may be formed from water or moisture, creating the necessary conducting solution from ingredients in the atmosphere.

For example, if steel (no. 3) and brass (no. 8) are in contact, the steel will corrode. A similar tendency exists between tin (no. 6) and copper (no. 9), but here the potential for corrosion is considerably smaller. This phenomenon must be kept in mind wherever metals are used in construction.

A very important factor in galvanic corrosion is the *area ratio*. A high cathode-to-anode ratio will often result in rapid corrosion, while a high anode-to-cathode ratio will cause slow corrosion, or none at all. For example, a small steel bolt will corrode when in contact with a large brass plate, while a small brass anchor will *not* corrode when in contact with a large steel plate.

Galvanic action increases in intensity as the metals are farther apart in the series.

Basically, to avoid galvanic corrosion, the electrical potential difference should be removed or minimized by one of the following methods:

1. *Choose metals that are close together in the galvanic series.*
2. *Provide separation by isolation,* using insulators such as vulcanized rubber, strips of sheet lead between the dissimilar metals, heavy tinning of ferrous metals, good-quality moistureproof building paper or felt, or

heavy coats of asphalt. In addition, in atmospheric exposures, sealants are often used to eliminate moisture penetration into the contact area.

3. Use *cathodic protection*, a method of eliminating galvanic corrosion that is lately gaining popularity. In simple terms, this method provides protection by reversing the electrochemical process. Either "sacrificial" anodes are incorporated into the structure or an auxiliary current is induced in the reverse direction. While "wet" cathodic protection in marine environments has been successfully used for many years, only recently has the "dry" method been introduced (Fig. 6.5). This latter method has met with partial success; it should be regarded as experimental but with good potential.

It must be understood that all cathodic protection methods must be accompanied by a well-planned and continuous maintenance program, without which they are doomed to failure.

In conclusion, the simplest way to eliminate galvanic corrosion is to use the same metal for all components (including fasteners) in the system. This solution, however, is not always practical, since many systems and structures are very complex.

The technology of how to avoid galvanic corrosion is well-understood by corrosion engineers. However, this knowledge has not been transmitted to

Figure 6.5 Cathodic protection prevents galvanic corrosion by reversing the electrochemical process. "Dry" cathodic protection, a relatively new method, should be considered experimental.

design engineers and architects, whose education in this respect will go a long way toward minimizing galvanic corrosion.

Stress corrosion is directly related to the combined effects of both stress and corrosion. Once localized corrosion occurs, pits and other small surface defects form, with resulting stress concentrations. These stresses, in turn, cause cracks and other existing defects to propagate. Stresses leading to stress corrosion may be either operating or residual stresses. They must, however, always be tensile stresses. The important fact to be understood is that stress-corrosion failures may occur at low tensile stresses, of a magnitude of 50 percent of the yield strength of the material and even lower.

Corrosion is common in:

1. Marine structures, particularly in saltwater environments, especially in the splash zone

2. Parking decks, bridges, and highway structures, especially those showered with deicing salts in the winter

3. Other exposed structures, such as transmission towers and sewage treatment plants

4. Structures exposed to chemical attack

5. Roof structures and steel embedments in the exterior envelope of buildings

6.1 Marine Structures

Structures situated along shorelines and fully or partially submerged in water often suffer heavy damage from corrosion. We have here the presence of seawater (chlorides in solution) or spray from seawater (oxygen). The heavy chloride concentration in seawater, coupled with the constant presence of winds and waves, makes the splash zone especially susceptible to corrosion. On the other hand, fully submerged concrete structures with well-embedded steel are rarely subject to corrosion.

The use of special weathering steels (covered under ASTM A-588)[3] has not proved to be a success. These steels develop oxide layers that are supposed to protect the base metal against additional corrosion. Thus, they are considered to be "maintenance-free." Unfortunately, when these outer layers are lost by washouts, rain, and wave action, the steel is left unprotected again.

6.2 Bridges and Highway Structures

Steel bridges are totally exposed to weather and must be continuously painted as protection against corrosion.

The use of special weathering steels in such bridges has become increasingly common, in an attempt to save painting costs. However, the use of this "perfect" material has not always resulted in happy endings. In general, a correlation has been found between the dryness of a region and the effectiveness of the protective layer of rust. Some state highway departments have decided to partially paint these bridges, especially in and around critical joints and connections.

Concrete structures similarly deteriorate because of continuous exposure to environmental effects. The deterioration is often accelerated as a result of direct and indirect application of deicing salts and the exposure of the structure to moisture and extreme temperature changes.

The problem of reinforced-concrete structures seriously deteriorating from deicing salts is widespread. The deterioration process is that of spalling of concrete because of the rusting of embedded reinforcing bars. The mechanism is initiated by the absorption of chloride ions from deicing salts and the penetration of water into the pores of the concrete. Recent estimates place annual repair costs to highway bridges alone at more than a billion dollars, nationwide — and growing fast. The expense to the private sector must be a comparably high figure.

The National Association of Corrosion Engineers believes that the corrosion of reinforcing steel is the biggest industrial corrosion problem facing the nation today.

This enormous problem has come upon us, relatively speaking, overnight. Heavy use of deicing salts in the United States did not begin until the late 1950s or early 1960s, when states and cities began "bare pavement" policies to enhance winter road safety. The amount of salts used in those early years were far less than typically used today. Since the chloride concentration in the concrete may take 8 to 10 years to build up before it initiates serious steel corrosion, the problem was not fully recognized until the late 1960s and early 1970s. A positive interrelationship between corrosion and deicing salts was not established until later. In fact, despite the severity and economic impact of chloride-induced corrosion of reinforcing steel, the technology to fight the problem and, more particularly, the practical applications of the technology are still in their infancy. While the benefits of using linseed oil were recognized for many years, the use of inexpensive, effective sealers only recently became standard practice in the construction of bridges and parking decks. The application of epoxy coatings on rebars has been recently gaining enormous popularity as a corrosion protection measure. However, users of this method must be *warned* to increase bond development length for full rebar load transfer. The effectiveness of epoxy-coated rebars and silane-based surface sealers is not yet fully proved.

The length of time that a structure can remain serviceable by resisting salt and water is directly related to the following factors:

1. The quality of the concrete in terms of strength, air entrainment, and permeability. (The water/cement ratio is an important criterion.)

2. The placing of the reinforcing steel. The greater the depth of the steel from the surface of the concrete, known as *cover,* the longer it takes for salts and water to reach the steel.

3. The rate of application of deicing salts. Depth and rate of penetration into the concrete are functions of the surface concentration of the salts.

As a summary, *the ten steps in the development of chloride-induced corrosion in concrete-embedded steel are as follows:*

1. Application of chloride deicing agents either directly onto the structure or indirectly from salted meltwater and slush carried onto the structure by vehicle tires.

2. Penetration of chloride ions and water into the concrete through cracks and surface pores, reaching the level of the reinforcing steel.

3. Lowering of the pH of the concrete in the vicinity of the reinforcing steel and the removal of the protective oxide layer coating the steel.

4. Increase of the electrical conductivity of the concrete in the vicinity of the reinforcing steel.

5. Electrochemical corrosion of the unprotected steel in the presence of oxygen and electrolyte, creating iron oxide (rust).

6. Expansion of the iron oxide corrosion products, increase in volume, and buildup of high internal stresses.

7. Internal splitting or delamination of the concrete.

8. Initial separation takes place and a loose spall is formed.

9. Complete separation of the spall occurs by the action of traffic and the added pressure of freezing water.

10. Eventual loss of sufficient concrete and reduction of effective steel cross section until the damage is so great that the structure no longer properly performs its intended function.

6.3 Parking Decks

During the 1960s, many engineers had the notion that because steel is inherently protected by concrete within reinforced concrete and prestressed concrete, it will cause no harm to expose reinforced- and prestressed-concrete parking structures to the environment.

The bitter truth came to light approximately 10 years later, when suddenly distressed concrete parking decks started appearing throughout the cold regions of the country.

There were several basic flaws in most of these structures:

1. Insufficient concrete cover

2. Porous concrete

3. Insufficient drainage

4. Inadequate surface protection or no protection at all

The biggest factor overlooked by designers and constructors of concrete decks was the fact that deicing salts are carried from the street by cars and dropped on the top of the concrete decks.

Corrosion is typically most severe on those deck levels with the most traffic. Sometimes the corrosion even follows the pattern of tire marks. On a parking-deck project in Ohio, corrosion was heaviest on the fourth floor (*not* the roof), since the elevator entrance was on this level.

Corrosion of the embedded bars causes delamination and spalls. Sometimes an adequate solution may be a slab overlay, but, where the chloride penetration and damage are heavy, the best solution is to replace the deck.

The lessons to be learned from all these deck failures is that we must not forget the following four principles:

1. Adequate concrete cover [1½ to 2 in (38 to 51 mm) minimum]

2. Low water/cement ratio (0.35 to 0.40)

3. Positive drainage (adequate number of drains and adequate slopes)

4. Efficient surface sealers

6.4 Structures Exposed to Chemical Attack

Industrial buildings are exposed to constant attack from a variety of chemicals used or produced during manufacturing processes.

Chlorine-treated swimming pools and chlorine tanks within water pollution control projects subject steel, concrete, and embedments within concrete to chemical attack.

Chemical properties of concrete ingredients may also have damaging effects on the performance of structures. Many older concrete structures were constructed using cinder concrete aggregates. The cinders were the residue of coal furnaces. Cinders had the advantages of low cost and light weight. However, their sulfur content is frequently the source of deterioration, especially when the concrete is soaked with water or even only damp. Often the sulfuric acid resulting from sulfur decomposition in cinder concrete results in corrosion of the steel embedments, such as steel beams, rods, wire mesh, and even pipes. Most vulnerable are roof slabs and beams, spandrel beams, and wet areas such as toilets and kitchens. The entire flange of the structural steel beam shown in Fig. 6.6 was eaten away by corrosion. Keeping cinder concrete dry will greatly increase its life.

Using dense, high-quality concrete with low permeability is the best means to ensure resistance of concrete to chemical attack. The use of good-quality ingredients coupled with proper proportioning, placing, and curing will produce concrete which is resistant to attack in many different environments. However, in some severe environments where heavy doses of chemicals are applied to concrete, it may be necessary to resort to the addition of protection by special coatings (see ACI-515)[4] and membranes.

There are many coatings and barrier materials on the market today. Sheet rubber, lead sheets, and bituminous materials (asphalt or coal tar) have been successfully used for years. There are also many new materials without sufficient use experience. What we ought not to overlook is that we must *pretest* the protective barriers (by accelerated testing) for every different chemical attack. New materials may be safely used only after adequate testing and sufficient field experience.

6.5 Buildings

Within building structures, structural elements close to the envelope are often subject to corrosion. While brick covering and spray-on fireproof

Figure 6.6   The entire flange of this steel beam was consumed by corrosion. Holes can also be seen in the web. The beam was embedded in cinder concrete which was kept constantly wet by leaks from plumbing pipes in a prestigious New York City hotel.

coatings will serve as fire protection layers, there is an incorrect conception that these will provide corrosion defense as well. Unfortunately, they do not. Water penetrating through brick joints will accumulate and remain in constant contact with structural-steel elements, such as spandrel beams, columns within exterior walls, and roof structures, causing their eventual destruction. When structural steel has been covered by flashing or other water-tight membranes, this protection has proved to be a factor in resisting corrosion.

Often steel embedments such as pipes are constructed within masonry walls. When the clearance between the steel and the brick is inadequate, once the corrosion process starts, the expansion of the rust results in compression stresses which will push the brick out, causing it to crack. This situation occurred in an office building in downtown Manhattan when spalled brick was noticed (Fig. 6.7). Once we probed the wall, the culprit was found: a corroded pipe (Fig. 6.8).

Ideally, steel embedded in dense concrete will not corrode. Earlier AISC specifications[5] (1969) stated in Sec. 1.24.1: "Steelwork to be encased in concrete *shall* not be painted." However, the latest AISC specifications[6] (1989) state in Section M3.1: "Steelwork which . . . will be in contact with concrete *need* not be painted." (Emphases mine.)

Figure 6.7   Spalled brick in this office building in Manhattan had been patched many times. This repair was useless because the cause of the spall had not been determined.

Figure 6.8 After removal of the spalled brick, a corroded steel pipe which had been embedded in the wall was found. See black spot and Fig. 6.7 for correlation.

Figure 6.9 During alteration of a hospital laundry building, many main structural columns and beams were found to be heavily corroded.

6.5.1   Bellevue Hospital laundry room beam and column damage

During alteration and rehabilitation of a laundry building at a major New York hospital,[7] it was discovered that corrosion damage to the building's main structural columns and beams was of such magnitude (Fig. 6.9) that at several sections steel columns were reduced to nothing but iron oxide powder. In other sections the amount of metal loss varied from 20 percent to nearly 100 percent. There was also delamination of steel flanges (Fig. 6.10). In fact, the vertical loads of the five-story building were typically carried to the lower levels by the masonry brick walls. The cracked brick is shown in Fig. 6.11. In Fig. 6.12, the corroded column flange is shown immediately after the removal of the brick covering.

Beams and girders were also subjected to corrosive attack. Figure 6.13 shows a big hole in the beam web that necessitated removal of the corroded section. This beam was framed to a new steel column as shown in Fig. 6.14.

The investigative studies revealed:

1. Water penetration through cracks in the exterior brick walls. The roofing material was also defective and permitted water penetration from that source.

Figure 6.10 Severe delamination (arrow) of column flanges was common.

Figure 6.11 Cracked brick column covering prior to removal of brick and exposure of heavily corroded steel column.

Figure 6.12 Dangerously corroded column flange, after removal of brick covering. Load was actually carried by brick covering.

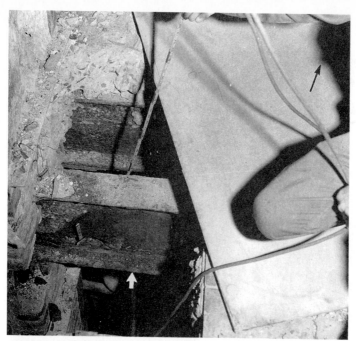

Figure 6.13 Big hole in beam web (arrow and carpenter's ruler) was a result of severe corrosion.

REMOVED CORRODED STEEL SECTION

Figure 6.14 New columns and beams were installed and welded in place to replace corroded steel structural members.

2. The steel was exposed to chemicals and brine from the laundry room.

3. There were also traces of ammonium nitrate, a corrosive agent.

The corrosion environment was so complex that it was not possible to determine the relative contribution of each of the factors. The fact that the exterior columns and beams were corroded to a greater extent than interior structural elements, however, indicated the important contribution of exterior water penetration.

When the actual extent of damage was discovered it was clear that the cost of repairs would outweigh the cost of replacement. However, the continuous use of this facility was an overriding factor, with cost-effectiveness secondary. The new steel members were protected with long-lasting epoxy paint and all cracks in the masonry enclosures were repaired and sealed.

Lessons

1. Cracks in exterior walls should not be left unattended for long periods of time. The corrosion process is time-dependent. Water penetrating such cracks, in the proper environment, will corrode embedded steel within the masonry.

2. Structural steel enclosed in masonry at the exterior envelope of a building must be either well-encased in concrete or protected by proper long-lasting coatings.

3. It is always advisable to conduct a thorough *condition survey* prior to launching a rehabilitation project. Such a survey is always an excellent investment for the owners, to enable them to make the right decisions.

6.5.2 Yonkers Racetrack stables

The concrete roadways connecting the stables at a New York racetrack showed signs of corrosion; extensive spalls were noticed at the bottom of the slabs. Figure 6.15 shows a view of the bottom of a roadway slab with spalls and rust lines coinciding with the rebars. Figure 6.16 shows the same view after concrete removal and exposure of the corroded bars. Several precast concrete rails also showed deterioration and corroded rebars (Fig. 6.17).

A close examination of the corrosion attack showed the majority of the bars with serious metal loss. No wonder, since they were in daily contact with a variety of strong chemicals. Chemical agents to which the tops of the slabs were subjected were:

1. Soap and water used for washing the race horses, daily

2. Horse urine and manure, daily

3. Salts used for deicing the roadways in winter

Figure 6.15 Bottom view of slab containing corroded rebars is shown *prior to* removal of concrete cover.

Figure 6.16 Bottom view of slab in Fig. 6.15 *after* removal of concrete cover exposed the corroded rebars.

Figure 6.17   Precast-concrete rails also showed severe corrosion of the embedded rebars.

The mechanism of corrosion was further complicated because we found that the original concrete, as constructed, contained calcium chloride additives. It was concluded in the survey that, while exposure conditions varied along the lengths of the roadways, the common denominator was always the deicing salts.

Lessons

1. Exterior concrete roadways and decks must be protected from environmental effects by protective coatings or membranes.

2. Concrete subjected to possible chemical attack, whether in exterior or interior use, must have similar protection.

3. Additives containing chlorides should not be used in concrete structures with reinforcing bars or other steel embedments.

6.5.3   Fort Lee balcony pop-outs*

Numerous pop-outs were observed at the soffits of many concrete balconies (Fig. 6.18) at a high-rise apartment building in Fort Lee, New Jersey. What was unique about these pop-outs was their conical shape. The

---

* Dov Kaminetzky, "Structural Systems-Corrosion," National Bureau of Standards Conference on "Corrosion and the Building Industry," Gaithersburg, Md., May 16–18, 1977.

Figure 6.18   Bottom view of balcony pop-outs at this New Jersey high-rise.

centers of these truncated cones (Fig. 6.19) corresponded to the centerlines of the aluminum balcony posts above (Fig. 6.20).

In some areas of the balcony, where the concrete had not popped out, circular cracks concentric with the posts were seen. These circles pointed to an imminent shear failure.

The consistent correlation of the spalls with the aluminum posts, each

Figure 6.19   Spalled pieces of concrete had a conical shape.

Figure 6.20 Bottom view of spall at balcony rail. The center of the railing post coincided with the center of the spall.

set into a steel sleeve atop a steel plate, suggested that corrosion of the posts may have been a possible cause for the distress.

No positive means of drainage was noted on the plans, hence rainwater which could collect in the post pocket could be expected to deteriorate the plate and sleeve forming the pocket and weaken the concrete below. The forensic engineer is constantly faced with the difficulty of identifying the exact failure mechanism because the most probable causes often create very similar stress patterns. In this case, water entering the pocket will freeze at low temperatures and expand. The circular ice block is then confined within the concrete and therefore will generate very high tensile stresses in the adjoining concrete. The resulting failure will be a side or corner breakout.

Similarly, the water penetrating the pocket will start the corrosion process of the steel sleeve. Water will enter the sleeve unless the most stringent waterproofing effort is made (Fig. 6.21). The corrosion product, which always requires a larger volume than the original base metal, will develop shear and tensile stresses in the concrete. When these stresses exceed the tensile strength of the concrete, it will fail in shear with a resulting conical pop-out.

In this case, it was clearly evident that the bottom popped out, which proved conclusively that corrosion was the culprit.

Scrapings we removed from the popped-out concrete at the bottom of the rail posts were analyzed chemically and found to contain lead, sulfur, and iron oxide. The lead and sulfur were ingredients of the pocket fill (grout used to fill the post pocket) as called for in the structural drawing. The iron

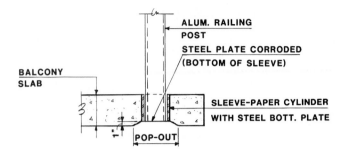

FAILURE CONDITION

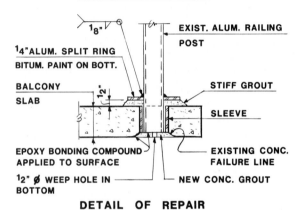

DETAIL OF REPAIR

Figure 6.21 Sketches of failure condition and of remedial work at balcony post.

oxide was the result of corrosion of the steel plate on which the post rested at the bottom of the pocket. No trace of bituminous material was found in the chemical analysis of the fill material, although it was called for in the architectural drawings.

The aluminum post in contact with the steel base would certainly be expected to cause a galvanic reaction. The result would then be a deterioration of the surrounding material. The laboratory report, however, indicated the presence of iron oxide but no aluminum oxide. This pointed to steel plate corrosion as the true cause because a galvanic reaction between steel and aluminum would have caused the aluminum to corrode, resulting in formation of white aluminum oxide.

Lessons

1. Balcony rail pockets must be properly sealed to prevent water penetration and accumulation.

2. Direct contact between steel sleeves and plates and aluminum posts

should be avoided. Isolation by bituminous coatings or other separators is strongly recommended.

3. Where possible, balconies should be sloped away from rail locations to minimize possible water flow into pockets.

6.5.4 Fordham University slab distress— Aluminum conduits

We witnessed cases of aluminum corrosion at the soffit of a slab at Fordham University where aluminum conduits were embedded as casings for electrical wires. The conduits corroded as a result of the galvanic action between steel rebars in the slab and the aluminum conduits. What made matters worse was that the concrete had high chloride content (from calcium chloride or other soluble chlorides which served as a catalyst), which accelerated the corrosion process. The formation of the oxide, which expanded, caused cracking and spalling of the slab (Fig. 6.22). These cracks paralleled the location of the conduits. When we removed the concrete cover, white powder (aluminum oxide corrosion product) could be seen (Fig. 6.23).

Later studies and tests have concluded that without the presence of chlorides, which act as corrosion accelerators, the corrosion process of aluminum embedded in concrete is very slow.

Lessons

1. Avoid direct contact between aluminum and steel.

2. Do not use any aluminum embedments in concrete unless absolutely necessary. In such cases, isolate the aluminum by special separators, such as bituminous coatings or waterproof membranes.

3. The presence of calcium chlorides in the concrete is the most important factor causing corrosion of aluminum in concrete.

6.5.5 Corrosion of masonry anchors—79th Street apartments

Masonry anchors or ties are generally used in wall systems for tying brick facings to block or concrete backup elements. These ties are often made of galvanized steel straps. Their use is most common in cavity walls where the ties are designed to resist:

1. Wind or suction loads acting perpendicularly to the surface of the wall

2. Buckling of the wall by tying two slender elements into one rigid system

Thus, the omission of these anchors or their loss as a result of corrosion will render such cavity-wall systems ineffective and make them prone to failure under the loads described above.

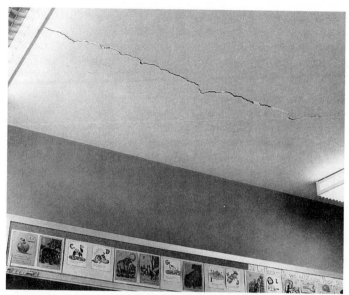

Figure 6.22   This concrete slab at Fordham University cracked along the location of embedded aluminum electrical conduits.

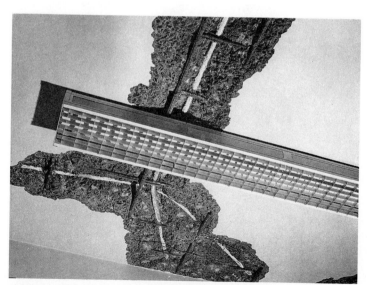

Figure 6.23   White aluminum oxide powder can be seen on the exposed conduits. The steel rebars were unaffected.

At this Manhattan apartment building, the masonry cracking (Fig. 6.24) was first initiated by excessive compression stresses. The cavity became exposed to the required environmental conditions for corrosion — namely air and water — with calcium chloride in the mortar acting as an accelerator.

The result of our inspection of the cavity showed significant corrosion of the metal ties (Figs. 6.25 and 6.26), rendering them useless and accelerating the deterioration of the masonry walls. We also found corrosion of the galvanized dovetail steel slots used to anchor the ties to the concrete columns or beams. These zinc-plated slots are often improperly installed in direct contact with steel rebars, thus creating a dissimilar-metal galvanic cell.

Lessons

1. The use of chloride additives in masonry mortar should be avoided.

2. Once masonry cracks appear in exterior walls, they should be repaired as soon as possible so as to prevent water penetration leading to possible corrosion and further deterioration.

6.5.6 High-bond mortar corrosion failures

A rash of unique masonry failures was reported in the late 1970s. The common denominator in all these cases was the corrosion of all steel

Figure 6.24   This 2-in-wide (51-mm) crack was the first sign of distress. Later, a brick missile released under compression landed in a baby carriage. Luckily, the carriage was empty!

Figure 6.25 Corroded metal ties were found in the mortar.

embedded within the masonry mortar. The similarities among these projects soon became evident. A special high-bond mortar used in these buildings contained a proprietary latex additive developed by a major chemical
company.

Around 1963, when high-bond mortar was introduced to the masonry

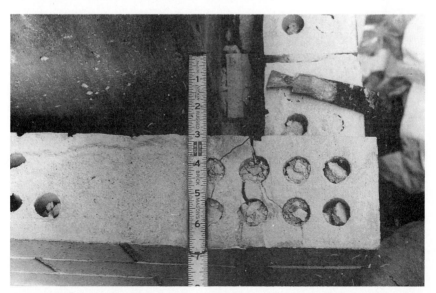

Figure 6.26 Corroded metal ties were extracted from mortar loaded with chlorides.

market, it was hailed as a breakthrough in masonry construction, and was quickly recommended by the Brick Institute of America.

The mortar additive was advertised as providing a bond strength at least 4 times greater than achieved with standard portland cement. Case histories described in the manufacturer's sales literature showed successful projects using the additive in mortar during the period 1961 to 1965.

Building departments in various cities approved the use of this mortar. Denver, Colorado, for example, approved the mortar in May 1962. Cleveland, Ohio, approved the mortar with certain limitations in May 1968.

The manufacturer's literature specified a minimum ultimate bond strength of 360 psi (2480 kPa) at 28 days. Actual tests showed values far above this [529 psi (3650 kPa), average]. The corresponding compression strength at 28 days was 4367 psi (30 MPa) average.

In April 1969 the manufacturer issued specifications, "Masonry Construction with High Bond Mortar." These specifications included the following:

1. *Suction.* Maximum of 5 grams/min per 30 in² (0.02 m²)

2. *Compression strength.* Average of 6000 psi (42 MPa); minimum of 5000 psi (35 MPa)

3. *Ultimate modulus of rupture of brick and mortar.* Minimum of 250 lb/in² (1723 kPa)

Because the developers of the mortar claimed that their new product had superior bond strength, it was particularly recommended for use in:

1. Prefabricated brick panels, built on the ground level and hoisted onto the facade with cranes.

2. Single wythe (no backup wall) brick-face construction.

The cost savings in time of construction and materials (no block or metal backup) attracted many builders, engineers, and architects, who started to use this material extensively.

At the time the bad news was beginning to spread, an article[8] stated that "The Chemical Company has notified the owners of about 100 buildings that may have masonry containing a mortar additive that the additive could cause corrosion of unprotected steel." What was at stake here was *the safety of at least 100 substantial structures throughout the U.S.*

The distress within the brick facing of these buildings appeared in the form of spalls, cracks, and rust stains (Figs. 6.27 and 6.28). Where reinforcing bars were used to increase the bending capacity of the walls, these rebars were encased in the special mortar. As free chloride ions were released and escaped, the corrosion process was on its way. Because the corrosion products of steel are greater in volume than the base metal by anywhere from 6 to 10 times, this expansion generated high tensile

Figure 6.27 Distress within the brick face of this parapet wall that used special high-bond mortar appeared in the form of rust stains. Steel joint reinforcing was embedded in the joints.

Figure 6.28 Distress within the brick face of this school built with special high-bond mortar appeared in the form of cracks and rust stains.

(a)

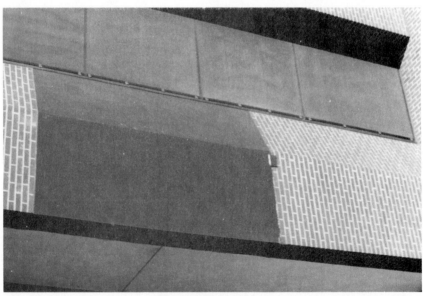

(b)

Figure 6.29 Some of the brick panels fabricated with high-bond-strength mortar actually fell to the street after their steel connectors had corroded (a). Brown plywood sheets (arrows) were used to temporarily cover the holes. Close-up view is shown in (b).

stresses within the brick. Where the confinement was tight, the stresses developed exceeded the strength of the brick itself, resulting in cracking of the brick.

Where panelized construction was used, anchorage to the structure was by steel clips. Once these clips corroded, the panels were left without adequate anchorage. In fact, in several of the buildings panels fell to the street (Fig. 6.29), endangering pedestrians. These failures obviously caused very serious concerns, and sidewalk protection bridges had to be erected.

Where steel was embedded within mortar which contained the additive, the steel corroded. The result was cracking and spalling caused by the expanding iron oxide and also surface rust staining, as can be seen in Fig. 6.30.

Lessons

1. New products must be tested by industry under realistic conditions and in the environment within which they are expected to perform.

2. Professionals proposing new materials for use in their projects should research the materials' records and, where necessary, supplement the available data with special additional testing of their own.

3. Users of new materials and, in particular, the owners of the structures

Figure 6.30   The rust emanating from the joints containing steel joint reinforcement embedded in the high-bond-strength mortar is evident. Note that corrosion was limited to five window-head courses.

containing them should be advised of the inherent risks involved with such new materials.

6.5.7 The Cleveland Bank Building

The Cleveland Bank building was the first to gain national attention as a result of the use of high-bond mortar in the construction of both its office high-rise tower and an adjacent garage.

In the tower the damage appeared in two areas:

1. The intermediate piers, where cracks formed below and above the steel shelf angles (Fig. 6.31).

2. The independent 8×8 in (203×203 mm) mechanical floor brick piers which contained vertical rebars within the brick cores (Fig. 6.32). Here, vertical cracks developed parallel to the vertical bars (Fig. 6.33).

The garage structure exhibited cracks similar to those described in 2 above (Figs. 6.34 and 6.35a). The corroded rebars which had been removed from the piers are shown in Fig. 6.35b).

Laboratory tests of mortar samples from the building face confirmed:

1. The existence of high-bond additive within the mortar.

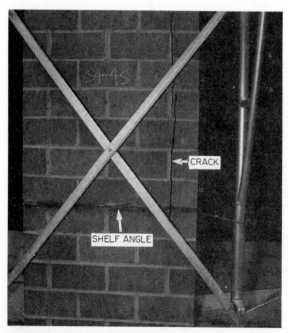

Figure 6.31 Cracks of brick piers developed where steel shelf angles were in contact with the high-bond-strength mortar. Building masonry piers cracked vertically and at the shelf angles.

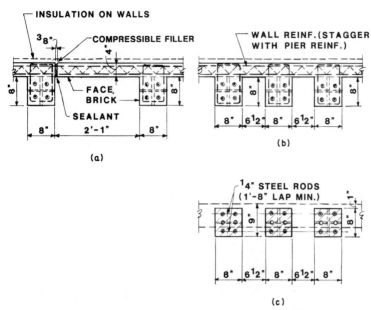

Figure 6.32 Drawing reproduced from the original plans for the Cleveland bank building shows the embedded bars within the brick piers (22d to 33d floors). (*a*) Between-pier reinforcing (with control joint), (*b*) pier reinforcing, (*c*) 8-in pier reinforcing.

2. The high chloride content of the high-bond mortar additive and its corrosive properties. Water-soluble chloride in large quantities ranging from 1.73 to 3.42 percent of the weight of mortar were found. These levels were *12 to 23 times* as great as the maximum level recommended by the ACI for prevention of corrosion of unprotected steel.

According to the laboratory report: "The amount of chlorides present is equivalent to between 8.5 and 9.5 percent calcium chloride dihydrate by weight of cement. Corrosion is significantly promoted when that value exceeds 2 percent; values 4 times that can be *disastrous.*"

One of the most important findings at the bank building, and the only consistent correlation, was that, *where there was a crack, there was steel in contact with mortar.* Cracks were found at points of maximum tensile stress and at points of minimum or no stress. There was no correlation between the damage pattern and a reasonable failure mechanism. The cracking intensity did not vary to any considerable degree as a function of the height or the location of the brick within the building, neither was there any consistent indication of substantially heavier damage on one specific elevation (south or north, for instance). We found the damage to the brickwork to be absolutely random and without any pattern. The only consistent factor was steel in contact with mortar. The delamination of the steel

Figure 6.33 Vertical cracks developed in building piers matched locations of rebars.

Figure 6.34   Vertical cracks in garage brick piers.

(a)

(b)

Figure 6.35 (*a*) Cracks in the brick piers emanate from vertical rebars. (*b*) Corroded rebars removed from brick piers are scattered on the ground.

shelf angles was intense (Fig. 6.36), with pieces of corrosion product (rust) falling to the touch (Fig. 6.37).

The damage and distress necessitated the removal of all the shelf angles in the intermediate piers in the high-rise and their replacement with new, properly coated shelf angles (two coats of epoxy paint). This required the removal of approximately four courses of brick above and below the shelf angles. The total replacement of the brick piers in the garage was the only reasonable solution. It was clearly concluded that any solution which would have involved leaving the existing steel rebars or shelf angles in contact with the corrosive mortar would have been totally inadequate and would have ultimately led to failure.

As in many investigations of facade failures, there were allegations regarding deficiencies such as inadequate soft joints, insufficient metal thickness of the shelf angles, inadequate painting of steel members, and improperly laid brickwork. High wind stresses (see Sec. 7.1.5, Case 4) and even acid rain effects were named as culprits. While some of these defects did exist, their contribution to the damage was insignificant.

Figure 6.36 Delaminated steel is seen hanging from heavily corroded shelf angle.

Figure 6.37 Corrosion products (pieces of rust) from delaminated steel shelf angle are the result of steel in contact with high-bond mortar.

In addition to the Cleveland bank case, I have investigated many others in cities such as Minneapolis, Minnesota; Albany, Syracuse, and Utica, New York; Cleveland, Ohio; and Boston, Massachusetts, in which the proprietary high-bond mortar had been used. Consistently, pachometer readings showed that behind every crack there was a steel rebar or other steel embedment. In many instances severe rust stains accompanied the distress. Figures 6.38, 6.39, and 6.40 show typical damage in various projects. The masonry cracking in many instances clearly emanated from a steel embedment in contact with the mortar (Figs. 6.41 and 6.42).

One of the side issues raised during this investigation was the question of whether steel members receiving spray-on fireproofing need to be painted. There has been a continuous seesaw regarding this question within the structural-steel industry.

AISC specifications[9] state that "unless otherwise specified, steelwork which will be concealed by interior building finish or will be in contract with concrete need not be painted." The term *building finish* has been interpreted to include spray-on fireproofing in the steel industry. The New York City Board of Standards and Appeals requires that "surfaces to receive spray-on-fireproofing shall be cleaned of dirt, grease, oil . . . and entirely covered with a rust inhibitive material." While it has been the practice in the industry to consider spray-on fireproofing a protective coating, it is my opinion that structural-steel sections requiring fireproof-

Figure 6.38 Typical rust stains and cracking caused by corroded steel in contact with high-bond-strength mortar.

Figure 6.39 Elevation of building showing severe cracking which coincided with steel embedded in high-bond-strength mortar.

Figure 6.40 Close-up of cracks emanating from corroded rebar embedded in high-bond-strength mortar. Note that caulked crack on face of brick matches location of rebar behind.

Figure 6.41 Crack leading from rebar to face of brick.

Figure 6.42 Heavily corroded steel embedment and brick cracks emanating from steel.

ing *must be painted* for corrosion protection prior to application of the spray-on. This is especially important for structural steel which is within the exterior enclosure.

At the Cleveland bank building, several columns were found to be totally unfireproofed and also unpainted.

Lessons

1. New products must be sufficiently tested under all conditions and environmental effects to confirm their adequacy. These tests must be performed prior to field use.

2. All new products should be used with caution, and initially only on small-scale projects, to confirm their reliability and effectiveness.

3. Construction materials or chemical additives for concrete or mortar containing chlorides should never be used where they may come in contact with steel elements.

4. Once a material is determined to be defective, it should be withdrawn from the market immediately by the manufacturer.

6.5.8 Southwest Florida condominium

In 1986, our firm participated in an investigation of severely corroded cantilevered balconies at a condominium in southwest Florida. The project consisted of two 12-story structures located on the beachfront. The environment was obviously a corrosive one, heavily loaded with airborne chlorides from driving winds off the ocean.

The distressed concrete balconies were prestressed with tendons running across the entire length of the building. The typical balcony was about 12 ft (3.7 m) wide overhanging 8 ft (2.4 m). The floor slabs were 4½ in (114 mm) thick, anchored at their ends with steel plates.

There was another group of cantilevered balconies on a different facade of the project that showed no distress; those were reinforced with conventional rebars.

The alarm was first sounded when it was discovered that two tendons had broken and could be seen hanging from the bottom of the exposed concrete floor/ceiling slabs. It was obvious that the stressed wires at the anchorages at the face of these balconies had failed. With all the necessary ingredients of the corrosion process — water, oxygen, and salts — there was no doubt in anyone's mind that the tendons were corroding rapidly.

The exposed edges of the balconies were spalled and heavily covered with rust stains (Figs. 6.43 and 6.44). Wide cracks were also visible from a distance, coinciding with the prestressing anchorages.

After removal of the concrete edges, the exposed anchorages could be

Figure 6.43 General view of corroded anchorages on the front face of cantilevered balcony slabs of Florida condominium located near the beach and exposed to airborne chlorides.

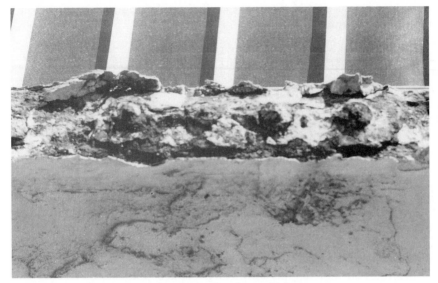

Figure 6.44   Close-up of corroded prestressed tendon anchorages.

inspected (Fig. 6.45). The projecting cut ends of the wires were heavily corroded, with an estimated area loss of up to 50 percent. The end anchorages were also found to be similarly corroded.

In cases of prestressed wire, any loss of cross-sectional area of the heavily stressed steel is accompanied by proportionate increase in the stress level. Failure in such cases is thus accelerated.

The actual level of corrosion activity was determined by the use of copper/copper-sulfate half-cell instruments. The tests were conducted in

Figure 6.45   Corroded anchorages and rebars after removal of concrete balcony slab.

accordance with ASTM Standard C876-80.[10] Among the recommendations of this standard:

    a. If potentials over an area are numerically less than − 200 millivolts CSE (Copper Sulfate Equivalent), there is a greater than 90% probability that no reinforcing steel corrosion is occurring in that area at the time of measurement.

    b. If potentials over an area are in the range of − 200 to − 350 millivolts CSE, corrosion activity of the reinforcing steel in that area is uncertain.

    c. If potentials over an area are numerically greater than − 350 millivolts CSE, there is greater than 90% probability that reinforcing steel corrosion is occurring in that area at the time of measurement.

While recognition of the need to provide adequate concrete cover for rebars is well-known, the same requirement to protect prestressed anchorages is often overlooked. The ACI-318 code, however, specifically recognizes the need for such protection.[11] In Sec 18.19.4, the code states that "Anchorages, couplers, and end fittings shall be permanently protected against corrosion."

Typical corrosion potentials taken at the balconies are shown in Fig. 6.46. Potentials as high as − 360 to − 380 mV had been recorded at the exterior edges of the balconies where the posttensioned anchorages were embedded. As expected, the corrosion potentials dropped at locations away from the edges, below − 200 mV.

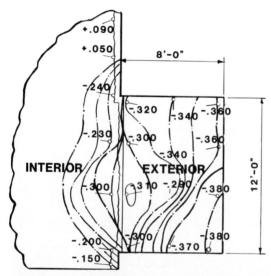

Figure 6.46 Diagram of high potentials outside and low potentials inside apartment. Potentials shown in copper sulfate equivalents millivolts.

Concrete samples were also removed from the affected balconies for chloride-content testing. These tests yielded chloride contents as high as 0.70 percent by weight of concrete, with the threshold level to support corrosion being 0.030 percent. Of 18 samples tested, only two were below the threshold level! Many were *10 to 20 times this value*. The anchorages at the interior of the slabs did not exhibit high chloride levels, since they were not directly exposed to salts.

It was evident that the corroded anchorages had to be replaced. This operation was complicated because each of the tendons carrying the weight of the floor slabs was under tension of 27 tons (240 kN), which had to be continuously maintained.

The repair. The corroded anchorages at the faces of the balconies were first removed; then the balconies themselves were demolished. New balconies 5 ft (1.5 m) wide were then cast with new posttensioned anchorages. Number 4 epoxy-coated rebar dowels were embedded into the slabs so that the balconies were extended to their original width. The interior bays within the buildings as well as the balconies themselves were fully shored. As the tendons were stretched, the elongations were measured constantly.

Repairing stressed tendons is a tricky, risky operation and must be performed by experienced personnel. It was reported that during the repair work, when a tendon was being released, it snapped, cracked, and rotated a part of one balcony slab, collapsing shoring and nearly killing a worker. Some corroded tendons were extended using a standard splice. A latex-modified concrete of over 5000-psi (34.5-MPa) strength was used. The exterior portions of the new slabs were cast-in-place after tensioning so that the tendons were properly protected.

Lessons

1. Prestressed end anchorages always must be protected against corrosion, especially in highly corrosive environments.

2. Concrete exposed to ocean environments should be protected by surface sealers to minimize salt penetration.

3. Buildings and other structures close to the ocean should be periodically inspected for early detection of corrosion problems.

4. When repairing prestressed tendons and anchorages, special care must be exercised so as not to stress-relieve existing prestress forces. *This is an extremely hazardous procedure which must be performed only by trained personnel.*

6.5.9 Manhattan Plaza eyebrows

When an architectural-award-winning structure exhibits corrosion only five years after its completion, "why" is repeatedly asked. This structure did suffer masonry spalls as a result of high compression (see Sec. 4.8.9), but deterioration of eyebrow slabs over the windows (Fig. 6.47) was clearly the result of corrosion. There was no possible contact with masonry here. The corroding rebars had been improperly placed too close to the exposed concrete faces and were subjected to constant rain-driven water and oxygen.

Lessons

1. Use side spacers in the concrete form to maintain a minimum 1½-in (38-mm) concrete cover over rebars exposed to weather.

2. Use clear surface sealers on exposed eyebrows.

Figure 6.47   Corroded reinforcing bars were common at the fascia of this Manhattan high-rise apartment building. Bars are shown after concrete cover spalled and fell to the ground.

6.5.10   Maryland parking deck

This 10-year-old three-level reinforced-concrete parking garage in Silver Spring, Maryland, showed serious distress and severe deterioration. Corroded bars were visible at both the top and the bottom of the slabs. The intensity of the corrosion was recognized after loose spalls had been removed from the bottom of the slab (Fig. 6.48). The corrosion was so heavy that in some locations more than 50 percent of the cross-sectional area of the rebars was lost. Metal ducts conveying electrical wiring were also corroded. For safety of the public, wooden shores were immediately placed at critical locations (Fig. 6.49).

Figure 6.48 Bottom of slabs spalled, exposing the corroded bars. Similarly, corroded bars were found at the top.

In November 1981, our firm was retained to evaluate and determine the structural adequacy of the parking deck. Nondestructive tests such as half-cell measurements were conducted and pachometer readings were taken. Concrete samples were removed for chloride tests. The half-cell readings indicated that over 18 percent of the upper deck was highly corroded and an additional 42 percent had possible corrosion. It was our estimate that over 50 percent of the upper deck was corroded. Concrete cover was found to be inadequate at several locations. An additional serious defect was found after an instrument survey of the deck (Fig. 6.50) was completed. Deflections as high as 2 to 2½ in (51 to 64 mm) were recorded. These low points created standing water puddles which eventually penetrated the deck.

Laboratory analyses of concrete core samples showed low pH levels (pH = 10) and very high chloride levels. An average of 6.0 lb of chloride per cubic yard of concrete (34.8 N/m^3) was found at the top of the core samples, while 1.6 lb/yd^3 (9.3 N/m^3) was found at the bottom.

Preliminary plans to repair the upper deck were prepared and cost estimates were obtained from contractors. Since these estimates indicated that a total replacement of the upper deck would be the most cost-effective solution, a new deck was cast as a replacement.

Figure 6.49 Temporary timber shores were added near the cracked columns on each side of the expansion joints, since the safety of the structure was at stake.

Lessons

1. Surface sealers are recommended in order to protect parking decks against corrosion accelerated by deicing salts.

2. Formwork cambers, proper slopes, and adequate drains must be provided in garage decks for effective drainage.

3. Proper concrete cover must be provided and a low water/cement ratio of the concrete mix must be maintained.

6.5.11   Tennessee high-rise — marble anchors

The failure of this stone facade, which was designed by one of the most prestigious and well-respected architectural firms in the country, is further evidence that nobody is failureproof.

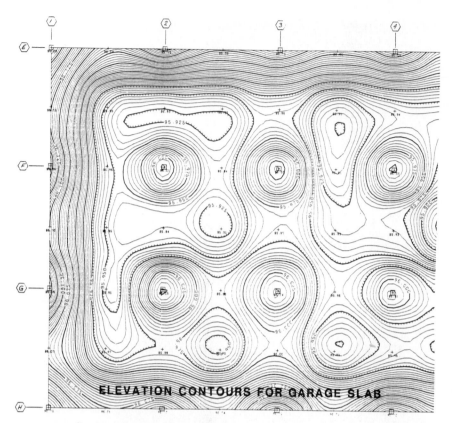

Figure 6.50 Contour map was drawn after instrument survey showed severe deflections. (Scale factors: $X = 5$ units/in, $Y = 5$ units/in.)

The concrete-framed building of 28 stories towers over a plaza. The 1¼-in-thick (32-mm) travertine marble panels were typically stacked in four units and were supported on brass bent-tee-shaped shelf angles (Fig. 6.51). The shelf angles were secured to the concrete columns with ¼-in-diameter (6.4-mm) expansion bolts. These bolts, or studs as they were commonly referred to, were the main issue of this failure, because many of them corroded (Fig. 6.52). Some 15 years after completion of construction, a large marble panel fell to the plaza level. An in-depth investigation followed.

While the expansion bolts attaching the brass shelves were intended to support the weight of the stone panels, the lateral wind resistance was provided by wind anchors. Not a single one of these wind anchors, which were made of stainless steel, corroded. Unintentionally, they became participants in supporting the weight of the stone panels when the expansion bolts were failing. And later, as these wind anchors began to bend, the stone split and spalled around their disks.

Figure 6.51 The stone support assembly, as constructed, consisted of brass angles and carbon-steel bolts. The bolts, which were altogether too small to support the load, corroded, rendering the entire support system useless (side view is shown).

Figure 6.52 Front view of brass shelf angle with two corroded mild-steel bolts. The stainless steel nuts and washers did not corrode, as is evident.

What made matters worse was the fact that many of the shelves were installed with excessive shims [up to 1⅛ in (28 mm)]. This further exacerbated the situation because the corroding expansion bolts were subjected to very high bending stresses. As a matter of fact, our analysis indicated that the ¼-in (6.4-mm) bolts were totally underdesigned and were doomed

Figure 6.53 The shape of the broken, corroded bolt (arrow) is characteristic of stress corrosion, indicating that the corroding bolts were also highly stressed.

to fail on the day of their installation (Fig. 6.53). They were only adequate in shear resistance, but totally overstressed in bending, even with standard shimming [⅜ in (9.5 mm) maximum].

The inadequate width of the expansion joints below the shelf angles [³⁄₁₆ in (4.8 mm) was used] resulted in closing of the joints, sealant failures, and shifting of loads within the height of the building.

Our investigation, which started in 1985, included removal of several stone panels, examination of the anchorage system, flatness measurements (Fig. 6.54), and laboratory tests of steel anchors and mortar samples. These tests showed that corrosion of the anchors resulted in cross-sectional loss that ranged up to 90 percent. It should be pointed out that when a threaded ¼-in-diameter (6.4-mm) bolt corrodes to ⅛-in (3.2-mm) diameter, it reflects a 75 percent loss of cross-sectional area (shear resistance) and an 88 percent loss of section modulus (bending resistance). The highest loss of section by corrosion was consistently associated with excessive shimming.We also found that those steel anchors that corroded were consistently those in contact with the brass shelf (Figs. 6.55 and 6.56).

The major conclusion of the investigation was that corrosion of the stone anchors resulted because low-carbon-steel expansion bolts were used for anchoring the brass shelf angles instead of the stainless-steel ones called for in the specifications. Brass support angles, stainless-steel washers and nuts, and carbon-steel bolts resulted in a fatal combination, creating an electrochemical corrosion cell.

The substitution of carbon-steel bolts for stainless ones escaped all the field inspections, and, unfortunately, the usual preconstruction sample

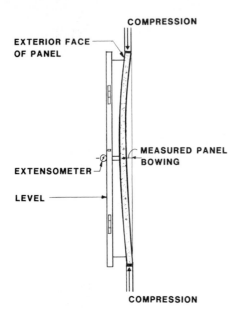

COMPRESSION

EXTERIOR FACE
OF PANEL

MEASURED PANEL
BOWING

EXTENSOMETER

LEVEL

COMPRESSION

Figure 6.54 This special instrument was built by our staff for measuring the flatness of the stone units. The extent of bowing is an indication of stress level within the stone.

approval was lacking. The amazing error escaped everybody involved in the construction, since the shop drawing referred to a "¼-in (6.4 mm) diameter stud w/S.S. nut and washer." The nuts and washers were indeed made of stainless steel. But the stud itself was made of ordinary carbon steel.

The repair.   The repair was accomplished by the installation of new ¾-in-diameter (19-mm) stainless steel anchors which were concealed by ¼-in-

Figure 6.55 The corrosion of the anchor bolts is clearly heavier away from the face of the concrete, at the contact with the brass angle.

Figure 6.56 The corroded bolt runs through an uncorroded stainless steel nut.

thick (6.4-mm) travertine plugs cut from the same stone (Figs. 6.57 and 6.58). The anchors were specially designed by our firm so that they could be proof-tested with their interior threading. Also, new wider ⅜-in (9.5-mm) expansion joints were saw-cut in the facade to permit movement.

The only other option was to remove and reset the travertine panels over the entire building, an option which was ruled out.

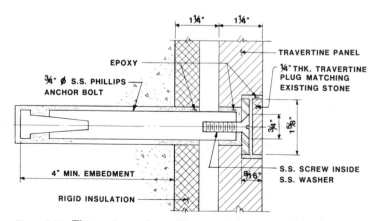

Figure 6.57 The repair was done with new stainless steel expansion bolts, installed through drilled holes in the stone panels. The bolt heads were concealed by thin stone plugs.

Figure 6.58 The finished repair was made aesthetically acceptable by use of stainless steel anchors and stone plugs (arrow).

Lessons

1. Avoid bimetallic connections, especially when the metal with the smaller contact area is lower in the electrochemical series. Particularly avoid metallic systems such as brass/steel, aluminum/steel, and stainless steel/mild steel.

2. Use consistent one-metal systems wherever possible. Stainless-steel support shelves and anchors are recommended.

3. Stone shelf angles should be supported by bolts with a minimum diameter of ⅜ in (9.5 mm).

4. Provide adequately wide soft joints [⅜-in, (9.5-mm,) minimum width is recommended].

5. Use a magnet to field-test stainless-steel components. Most construction-grade stainless steels are nonmagnetic.

6. Submission of material samples should be carefully reviewed by architects, engineers, and contractors for compliance with the project specifications.

7. The adequacy of the support system (shelves and anchors) must be reviewed by a structural engineer *prior to* installation.

8. Last, but not least, use unambiguous specifications and notes on drawings. The drawings in this example should have referred to "S.S. studs and S.S. nuts and washers."

References

1. "Building Code Requirements for Reinforced Concrete," American Concrete Institute Standard 318-89, Detroit, 1989.
2. "Corrosion of Metals in Concrete, American Concrete Institute Standard 220R-85, Detroit, 1985.
3. "Standard Specification for High Strength Low-Alloy Structural Steel with 50 ksi (345 MPa) Minimum Yield Point to 4 in (100 mm) Thick," American Society for Testing and Materials Standard A588-84a, Philadelphia, 1984.
4. "A Guide to the Use of Waterproofing, Damproofing, Protective and Decorative Barrier Systems for Concrete," American Concrete Institute Standard 515.1R-79, Detroit.
5. "Specification for the Design, Fabrication and Erection of Structural Steel for Buildings," American Institute of Steel Construction, New York, February 12, 1969.
6. "Specification for Structural Steel Buildings," American Institute of Steel Construction, New York, June 1, 1989.
7. Dov Kaminetzky, "Corrosion Problems in Structural Systems," National Association of Corrosion Engineers International Forum, Houston, March 6–10, 1978, p. 91.
8. "Mortar Additive May Corrode Steel," *Engineering News-Record,* May 3, 1979.
9. "Specification for the Design, Fabrication and Erection of Structural Steel for Buildings," *Engineering News-Record,* May 3, 1979.
10. "Half-Cell Potentials of Reinforcing Steel in Concrete," American Society for Testing and Materials Standard C876-80, Philadelphia, 1980.
11. "Building Code Requirements for Reinforced Concrete," American Concrete Institute Standard 318-89), Detroit, 1989.

7

Failure Investigation/ Forensic Engineering

7.1 Who? Why? When? What? and How?

The effectiveness of the investigator of any failure, whether a total collapse or minor distress, depends on the data obtained, the thoroughness of the analyses, and the team of experts selected to assist in the investigation.

The three essential elements are shown in the "investigative circle" (Fig. 7.1):

1. Act with the necessary *speed* to obtain data.

2. Determine the failure *pattern.*

3. Establish the *correlation* between analysis and field measurements.

The circle may be closed only when all elements are connected.[1]

7.1.1 Who?

Selecting the technical team. The lead forensic investigator is the captain of the team. This failure expert will recommend, as necessary, a testing laboratory, a field instrumentalist, surveyors, metallurgists, and other specialist consultants. The lead investigator should possess sufficient experience and know-how to collect all the necessary data to establish the correct cause of the failure.

An improper or inadequate investigation will jeopardize the development and preparation of a case — especially when other parties have well-organized teams and search programs!

During the course of the analysis and search for cause of failure, the lead investigator must construct a mental, mathematical, or even physical model and evaluate the probable forces that could cause the type or mode of failure observed.

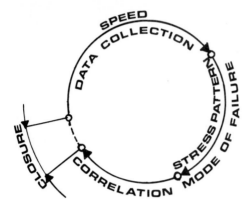

Figure 7.1   The three essential elements in the investigative circle: act with *speed* to obtain data, determine failure *pattern*, establish *correlation* between analysis and field measurements.

7.1.2   Why?

To find answers to the "why" of failures, the entire design and construction process of the failed structure must be reviewed. The flow chart of failure investigation is shown in detail in Fig. 7.2. Alternative paths of travel of loads must be considered.

To establish the cause of failure, the following factors will have to be determined:

1. Assumed strength (design)
2. Actual strength (construction "as is", before failure)
3. Assumed loadings
4. Actual loadings
5. Failure pattern: relative location of the failed section
6. Type of main failure: tension, compression, shear, torsion, or stability

There is no way the investigator can arrive at a solution without knowing *all* the loads acting on the structure.

A serious failure of a sheet pile cofferdam which was constructed near the approach to the Whitestone Bridge in New York City is a case in point. Under most probable loading conditions, which could not be confirmed, the failure would have been caused by a welding defect. "Would have been," but not quite, since three months after the failure of the cofferdam (Fig. 7.3), it was brought to light that a heavy barge tied to the cofferdam departed by tug without being disconnected from the cofferdam!

Once the failure mode has been determined, the equations are basically solved. However, this feat is not always easy, and sometimes is even impossible to perform. The investigation may end with more unknowns than equations (Fig. 7.2), such as five equations with seven unknowns, in which

FAILURE   INVESTIGATION   FLOW-CHART

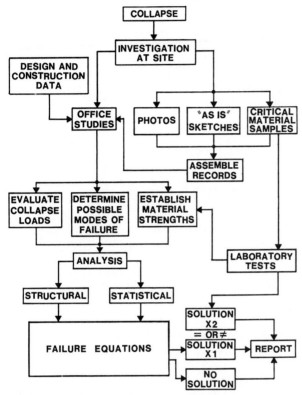

Figure 7.2 Failure investigation may end up with more un-knowns than equations, requiring resort to statistical methods of evaluation.[2]

case it is necessary to resort to statistical methods of evaluation. For the solutions we may end with:

1. A most probable mode

2. A probable mode

3. A least probable mode

In cases of progressive collapses — which are some of the most dramatic failures — it is most important that the trigger be established. The *trigger* is the load or induced deformation which causes the failure to occur at the *particular time*. The trigger may often be found by studying the funnel (Figs. 2.142, 2.143, and 4.137) created by the debris of failed elements in which a defective part may be found. However, where excessive loads are the trigger, such loads sometimes cannot be confirmed after the failure.

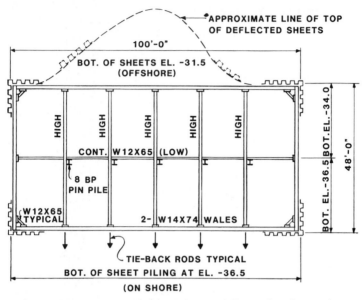

Figure 7.3 Plan view of cofferdam failure initially attributed to inadequate welding. The sheet piles were inadvertently pulled by a construction barge.

Frequently, a deficiency in strength is identified. This may be in the form of understrength material or a defective structural member, a deficient weld or connection material in structural-steel frames, or missing or misplaced rebars in concrete structures. In these instances, solution X2 (Fig. 7.2) emerges directly. However, the investigator *must* follow all of the steps of solving the equations, since the possibilities of other contributory causes *may* exist and they *must* be found. Once all solutions (X1, X2) are established, an effort should be made to determine whether the contribution of each cause can be quantitatively determined.

It is prudent to follow *all* the steps. Elimination of one seemingly routine check may result in missing the cause completely. The following case will illustrate this point of view.

Shortly after erection of precast-concrete beams for a health-care facility, a consistent crack pattern appeared in many units. The distress was evident in the form of vertical tensile bending cracks at the bottom of the beams. Loss of bond was also evident in many instances. Other deficiencies were found at the supporting ends which reduced the ultimate capacity of the beams. To confirm the actual load capacity of the beams, several beams were sent to a laboratory for full-scale load testing (Fig. 7.4). These tests (to failure) confirmed the low load-carrying capacity of the beams, and the cause of the failure was determined to be solely the end-support defects. It was not until later, at the end of the investigation, that a routine laboratory

Figure 7.4 Full-scale laboratory load test of precast beams led to routine test of rebar tensile properties that established (unexpectedly!) that beam cracking was a result of understrength rebars. [Low-strength 40-ksi-grade (275 MPa) rebars were placed by error.]

test of tensile properties of the rebars determined that the yield point of the main reinforcing steel was 40 ksi (275 MPa), rather than the 60 ksi (414 MPa) specified by the structural engineer.

7.1.3 When?

Speed. The more data at the investigator's disposal, the greater the possibility for in-depth analysis of the various failure causes. *All* possible causes are listed and studied carefully. The process of elimination is used thereafter. The reasons for initial elimination of causes are recorded. Additional studies are then performed to eliminate as many of the probable causes as possible. Fast response will give the investigator a tremendous advantage. For example, the investigator should never hesitate to go back to the site and collect additional data before rubble and remnants are carted away and disposed of. Laboratories that can render rapid and efficient service add advantage to the team. (Unfortunately, there are too few of these!)

I recall several instances when the engineering team arrived at the failure scene *many days* after a collapse (due to no fault of its own). At this time, a few *critical* samples relating to the collapse either had been discarded and thrown away or had already been removed by one of the parties. This puts investigators at a great disadvantage: they may never see important data at all, or when they finally obtain these data, the investigators may have already formed an opinion based on earlier and incomplete information. It may be quite damaging to change an opinion at this juncture, and this obviously places the entire investigation and the attorney in charge at a disadvantage.

It is essential to determine the causes of the failure as soon as possible for additional reasons. If the *correct* cause and mode of failure is determined early in the investigation, additional damage or even loss of life in other areas of the complex having similar structural framing, conditions, and loadings can be avoided.

At that time, the investigator may determine that because of overloads or insufficient strength, temporary shoring is required to avoid further property damage or loss of life.

The distress or potential hazard may be eliminated also by removal of actual loads, even simple overloads in case of accumulated, overstocked, and piled-up storage. The same effect may be produced by elimination of vibrating equipment causing damage to the structure.

7.1.4 What?

Establish the pattern. Get the mode *while* investigating. Concurrent field and office work is essential. Working simultaneously at both will assist in identifying the essential data which should be obtained.

Accurate *photography,* with clear identifications, and supplementary field *sketching* are critical.

To ascertain the load-versus-strength relations, the possible acting loads and the material strength should be obtained.

Establish the loads that could reasonably have acted. These may be dead loads, live loads, or temperature and vibration effects. The existence of unusual loads and accidental loads such as gas explosions or vehicle impacts should be confirmed and their magnitude determined.

To establish material strength, samples should be tested *in situ* or removed for laboratory testing. The number of tests should be based on the coefficient of standard deviation of the existing materials or details and on the maximum allowable error in variation of the results (as per ASTM C-122-76).[3]

For *in situ* tests, field instruments such as corrosion and delamination testers, half-cell testers, magnetic locators, and pachometers may be used as well as such methods as in-place stress testing, penetration tests, and other nondestructive techniques.

Nondestructive testing using penetration probes or sonic tests (Fig. 7.5) may be used effectively to establish material strength and continuity.

Figure 7.5 Nondestructive sonic testing is a common way to check for voids in concrete or mortar. Here, mortar below base plate is checked for voids.

However, the accuracy and validity of these tests must be confirmed by correlation with a sufficient number of companion destructive tests. This was not followed in the case shown in Fig. 7.5. Core samples taken later proved the unreliability of the nondestructive tests in this instance.

Movements may be accurately measured in the field. Continuous monitoring of the widths of expansion joints of a large structure (Fig. 7.6) and their variation with temperature was conducted over a 1-year period. Thus, the structure was subjected to a whole range of temperatures while the joint width was being measured. The adjacent cracks were also measured; the measurements confirmed that the planned expansion joints were actually locked, while the undesirable cracks which had developed adjacent to the joints were changing width with the temperature in the same manner as well-performing expansion joints should have been.

7.1.5 How?

The correlation. Correlation is a powerful tool for finding the truth, and later for convincing other individuals—be they adversaries, judges, or juries. The *correlation* between (1) computed stresses and deformations obtained by sophisticated analysis and (2) stresses and deformations obtained by actual measurements in the field is a proof which cannot be easily denied or destroyed.

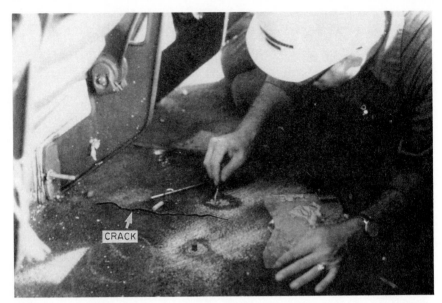

Figure 7.6   Crack-width monitoring proved correlation between expansion-joint failure and temperature variation.

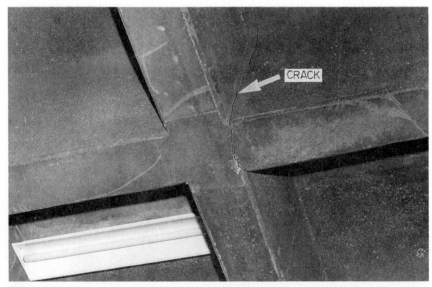

Figure 7.7   Problem of vertical cracking in a two-way joist system supporting a roof was resolved by destructive test of the concrete that exposed incorrectly placed rebars.

The following cases will demonstrate the validity of these statements.

Case 1. A cracking condition (Fig. 7.7) of a reinforced-concrete two-way joist at a university roof was resolved by a destructive test. A probe was cut into a joist by removing concrete and exposing the rebars. It was determined that the rebars had not been lapped at the splice location as required. The bars were placed erroneously with a gap (Fig. 7.8) instead of a lap (Fig. 7.9), thereby losing the necessary transfer of load by concrete bond.

Case 2. Another example where correlation played an important role was the establishment of a clear-cut relationship between brick units with a very high coefficient of saturation and the degree of damage. The roof parapets of this building had damaged brick units with their faces spalled (Fig. 7.10). This relationship was proved quantitatively because the number of cracked units per building was directly related to the coefficient of saturation of the brick.

Case 3. Several high-rise, flat-plate concrete structures in a residential complex exhibited distress in the form of vertical compression cracks in the slab faces. The condition had deteriorated to such an extent that sections of concrete had spalled and separated from the slabs. Stress-relief tests were performed at several floor levels (Fig. 7.11). These tests showed

Figure 7.8 Critical gap in rebar placement was discovered by destructive probe.

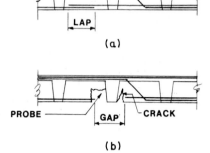

(a)

PROBE

GAP   CRACK

(b)

Figure 7.9 Reinforcement details show design intention, (a), namely overlapping of rebars at splice, and as-found condition, (b), proving gap was cause of cracks.

unusually high vertical compression stresses in both the brick masonry and in the concrete slabs. It was also determined that these stresses increased linearly from the top of the building going downward. A computer model was analyzed using concrete creep as a major input and verified the field-test results (Fig. 4.92). Since the measured stress and damage pattern also confirmed the pattern of stress and mode of failure, the investigative circle (Fig. 7.1) was closed.

Figure 7.10 Investigation of spalling of brick parapet established a clear-cut relationship between the distress and brick units with very high coefficient of saturation.

Figure 7.11 Electrical strain gauge installation used for stress-relief test proved correlation of damage with creep.

Case 4. A steel frame high-rise building that exhibited distress was investigated to establish the cause. The distress was in the form of corroding steel embedments and cracked brick units. Accordingly, a wind analysis was performed by an investigator. He obtained the highest wind speed recorded by a nearby meteorological station during the period in question and then calculated the corresponding wind load. He then computed the wind sway and the stresses in the masonry elements of the structure which were alleged to be high enough to cause the cracking and damage. Our firm subsequently performed an independent investigation. We, contrary to the earlier investigators, performed *in-place* wind speed measurements and also measured the *actual* sway of the building (Fig. 7.12) instead of relying on a theoretical computer analysis. The measured sway was considerably *less* than the one originally computed, which proved that the computer-model stresses were erroneous.

Concrete cores drilled *through* cracks (Figs. 7.13 and 7.14) are very useful in establishing:

1. The existence of reinforcing bars extending through the cracks, or the lack thereof.

Figure 7.12 Wind analysis was performed (by others) by obtaining wind load from highest wind speed *recorded* near building location during the period in question. Wind sway and stresses in structure were then computed and alleged to be high enough to cause masonry distress. Photograph shows in-place instrumentation for measuring wind sway performed by our firm, which disproved the baseless analysis by others.

2. Approximate age of cracks by microscopic surface examination. The fact that cracks existed prior to the actual date of the collapse may alter the whole direction of a failure investigation.

Load tests and models.   Full-scale, in-place load tests are a powerful tool to confirm the strength of a structural element or frame. When excessive deflections, cracks, or complete collapse are the outcome of such tests, they constitute undeniable proof of the insufficient strength of the member or structure. However, when these tests are used for comparison with failed sections, great caution must be exercised to ascertain that the tested and failed sections are identical, and, when they vary, the effect of the variables must be carefully considered.

It is often useful to build a small-scale model of the failed structure. This is done for two purposes:

1. To load-test the small model and thus extrapolate the behavior of the prototype.

2. To study the model so as to better understand the collapse mechanism. For example, during the investigation of the collapse of the formwork of

Figure 7.13 Concrete cores drilled through cracks in slabs are useful test specimens for establishing existence of rebars. Approximate age of cracks can also be determined by microscopic surface examination.

(a)

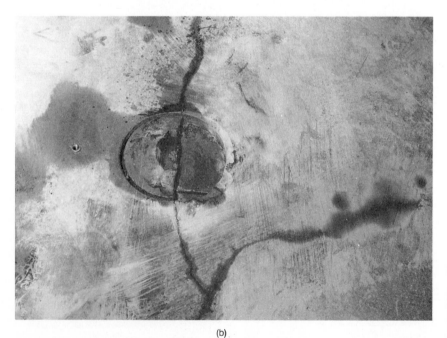

(b)

Figure 7.14 (a) Extensive cracking of floor slab was investigated by extracting drilled core specimens. (b) Enlargement of core through crack.

the W. T. Love Dam, the small-scale model shown in Fig. 7.15 was built and used extensively.

Escape hatches. Once the cause of failure has been determined, some "escape hatches" are open for the designer and constructor. They are:

1. "Act of God"
2. Service overloading
3. "State-of-the-art"
4. Poor maintenance

The first is applicable when the actual load has greatly exceeded the assumed value, as in the case of a hurricane, when failure may occur even if both the design and the construction have been blameless.

Service overloading and poor maintenance will be considered as abuse of the structure by the user or owner.

Invoking state-of-the-art can be justified when the profession as a whole is not aware of the particular failure or defect at the time of the event.

The solution as to the actual cause of a failure is not always clear-cut and may not be what it appears to be at first glance. Nothing should be taken for granted!

Erroneous assumption. A typical example is the soil-loaded roof of a parking deck which collapsed in Nyack, New York (Sec. 3.8.2). A broken gas pipe (Fig. 7.16) was found in the rubble, and it was first thought that the collapse was triggered by a gas-main explosion. When a second section of the deck collapsed several days later, it was clear that the broken gas main had nothing to do with the original failure.

Figure 7.15   Scale model of cantilevered formwork for dam construction was constructed for studies into the causes of fatal formwork collapse.

Figure 7.16   The misleading clue — the broken gas pipe!

Figure 7.17   Failures are not necessarily what they seem to be at first glance. This photograph shows the front view of a building facade which appears to have its center section collapsed.

Similarly, looking quickly at one of two photographs of a collapse in midtown Manhattan (Fig. 7.17), the investigator might conclude that the center portion of the facade of this building had collapsed, leaving the interior in full view.

The second photograph of the same collapse, from a different angle (Fig. 7.18), tells a much different story. In fact, another building located in front had collapsed, leaving the larger building behind intact.

7.2 Conclusion

The results of failure investigations shed light on the behavior of structures and indicate the importance of providing effective feedback for future competent designs.

As a by-product of the findings which the failure investigator reaches after reviewing the abundant and complex information available, the safety of structures will be enhanced.

Knowing which particular structures are susceptible to damage, we can

Figure 7.18 This view from another angle clearly shows the building in Fig. 7.17 to be totally intact. The collapsed building was, in fact, another structure which was in front.

arrive at a more accurate estimate of the safety of existing structures, improve future designs, and minimize failures.

References

1. Dov Kaminetzky, "Failure Investigations," Construction Failures: Legal and Engineering Perspectives Joint Conference, American Bar Association and American Society of Civil Engineers, Houston, Oct. 14–15, 1983.
2. Dov Kaminetzky, "Anatomy of Failures of Roofs and Other Structures," American Society of Civil Engineers Reprint 81-167, ASCE Conference, New York, May 1981.
3. "Standard Recommended Practice for Choice of Sample Size to Estimate the Average Quality of a Lot or Process," American Society for Testing and Materials Standard C-122, Philadelphia, 1976.

The Expert Witness

Experts in construction litigation are often engineers, but others, such as chemists, estimators, contractors, and architects also serve as expert witnesses within their specialty field.

Engineers may be called to testify on engineering matters after an investigation of a serious collapse involving the loss of life, or sometimes to render an opinion regarding less important cases such as, for example, the excessive dusting of an industrial floor, surface cracking or crazing of a concrete slab on grade, or even the proper costs of a structural repair. Sometimes the expert is requested to give testimony as a fact witness (rather than an expert witness) which relates to facts within the scope of the witness's knowledge which do not reflect an opinion.

The testimonies given by such experts often weigh heavily in court cases where judges or juries have to rule on the serious liability of individuals or corporations involved in construction when a dispute arises as to responsibility. Sometimes these testimonies are given in arbitration proceedings, hearing boards, or depositions, also referred to as *EBTs* (examinations before trial). Because, in fact, most construction disputes are settled out of court, the part played by the expert witness is very important to the resolution of these disputes.

The engineer is typically requested to serve as an expert in:

1. Failures and collapses

2. Delay claims

3. Interpretation of drawings and specifications

4. Defective design and construction

5. Safety during construction

6. Interpretation of codes

The single most important requirement of an expert witness is expertise in a particular field, earned as a result of professional experience, skill, education, and research. Further, the engineer obviously cannot offer significant help to an attorney unless he has, in addition, specific knowledge of the subject that is the main issue in the dispute. Understanding legal terminology and procedures will be helpful, but is not a necessity. The expert witness must, however, be ready and prepared to appear in court when requested.

In my many years of work as an engineer engaged in the study and investigation of distressed structures, I have been called on frequently to give testimony in support of findings, opinions, and conclusions contained in my reports and studies. On many occasions at the conclusion of our investigations, my services as an expert witness were *not* requested, because the opinions I expressed or the reports I issued were not supportive of the client's legal position. It goes without saying that an ethical engineer will *never* modify, fashion, or tailor a report only to satisfy a client's position. As an engineer, you must base your opinion solely on the facts as you understand them to be and support it by analysis, review of drawings and specifications, field and laboratory tests, and expert testimony by others.

I referred earlier to opinions based on accurate findings of the facts. *No half-truths can be accepted as facts.*

As a vivid example, a landmark case in Ohio comes to mind. I gave a testimony in a major litigation involving a mortar additive which was alleged to have caused catastrophic corrosion in many structures throughout the United States. In defense of this material, a photograph was introduced as evidence showing masonry buildings where *supposedly* this additive was used successfully without distress (Fig. 8.1). The purpose of this photograph was obviously to prove that the material was not defective. The reaction on the faces of the jurors was shock and disbelief when the opposing attorney introduced as evidence *the identical photograph, but complete and uncropped* (Fig. 8.2) showing unmistakable cracks at the top of the

Figure 8.1   Cropped photo showed no cracks.

Figure 8.2   Full-image photo showed cracks at top resulting from corrosion of embedded steel.

print! It was clear to the jurors that the photographs introduced earlier as evidence *had been cropped to eliminate the damning evidence of corrosion!*

Because most courts will not accept an engineering report as evidence without the presence of the person who wrote the report, the engineer has to appear in court to support the opinions expressed in the report. Thus, the engineer's technical knowledge, credibility, and integrity are often put to the test.

It is very important that while reviewing the data with the attorney, the expert witness should state any findings and opinions well before both arrive in court. There should not be any surprises. The attorney may advise the client to settle the case if the opinions of the witness are not favorable to the client's position. If there is any doubt in the attorney's mind as to the position of the expert witness, the attorney usually will try to avoid the particular subject.

One of the acts of an expert witness which is most disturbing to attorneys is the expert who produces from his brief case new documents in court. It is vitally important that all documents be reviewed prior to testimony, and only material requested by counsel be brought to court by the witness.

Most of all, do not become an advocate. Do not testify *for* a client position or *against* the opposing party. Testify to the facts while giving your opinion "as it is," straight and to the point. You will score more points this way and your good reputation will be your reward.

It is of paramount importance that testimonies given by engineers be accurate, based solely on the facts. I have seen embarrassed faces of expert witnesses when confronted by their own letters, reports, or other documents in which they have given contradictory opinions on *the same issue in the same case.*

It is for this reason that many attorneys advise engineers conducting an investigation into a collapse, distressed structure, or other type of engineering problem *to refrain from giving engineering opinions in writing.* This requirement is essential, especially at the early stages of the investigation, when all the facts are not known. These early opinions are often used by sharp opposing attorneys to their advantage.

During investigations of major collapses I have often been confronted with groups of inexperienced so-called investigators loudly discussing the case and immediately giving their opinions as to the causes for the collapse: "It is a punching shear failure!" or "It is a simple welding failure!" and so on. Once, when investigating a school complex with classrooms full of cracked double-tees (see Sec. 2.15.8), I was even given, gratis, another "investigator's" chalk sketch, left on the blackboard illustrating the failure's cause (Fig. 8.3).

Witnesses who write books or magazine articles should review their earlier writings. Thus, they will be aware of the reasons for differing opinions in the past. Diligent attorneys do a great deal of research into the backgrounds of important expert witnesses.

A most common technique used by some attorneys, especially in depositions, is first to get the witness into a "very comfortable state of mind." While the witness is relaxed and self-satisfied, they will proceed with their cross-examination as if it were a friendly conversation. The witness is then inclined to "fill any void with the sound of his voice," and talks, talks, and talks — sometimes contradicting earlier testimony, sometimes falling into other traps.

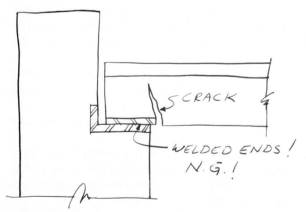

Figure 8.3 Sketch of school precast-concrete failure showing the damaging weld.

Depositions are often used as a test of strength of one's own case. Not infrequently, after hearing testimonies given in EBTs, attorneys may advise their clients to settle then for less than they might have to pay later.

A very common advice attorneys give their witnesses prior to such examinations is "Anything you say can be used against you, nothing you say will be used for your benefit. Therefore, don't volunteer information and be as brief as possible."

A witness must listen carefully to each question. Some attorneys will use double and triple negatives in their questions to try to confuse the witness. If you do not understand the question, do not hesitate to ask for a clarification or ask to have the question repeated. Incidentally, you can use the pause to collect your thoughts, and thus be able to respond more effectively to the question. *After* testifying, you have a chance only to review the written record of your testimony, to correct any errors in the transcript, *but not to change its content!*

Beware of a crafty attorney who may show you a document from a distance and ask you to comment about it without your really seeing it. Ask to examine the document, and take as long as you need to study it carefully.

Where possible, an expert witness, when testifying in front of juries, should translate complex engineering and technical information into simple language to be understood by lay people. My late partner, Dr. Jacob Feld, was testifying in Philadelphia, Pennsylvania, one day when he was asked by the opposing attorney to explain why there was a variation in the strength of the various concrete samples. He answered it this way: "When you slice a loaf of bread, do you expect each slice to be exactly the same thickness?" His point came across to the jury loudly, clearly, and unambiguously.

There is absolutely nothing wrong in disclosing that a witness discussed the case with counsel prior to testifying, and reviewed various documents.

There is also absolutely nothing wrong in disclosing that you have an understanding to be paid for the time you spent in preparing for testifying and during the testimony itself. You should, however, be aware that smart lawyers will try to play your fee arrangements to the hilt *in front of juries.* This is so because the average juror, while aware that nobody should be paid *for testimony,* or paid *to testify one way or another,* is unaware that it is perfectly proper for expert witnesses to be paid *for their time.*

The following, then, are the golden rules for an expert witness that I have obtained from attorneys over the years. It is my belief that these "thirteen commandments" should be extremely helpful to an engineer venturing into the sometimes perilous water of expert testimony, whether in court or in EBT:

1. Take your time before answering. Allow yourself sufficient time to think and your counsel to object.

2. Answer only the questions asked. Do not answer a question unless you fully understand it.

3. If you do not remember, do not guess. Guesses are often wrong.

4. If you are not sure, say so.

5. Be well-prepared. Review the facts prior to testifying. During testimony, if you are not sure, ask to see the record, reports, or other documents.

6. Never volunteer any information.

7. A deposition is not a conversation. When testifying in depositions, do not volunteer or guess mountains of information such as dates, figures, or names of people present at meetings, as they can be used later in court to destroy your credibility, the reliability of your memory, and even your integrity.

8. Never argue with the opposing attorney. Do not show any hostility to anyone. Be calm, and answer the best you can.

9. If the opposing attorney badgers you, answer him using *highly technical terms*. He is sure to let go, as he will not understand your answers. However, wherever possible give judges and juries a simple translation of complex engineering problems, so that they understand the facts. Use examples for illustration. (Make sure you try them on for size first for the attorney to evaluate *prior to trial.*)

10. When possible, refer in your testimony to your written reports. Thus, you will reduce the chance of contradiction.

11. Beware of the examiner who shows or waves in your face only a small portion of the document. Ask to see the entire document!

12. You do not learn how to become a good expert witness in college. Reading articles and getting tips from others will certainly help. However, there is no substitute for experience "in the dock."

13. And most importantly, always tell the truth. Tell it politely and courteously and never lose your temper.

9

Responsibilities

It is no secret that misunderstanding of responsibilities among the various members of the construction team can lead to failures. In particular, we often see how legal conflicts arise from unclear definitions of professional or contractual responsibilities.

I have been preaching for years on the importance of the following flow chart:

BETTER DEFINITION OF RESPONSIBILITY

BETTER QUALITY CONTROL

MINIMIZE FAILURES

Unfortunately, all the participants in a construction project have their own ideas about their responsibilities and roles. Contractors think that any checking or review of their shop drawings, which results in an approval by the engineer, totally relieves them of any responsibility for the construction covered by these drawings. Some contractors also think that since there is inspection on the project, this also relieves them from responsibility, i.e., they believe that the presence of an inspector on the site absolves them from their contractual responsibility.

We find out what each person involved considers a personal responsibility only after the occurrence of a collapse or other failure. And this is when the fingers begin to point in all directions — always at someone else. The simple fact is that there is rarely clear understanding of responsibility. Some of the most common areas of ambiguity of understanding are:

- Shop drawings

- Cladding

- Expansion joints

- Inspection

- Formwork and shoring

- Precast- and prestressed-concrete design

The American Society of Civil Engineers has recently issued a guideline for owners, designers, and constructors, under the name of "Quality in the Constructed Project."[1] This manual attempts to define the role of each member of the construction team: exactly what each is supposed to do and what not to do. One of the questions that has not been answered in this manual — and should have been — is the meaning of *shop drawings:* What are they, and who is responsible for their content? This is something that is still not clear to many in our industry.

Who is responsible for "design" is another question. But then, what is "design," and how is it defined? How is "design" different from a "shop drawing," or a "placement drawing," or a "detail drawing?"

The introduction to the American Concrete Institute's building code (ACI-318-89) says, specifically, ". . . the Code and Commentary are not intended for use in settling disputes between the Owner, Engineer, Architect, Contractor . . ." and that "therefore, the code cannot define the contract responsibility of each of the parties . . ."[2] The code also says that references to compliance with the code should not be put into job specifications, since "the contractor is rarely in a position to accept responsibility for design details or construction requirements that depend upon a detailed knowledge of the design."

I do not think that the contractor should ever be responsible for design! That should be clearly defined, but unfortunately, rarely is.

Let us take *precast concrete,** for example. An engineer, in drawings, shows a double tee and specifies floor load and depth, but then leaves the rest to the contractor to develop in a shop drawing. Is this drawing then, in fact, a true shop drawing? I do not believe it is. This drawing is indeed a design. The engineer is asking the contractor to design part of the structure. The engineer is asking the contractor to draw, compute, apply loads

---

* Dov Kaminetzky, "The Three Orphans," *Concrete International: Design and Construction,* American Concrete Institute, Detroit, Mich., June 1988, pp. 47–50.

and prestressing forces, evaluate stresses, and put everything together. This is an engineering process. Thus, the engineer who designs in such a way is delegating a personal responsibility for the building to others and asking the contractor to do the work that should properly be done by a structural consultant and not by a contractor.

One such example was a big precast, prestressed project in upstate New York, where a girder spanning 147 ft (45 m) collapsed (Fig. 9.1). The contractor actually designed the girder. After the collapse he claimed: "I didn't design that girder, I just prepared the shop drawings." Obviously he was pointing the finger at someone else as the responsible party.

There are presently a great number of ambiguous areas of responsibility. In Europe, engineers produce a complete set of drawings — every rebar is indicated, every rebar bend is specified. What the contractor gets is the complete story, allowing no deviation. We in the United States, on the other hand, do not have such a system. Our construction philosophy is that the innovative contractor be given some edge for competition. If contractors can come up with a more effective solution, then it is to their benefit — and in turn, the owner's — but they must not actually be doing the design. Once the contractors are developing something which is not spelled out in the design, then they are, in fact, designing, and that is where the trouble starts.

It is important to recognize that contractors should only interpret a design — determining the direction of the design, but not actually design-

Figure 9.1 A long-span precast, prestressed girder collapsed during construction of an ice skating rink. Both design and construction were found to be at fault.

ing. If they are designing, then they are taking the responsibility for something that is not their job.

Another area of conflict is the *inspection* of construction. What is inspection? Actually, it is a process of verifying that somebody is fulfilling a contractual obligation. The inspector does not have the authority to direct the construction operation, and definitely does not have the authority to stop the job. Is the inspection "periodic" or "complete"? What does the inspector check against — the original contract drawing or the contractor's shop drawings? If the latter are okay, then how does the inspector know that these shop drawings are the true intent of the design? While shop drawings are the contractor's "convenience" drawings, they are not design drawings. Some public agencies insist in their contracts that the inspector check the installed construction against the original design drawings. If the completed construction is inspected and checked against shop drawings, no one is really checking against anything!

An area of frequent conflict between architect and engineer is the *design of cladding,* which often becomes an orphan that nobody wants to adopt. Expansion joints constitute another area where disputes often arise, especially about the size of the joints and their location, which quite often are in error. Who is responsible for the design of expansion joints? This question has not yet been answered satisfactorily.

Formwork and shoring is another difficult point of contention. The introduction to the ACI Building Code (ACI318-89) states on one hand that ". . . the code cannot define the contract responsibility of each of the parties. . . ." On the other hand, it also states that shoring and formwork is still the responsibility of the contractor. According to Sec. R1.3.2, ". . . it remains the *contractor's responsibility to design* and build adequate forms. . . . " (Emphasis mine.) There is a variety of opinions in the concrete industry on this matter: Who should *design* formwork and who should be responsible?

But what is the definition of "design?" There are numerous definitions, including "make a preliminary sketch" and "create, fashion, execute, or construct." Ironically, one definition of the plural, *designs,* is "aggressive or evil intent." Most contractor's insurance policies contain exclusions prohibiting them from doing any design. I can recall one collapse case in Philadelphia (see Sec. 2.20.12) in which I gave expert testimony, where several days were spent in court trying to determine whether the selection by the contractor of the number of pumps required for drainage in temporary construction was, indeed, a design.

Assignment of responsibility in precast- and prestressed-concrete design is an area that quite often leads to litigation. The problem is with the shop drawings; who is actually doing the design? Many precast elements crack because of welded connections where those details were developed by the contractor and indicated on shop drawings. The engineer may only review and not actually approve these drawings. Even if the engineer only looks at them, it can be interpreted as approval; the engineer will be involved and can be blamed. The engineer of record should prepare the design personally. If not, then the shop drawings, which become the design, should be prepared by an engineering consultant, engaged by the engineer of record, who is willing to take the responsibility. This is preferable to assigning the responsibility to the contractor for the design of that part of the construction.

A better and clearer definition of responsibility will lead to improved quality control, because all will know what they are expected to do before the problem arises. That will minimize failure and conflict within the construction process itself.

Failures of understanding.   Take the case of a slab failure that we investigated. This slab outside a big residential project in a suburb of Boston had settled and cracked. On one side it was supported and on the other side it was on fill that was supposed to have been compacted (Fig. 9.2). The specifications spelled out that the fill should be compacted to 95 Proctor, sampled, and tested. Unfortunately, the slab settled because the soil was not properly compacted, despite being inspected by a laboratory that was contracted by the owner. Therefore, the question is asked: Is the slab failure thus the owner's responsibility?

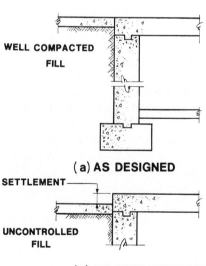

WELL COMPACTED FILL

(a) AS DESIGNED

SETTLEMENT

UNCONTROLLED FILL

(b) AS CONSTRUCTED

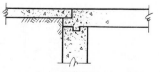

(c) PROPER DESIGN

Figure 9.2   (a) Slab supported on properly compacted soil. (b) Slab after failure on uncontrolled fill. (c) How slab should have been designed.

During the court case, the jury decided to go to the site to see what had really happened and the foreman of the jury tripped over the slab drop (Fig. 9.3)! The case was settled shortly afterward. Nevertheless, this was a case of "mixed" responsibility, and court decisions can go either way. The contractor failed to compact properly and the owner's testing laboratory failed to detect the defect. I do not believe that inspection, by itself, relieves the contractor of responsibility. Contractors have a contract to construct a good-quality job, and just because an inspector does not catch them, they are not relieved of their obligation.

While investigating the collapse of a whole floor of a parking deck near Washington, D.C. (Fig. 9.4), we found that the dowels were installed on the bottom of the slab at the support (Fig. 9.5). They obviously could not support anything unless they were placed on the top as required by the design. The steel was simply set in the wrong place, which was one of many problems on the job.[3] I have often seen a collection of defects on a single project: stirrups misaligned, slabs too thin, steel misplaced or even missing. But what is the rest of the story? The expansion joint did not work properly because all the designer specified was building paper at the sliding surface, which was insufficient to produce a free-moving slab, and the joint locked. Was this a design error? Yes, it was. But it was also a construction error and an inspection error.

A court will want to know if there is one party who is *most* responsible and what portion of the damage may be assigned to each party. Is an inspector who does not detect a construction error therefore solely respon-

SLAB DROP

Figure 9.3 The slab after it had dropped.

Figure 9.4 The collapse of an entire floor of a parking deck was traced to dowels which had been installed at the bottom of the slab. The dowels obviously could not support anything unless they were placed on top.

Figure 9.5 The misplaced dowels.

sible? No; the prime responsibility still should be with the contractor, but the inspector has a secondary responsibility. Also, the designer may also have to assume some of the responsibility. It is not always possible to determine one fault which by itself is sufficient to have caused the failure.

A building in Florida developed cracks along one corner (Fig. 9.6). First, there was a design deficiency because the analysis determined that the stresses were too high at this corner. Also, there was insufficient height of the reinforcing bars, coupled with premature removal of the shoring. Therefore, three different factors contributed to the failure: one was a design deficiency and the other two were the contractor's errors. So, who was responsible? How much responsibility is allocated to each party depends on the law of the state. But there is obviously joint responsibility. Who pays? That is up to the courts.

A major collapse which we investigated in Boston several years ago (Fig. 9.7) killed several workers. The mayor of Boston assigned an investigating commission that listed a catalog of irregularities and deficiencies. Incredibly, the project had started without a building permit! Problems were then identified in the structural design: there was lack of inspection, reinforcing steel was misplaced, there was insufficient concrete strength, shoring was removed prematurely, and on and on. Nearly every step of the construction sequence was flawed. That the myriad problems in combination caused the failure was easy to define; assigning specific responsibility, on the other hand, was difficult.

Figure 9.6 Three different deficiencies contributed to the cracking of floor slabs in this Florida development — one design flaw and two contractor errors. Who is responsible?

Figure 9.7 A major collapse in Boston killed several workers. An incredible series of irregularities and deficiencies was uncovered, including lack of a building permit, lax inspection, misplaced rebars, weak concrete, and prematurely removed shoring.

In another case, the Bailey's Crossroads project near Washington, D.C. (Fig. 9.8), there was a failure which resulted in total collapse of a high-rise concrete building during construction. The cause of the collapse was determined to be the premature removal of shores. The engineer did not have a contract to perform engineering inspection services during construction. Still, the engineer was found liable because several codes, the ACI code included, state that ". . . Concrete construction shall be inspected throughout the various work stages by a competent Engineer or Architect . . ." (ACI code Sec. 1.3.1).[2] But is failure to inspect really the engineer's fault, or is it the owner's fault? Is it the owner's responsibility to know that inspection is needed, or should the engineer inform the owner of this need?

Figure 9.8 Failure of this high-rise building was traced to premature removal of shores. The engineer was not contracted to perform inspection during construction, yet was found liable. Is failure to inspect the engineer's fault, or the owner's?

In the Bailey's Crossroads case, though, the engineer had signed inspection reports even though he had no responsibility, solely because the owner had asked him to. The engineer was therefore found to be responsible for inspection even though nowhere was he contractually bound to perform and be responsible for inspection.

In another case, a Newark, New Jersey, parking garage constructed of concrete floors collapsed during slab screeding (Fig. 9.9). The concrete was supposed to have been high-early-strength concrete. The contractor ordered high-early-strength concrete and thought he was getting it. A laboratory hired by the owner checked the concrete strength during construction and advised that the strength was fine. Unfortunately, the laboratory was wrong. How they arrived at their results — or even whether the tests were performed at all — is unknown.

So the contractor blamed the owner and the laboratory for failing to inform him that the concrete was deficient. The court, however, found that the contractor was still responsible because he had a contract to supply high-early-strength concrete — and he had not done so. Regardless of the erroneous laboratory report, regardless of the fact that the supplier delivered the wrong material, the contractor was still responsible. Does a failure of a testing agency retained by the owner relieve the contractor from the obligation to supply the proper materials and to do the proper work? The court's answer in this case was in the negative.

Shores, reshores, and formwork are another common problem. A failure in Alexandria, Virginia, that resulted in collapse (Fig. 9.10) and the deaths of several workers brought up the old question of who is supposed to design

Figure 9.9 Concrete parking garage in Newark, New Jersey, collapsed during slab screeding. Investigation showed that high-early-strength concrete had been specified and ordered by the contractor, and that the laboratory supposedly tested the concrete strength. Does failure of a testing laboratory retained by the owner relieve the contractor from the obligation to supply proper materials?

Figure 9.10 Collapse of formwork during construction near Alexandria, Virginia brought up old questions about who is supposed to design shores and formwork, who is to determine when to strip forms, and who is responsible for designing and inspecting bracing for shores and reshores.

the shores and formwork, who is supposed to determine when to strip the forms, and who is responsible for designing bracing of the shores and reshores. These responsibilities continue to remain a fuzzy area.

Conclusion. We must have a clear definition of responsibility; a clear definition of design; a clear definition of the purpose, required completeness, and extent of inspection. There must be an understanding among all members of the construction team. If there is not a clear definition of responsibilities, failures result and we end up with litigation. Better definition of responsibilities will lead to better quality control which, in turn, will minimize failures and legal conflicts.

References

1. "Quality in the Constructed Project," Preliminary Edition, American Society of Civil Engineers, New York, 1988.
2. "Building Code Requirements for Reinforced Concrete" American Concrete Institute Standard 318-89, Detroit, 1989.
3. Dov Kaminetzky, "Failures and Responsibilities," Concrete International: Design and Construction, Published by American Concrete Institute, August 1987.

Chapter 1

Recommendations for Design and Construction Improvements

10.1 Common Pitfalls

10.1.1 General

1. New, unusual materials, without sufficient track record
2. New, untried methods of construction, especially on large-scale projects
3. Doing somebody else's job and ending up with someone else's responsibilities
4. Complicated design and construction methods and techniques
5. Work outside your area of expertise

10.1.2 Concrete

1. Flat-slab punching shear failure
   a. Large openings directly at the column face
   b. Small columns supporting large areas
   c. Few or no reinforcing bars passing through the column periphery
   d. Unsupported reentrant corners, especially at slabs without spandrels
2. Flexure failure
   a. Very shallow beam supporting long spans
   b. Deep girders framing into small columns
   c. Deep girders supported by shallow beams
   d. Girders with less than ½ percent reinforcing
   e. Cutoff of reinforcing bars in the tension zone
3. Bearing failure
   a. Small bearing areas (especially precast and prestressed elements)
   b. Bearing on nonconfined brackets and corbels
   c. Bearing at the edge of brackets, corbels, and beam seats

    d. Improper sliding and rotation details (especially at expansion joints), missing bearing pads

    e. Unreinforced and/or shallow brackets or corbels

4. Formwork

    a. Improper shoring and reshoring of high-rise construction, especially during winter months

    b. Use of double-tier shores

    c. Early stripping and premature removal of shores

    d. Insufficient cross-bracing

    e. Use of long, unbraced shores

    f. Use of warped and/or bent shores

    g. Inadequate mudsills

5. Deterioration

    a. Lack of air entrainment

    b. Accelerated corrosion of steel in presence of chlorides

    c. Aluminum embedded in concrete

10.1.3 Prestressed and precast concrete

1. Lack of detailed, step-by-step tensioning procedures

2. Lack of allowances for posttension movements

3. Welding of ends of precast elements

4. Thin, unreinforced panels

5. Hanging utilities from bottom of precast webs

6. Jacking or supporting prestressed elements at midspan

7. Temperature differentials between interior precast panels and exterior columns and beams

10.1.4 Structural steel

1. Nonredundant elements, such as single tension members, hanging rods, eye bolts, single anchors.

2. Deep blocks and cuts of beam ends, especially when not strengthened with plates or angles.

3. Steel frames (columns and beams) which are designed to be braced by *future* slabs or walls.

4. Insufficient drainage of large, flat areas (ponding failures).

5. Slotted holes used as expansion joints.

6. Welding of thick plates.

7. Stress raisers, abrupt changes in section.

8. Exposure of steel to impact and low temperatures, leading to brittle fracture.

9. Lack of weld returns.

10. Unbraced compression chords of trusses.

11. Excessive deflections of low, shallow girders; no camber and/or clearance for doors and windows.

10.1.5 Masonry

1. Thin, high, unbraced walls

2. Walls with many openings (should be reinforced with piers)

3. Missing or inadequate vertical and horizontal soft joints

4. Locked soft joints — full of mortar

5. Inadequate masonry anchorage (design and/or construction)

6. Corrosion of masonry anchors

7. Brick with high coefficient of saturation (greater than 0.78)

8. Lip brick, soap brick

9. Parapet walls lacking vertical expansion joints

10. Excessive shimming of shelf angles

11. Paper or wood in mortar joints

12. Inadequate flashing: no laps, loose laps, no reglets, no mastic bed, damaged flashing

13. Stacked, thin, stone panels

14. Glue-laminated stone liner plates for support

15. Cavities filled with mortar

16. Shelf angles not supported close enough to end of span, not carried around corners

17. Caulked joints not properly dimensioned, caulking bonded on three sides

18. Insufficient or blocked weep holes

19. Walls not rigid enough to prevent cracking due to lateral loads, especially masonry with wood or metal studs backup

20. Improper mortar for durability or strength

21. Inadequate foundation or spandrel beam support

22. Spandrel beam improperly braced to prevent rotation

10.1.6 Foundations

1. One structure supported on two different foundation systems, i.e., piles and spread footings

2. Inadequate drainage for retaining and basement walls, leading to buildup of hydrostatic pressure

3. Inadequate lateral stability of structures with one-sided soil pressure

4. Corrosion of piles and sheet piles in the splash zone

5. Lack of waterstops in foundation walls and pressure slabs

6. Ignoring negative friction effect on piles in unconsolidated soils

7. Lack of piles to support slabs on grade where soils are of poor bearing capacity

8. Insufficient or no borings

9. Borings not taken deep enough

10. No consideration of groundwater level

11. Failure to design for uniform settlement

12. Backfilling before foundation walls are adequately braced by floor framing

13. Inadequate depth of foundations subjected to scouring

14. Compacting upper-10-in (25.4-mm) layer of loose, unconsolidated soil and fills

10.1.7 Miscellaneous

1. Expansion joints bearing on isolated brackets or corbels.

2. Decks supporting heavy earth fill.

3. Long cantilevers with short backup span.

4. Long structures without expansion joints, or locked expansion joints.

5. Support of heavy, concentrated loads.

6. Very flexible members used to support vibrating equipment or gyms and dance halls, where vigorous activities generate high levels of vibration.

7. Hybrid construction systems such as reinforced concrete and structural steel.

8. Overloading of structures with construction materials during construction.

9. Inadequate number of roof drains; drains located at high points near columns, leading to ponding.

10. Bimetallic corrosion of dissimilar metals in direct contact.

See Lessons at the end of case histories for additional pitfalls.

10.2 Checklists/Guidelines

The following items should be checked by the design and construction teams prior to and during construction of the project

10.2.1 General

1. Location of the project
2. Qualifications of the design team
3. Usage of the structure
4. Available documents, surveys, borings, reports, etc.
5. Governing codes, standards, regulations, etc.
6. Live loads, wind loads, earthquake requirements, impact loads, alternating loads, vibrations, storage loads, equipment, and piping loads
7. Existing structures, utilities, and conduits on the site
8. Special local environmental conditions
9. Comparisons with similar projects
10. Expansion joints (number and width of joints and sealants)
11. Protective coatings and sealers
12. Complete list of the participants of the team
13. Responsibilities of the participants
14. Information flow
15. Preconstruction conference

10.2.2 Structural design

1. Selected structural system
2. Compatibility of materials
3. Connections between components
4. Magnitude of loads
5. Direction of loads
6. Load combinations
7. Equilibrium of loads
8. Stability of the entire structure and individual members

9. Bracing of the structure

10. Effect of temperature, shrinkage, and creep, especially on long structures

11. Deformations and angular rotations

12. Transfer of reaction forces

13. Assumed structural model

14. Geometry, dimensions, and spans

15. Selected foundation system

16. Interface between the structure and the soil

17. Possible settlements, effect of differential settlements on the structure

18. Allowable stresses

19. Effects of adjacent structures

20. Constructibility and tolerances

21. Availability of materials, equipment, and manpower

22. Compatibility of cost estimate and budget

23. Compatibility of quality of field inspection and testing services with the level of complexity and sophistication of the design

24. Verification of blind work

25. Coordination of specifications with the drawings

I dare to make the following statement regarding failure prevention: If a structural engineer is to remember at all times only two things, these should be:

1. Σ of all forces and moments = 0.

2. For every load there is a deformation. Remembering this by itself will go a long way to minimize the possibility of failure.

10.2.3 Construction

1. Qualification of contractors and suppliers

2. Written excavation procedures

3. Written erection procedures

4. Written safety procedures, designation of safety coordinator

5. Permits

6. Material approval

7. Field and laboratory material testing procedures

8. Planned quality control and inspections
9. Verification of existing conditions, including utilities
10. Shop drawing approval
11. Check actual construction loads, mechanical equipment loads, etc., against design loads
12. Provide temporary bracing during construction
13. Retain an engineer to design temporary structures

10.2.4 Reinforced concrete

1. Reinforcing-steel grade
2. Strength of concrete of all elements
3. Concrete mix, water/cement ratio, and admixtures review
4. Dimensions of all concrete members (slabs, beams, columns, and girders)
5. Concrete cover (over reinforcing)
6. Rebar details:
   a. All components and zones reinforced
   b. Minimum reinforcement provided
   c. Length, size, and location of rebars
   d. Spacers, bar supports, and chairs
   e. Ties, hoops, and stirrups
   f. Additional reinforcing at openings
   g. Effective depth of top bars (field verification)
   h. Integrity reinforcing, i.e., flat-slab top and bottom bars running continuously through column periphery
7. Construction joints, reviewed by engineer of record
8. Brackets, corbels, and daps reinforcing
9. Expansion joints: movement, spacing and width
10. Concrete vibration requirements
11. Dimensional tolerances
12. Chloride threshold and air-entrainment content
13. Curing requirements
14. Waterstops in expansion and construction joints
15. Coordinate embedments, inserts, and box-out requirements with rebar location
16. Flat-slab punching shear at columns, especially near openings and slab edges

10.2.5 Formwork design and construction

1. Prepare formwork shop drawings
2. Prepare shoring and reshoring sequence and procedure

3. Verify that actual rate of casting and temperature is that used in form design

4. Provide cross-bracing against lateral loads

5. Check pull-out capacity of anchors and expansion bolts embedded in concrete

6. Check lumber and plywood grades

7. Check effects of uplift

8. Check wind resistance of formwork

9. Brace high, flexible shores

10. Check concrete strength prior to stripping by cylinders or pull-out tests

11. Check for proper bearing on mudsills

12. Check plumbness of shoring

13. Be sure all shoring is resting on plate or sill, not hanging loose from deckwork

14. Check bracing locations and see that all bracing is connected

15. Check shoring against layouts for correct positioning

16. Check for proper grade adjustment

17. Check all locking devices

18. Be sure shoring is properly secured to deckwork above

19. Check layouts and shop drawings to see that correct sizes have been used and that nothing is missing

20. Provide cambers and chamfers as required

21. Double-check spans so that overloading will not occur

22. Check span and deflection of metal deck used as forms

23. Check shoring and reshoring sequence

10.2.6   Precast concrete (see also Reinforced concrete)

1. Connections should not create 100 percent rigidity; avoid welding; provide sliding and rotation pads.

2. Connections designed for all forces.

3. Dimensional tolerances.

4. Written erection procedures.

5. Bracing during erection (wind, crane impact); check for stability and erection stresses.

6. Transportation and lifting reinforcement.

7. Special reinforcement at brackets, corbels, and daps.

8. Special shrinkage and temperature reinforcement.

9. Protection sealers.

10. Air-entraining agents for exposed precast panels.

11. Concrete reinforcing cover.

12. Rebar protection (galvanization or epoxy coating).

13. Form-release agents.

14. Inserts for hangers and lifting hooks.

15. Insert pull-out capacity and edge distance.

16. Surface finish.

10.2.7 Prestressed concrete (see also Reinforced concrete; Precast concrete)

1. Type of tendons

2. Number of wires

3. Prestressing force

4. Tie-down clamps

5. Tendon sheaths

6. Free movement of tendons inside the sheaths

7. Anchors and couplers

8. Corrosion protection of wires and anchors

9. Additional reinforcing at couplers, tie-down cables, daps, corbels, and reentrant corners

10. Written prestressing sequence

11. Check tendon elongations for consistency with prestress force

12. Partial prestressing: friction losses, cambers

13. Grouting of tendons and corrosion protection

14. Concrete strength at time of destressing

15. Deduct area of holes created by couplers and sheaths from concrete compression area

10.2.8 Structural steel

1. Structural-steel grade and yield point

2. For high-strength steels, especially thick sections, review need for minimum toughness requirement

3. Size, length, location, and elevation of all members

4. Connections, bolts, rivets, and welds (design adequacy and field conformance)

5. Buckling of columns, plates, and shells

6. Stiffeners at load concentrations and reactions

7. Reinforcement at excessive blocks and cuts

8. Lateral support at all compression flanges (especially cantilevers) and compression chords in trusses

9. All columns are braced in two directions

10. Anchor bolts at all bearings on concrete (columns and beams)

11. Cambers

12. Fireproofing

13. Protection against corrosion (coatings, membranes, or concrete cover)

14. Field and laboratory testing of connections

15. Stability and bracings

16. Members and/or connections subjected to fatigue

17. Members and/or connections subjected to impact and/or vibrations

18. Written erection procedures

19. Temporary bracing

20. Approved shop drawings

21. Ponding of flexible, light roofs

22. Eccentric connections

23. Preheat welding requirements

24. Avoid bending stresses in tension members

10.2.9 Masonry

1. Brick properties: compressive strength, coefficient of saturation, water absorption and initial rate of absorption, temperature coefficient of expansion, severe weathering exposure

2. Stone properties: tensile and compressive strength, elastic modulus and its variation with environmental exposure, temperature coefficient of expansion

3. Block properties: tensile and compressive strength, shrinkage, water absorption

4. Mortar properties: tensile strength, compressive strength, bond strength, no chlorides in mortar

5. Selection of proper mortar type for actual environmental exposure

6. Compatibility between mortar and masonry unit (brick, block, or stone)

7. Efflorescence

8. Expansion joints (both vertical and horizontal)

9. Anchorage to backup walls, beams, and columns

10. Stability of walls

11. Height/width (slenderness) ratio

12. Drainage of walls (weep holes, flashing, sloped copings, etc.)

13. Corrosion resistance of shelf angles, ties, anchors, and other metal embedments

14. Continuity of flashing system (laps and reglets)

15. Positive anchorage at top of backup walls

16. Wind resistance of masonry system

10.2.10 Foundations

1. Visit site to observe existing structures which may be affected by new construction. Find out type of foundation used for nearby existing buildings.

2. Take proper borings, of sufficient number and adequate depth. Verify groundwater table.

3. Choose appropriate type of foundation. Consult geotechnical engineer.

4. Design for uniform bearing and settlement or provide for any differential settlements which may occur between parts of the structure, by use of expansion joints.

5. Brace foundation walls before backfilling. Do not load unbraced foundations with material or equipment surcharge.

6. Provide drainage at retaining or basement walls.

7. Verify that actual soil is as assumed in design, both for vertical support and for lateral load against walls below grade.

8. Check for effect that pumping and/or vibratory equipment may have on existing structures.

9. Design foundation for all directional loads and within allowable stresses and settlement tolerances.

10. Support slabs on piles when slabs are bearing on compressible or otherwise inadequate soils.

11. Provide step-by-step construction procedure when temporary support, underpinning, shoring, or bracing is required during installation of new foundation.

12. Tremie placing of concrete should always be performed in still water. Do not pump water out of the shell during placement operations.

13. For drilled pier and caisson installations, extract full-depth concrete test cores from the first five to verify integrity of the concrete and adequacy of bottom bearing.

14. Monitor groundwater level during construction , especially if pumping is required, since it may affect existing buildings and the bearing capacity for new structures.

15. Establish vibration criteria for effect on existing structures during pile-driving operations. Special care is required for old and historical landmark buildings.

16. Provide batter piles to resist lateral loads.

17. Avoid bending moments and eccentricities on footings.

18. Check pile caps for eccentricities of as-driven pile locations.

10.2.11 Safety during construction

1. Check the stability of the structure during construction in partially completed stage. Is there a need for temporary bracing and guying?

2. Prepare written construction sequence. Is it possible to advance construction of certain components that will enhance safety during construction? An example is installing a staircase at an early stage to avoid the use of ladders.

3. Check number, strength, and safety of lifting hooks for elements such as precast panels and planks.

4. Provide inserts for temporary props, to be used prior to installation in final position.

5. Provide sockets for temporary guards, handrails, toeboards, etc.

6. Check strength of temporary handrails to resist lateral loads.

7. Check for temporary covers over all openings in floors and roofs.

8. Where temporary handrails and guardrails are not possible, check for safety belts and safety nets.

9. Provide an adequate number of access ladders. Check strength of ladders. They must be secured in position at all times and rechecked when moved to new locations.

10. Appoint a safety supervisor for the job site.

11. Keep a neat and organized site, with materials stored outside accessways. This will minimize construction accidents.

12. Arrange scheduled safety inspection visits.

13. Have regular safety training sessions with personnel.

14. Provide first-aid equipment and plan for medical care in case of accident.

Index

ABOUT THE AUTHOR

Dov Kaminetzky, P.E., is a principal and founding partner
of the distinguished consulting engineering firm of Feld,
Kaminetzky & Cohen, P.C., in New York City. He is
widely recognized as one of the pioneers and leaders in the
field of forensic engineering—the investigation and
analysis of the causes of structural distress and failure. In
addition to his studies of failed structures, Mr. Kaminetzky
has to his credit the structural design of numerous
successful projects, such as the Guggenheim Museum, the
North River pollution control plant, and the Yonkers
Raceway suspension roof.

Mr. Kaminetzky lectures around the world to practicing
engineers, designers, architects, contractors, attorneys, and
students on such subjects as design and construction
failures, and repair and rehabilitation of structures. He has
been an adjunct professor at the Graduate School of
Engineering of the City University of New York. He is a
Fellow of the American Society of Civil Engineers, a
Fellow of the American Concrete Institute, a member of
the Moles, and a licensed Professional Engineer in many
states.